About Island Press

Since 1984, the nonprofit Island Press has been stimulating, shaping, and communicating the ideas that are essential for solving environmental problems worldwide. With more than 800 titles in print and some 40 new releases each year, we are the nation's leading publisher on environmental issues. We identify innovative thinkers and emerging trends in the environmental field. We work with world-renowned experts and authors to develop cross-disciplinary solutions to environmental challenges.

Island Press designs and implements coordinated book publication campaigns in order to communicate our critical messages in print, in person, and online using the latest technologies, programs, and the media. Our goal: to reach targeted audiences—scientists, policymakers, environmental advocates, the media, and concerned citizens—who can and will take action to protect the plants and animals that enrich our world, the ecosystems we need to survive, the water we drink, and the air we breathe.

Island Press gratefully acknowledges the support of its work by the Agua Fund, Inc., The Margaret A. Cargill Foundation, Betsy and Jesse Fink Foundation, The William and Flora Hewlett Foundation, The Kresge Foundation, The Forrest and Frances Lattner Foundation, The Andrew W. Mellon Foundation, The Curtis and Edith Munson Foundation, The Overbrook Foundation, The David and Lucile Packard Foundation, The Summit Foundation, Trust for Architectural Easements, The Winslow Foundation, and other generous donors.

The opinions expressed in this book are those of the author(s) and do not necessarily reflect the views of our donors.

Ecological Economics

To Andrea and Marcia
And to the next generation, especially Liam, Yasmin, Anna, Will, and Isabel

"The human mind, so frail, so perishable, so full of inexhaustible dreams and hungers,
burns by the power of a leaf."

—*Loren Eisley*

Library of Congress Cataloging-in-Publication Data.

Daly, Herman E.
  Ecological economics : principles and applications / Herman E. Daly and Joshua Farley. — 2nd ed.
    p. cm.
  Includes bibliographical references and index.
  ISBN-13: 978-1-59726-681-9 (cloth : alk. paper)
  ISBN-10: 1-59726-681-7 (cloth : alk. paper)
1. Environmental economics. I. Farley, Joshua C., 1963– II. Title.
  HD75.6.E348 2010
  333.7—dc22
                                        2010012739
British Cataloguing-in-Publication Data available.

Design by Mary McKeon

Printed on recycled, acid-free paper ✪

Manufactured in the United States of America
10 9 8 7 6 5 4 3 2 1

Ecological Economics

Principles and Applications

Second Edition

Herman E. Daly and Joshua Farley

Washington | Covelo | London

Contents

Acknowledgments

W e are grateful to our many colleagues in the International Society for Ecological Economics for their intellectual contributions and the community of scholarship and support they provide. We especially wish to acknowledge Robert Costanza, and the faculty at the Gund Institute for Ecological Economics at the University of Vermont, as well as our colleagues at the University of Maryland School of Public Affairs. We are also immensely grateful to the Santa Barbara Family Foundation for financial support and to Jack Santa Barbara for encouragement and substantive help far beyond the financial. Our editor at Island Press, Todd Baldwin, provided many helpful suggestions, in addition to guiding the whole process from idea to published book.

A Note to Instructors

A textbook is usually a pedagogically efficient presentation of the accepted concepts and propositions of a well-defined academic discipline. Although we try to be pedagogically efficient, this book is not a textbook in the above sense, because ecological economics is not a discipline, nor does it aspire to become one. For lack of a better term, we call it a transdiscipline. We think that the disciplinary structure of knowledge is a problem of fragmentation, a difficulty to be overcome rather than a criterion to be met. Real problems in complex systems do not respect academic boundaries. We certainly believe that thinking should be "disciplined" in the sense of respecting logic and facts but not "disciplinary" in the sense of limiting itself to traditional methods and tools that have become enshrined in the academic departments of neoclassical economics. Furthermore, ecological economics is still "under construction," and therefore no fully accepted methodologies and tools exist. Instead, its practitioners draw on methods and tools from various disciplines to address a specific problem.

Much of what we present is more contentious and less cut-and-dried than what you would find in a standard economic principles text. While we are especially critical of standard economics' excessive commitment to GNP growth and its neglect of the biophysical system in which the economy is embedded, we also recognize that much environmental destruction and other forms of misery are caused by insufficient attention to standard economics. For example, subsidized prices for natural resources, neglect of external costs and benefits, and political unwillingness to respect the basic notions of scarcity and opportunity cost are problems we join standard economists in decrying.

As will be clear to any economist, the sections presenting basic micro- and macroeconomics, as well as other parts discussing distribution and trade, are based on standard economics. We want to be very clear: We are not claiming that ecological economists invented supply and demand, or national income accounting, or comparative advantage. Although it may be unnecessary to state this, experience has taught us to be very explicit in recognizing the origins of certain economic concepts, even if they are now so accepted as to be in the public domain. There are enough real points of contention between standard and ecological economics that we don't need to add any fictitious ones! On contentious issues we do not shy away from controversy, but we do try to avoid the temptation to fan the flames unnecessarily and to remember that the conflict is primarily between ideas, rather than the people holding them. It would be dishonest, however, not to admit that we *all* hold certain of our ideas with passion. If that were not the case, then no one would stand for anything, and studying economics would be very boring indeed. But among our passions should be a commitment to fairness—first in considering the views of others, and second in demanding equal treatment for our own views.

We the authors are both economists trained in the standard neoclassical Ph.D. programs that one finds in nearly all American universities. Between us we have taught and practiced economics for over 60 years in universities and development institutions in various countries. We are not "non-economists," nor do we consider that the epithet describes an irredeemably fallen state. After all, most policy makers are "non-economists," a fact for which we are sometimes grateful. We accept more of traditional economics than we reject, although we certainly do reject some of the things we were taught. We have little patience with anti-economists who want to abolish money, who consider all scarcity to be an artificial social construct, or who think that all of nature's services should be free. On the other hand, we do not share the view of many of our economist colleagues that growth will solve the economic problem, that narrow self-interest is the only dependable human motive, that technology will always find a substitute for any depleted resource, that the market can efficiently allocate all types of goods, that free markets always lead to an equilibrium balancing supply and demand, or that the laws of thermodynamics are irrelevant to economics. Precisely because ecological economists have some basic disagreements with standard economics, it is necessary to emphasize that these divergences are branchings from a common historical trunk, not the felling of that common trunk.

We have provided a workbook, Farley, Erickson, Daly, *Ecological Economics: A Workbook for Problem-Based Learning*, to supplement this textbook. We emphasize that the textbook is self-contained and in no way depends on the workbook. Nevertheless, some instructors and students will find the workbook helpful, especially regarding systems thinking, case studies, applications, and design of class projects.

Introduction

Probably the best introduction to our book is the conclusion of another book. The other book is *Something New Under the Sun* by historian J. R. McNeill.[1] McNeill argues that the Preacher in Ecclesiastes remains mostly but not completely right—there is indeed "nothing new under the sun" in the realm of vanity and wickedness. But the place of humankind within the natural world is not what it was. The enormity and devastating impact of the human scale on the rest of creation really is a new thing under the sun. And it greatly amplifies the consequences of vanity and wickedness. McNeill's findings help to place ecological economics in historical context and to explain why it is important.[2] His conclusions are worth quoting at length:

> Communism aspired to become the universal creed of the twentieth century, but a more flexible and seductive religion succeeded where communism failed: the quest for economic growth. Capitalists, nationalists—indeed almost everyone, communists included—worshipped at this same altar because economic growth disguised a multitude of sins. Indonesians and Japanese tolerated endless corruption as long as economic growth lasted. Russians and eastern Europeans put up with clumsy surveillance states. Americans and Brazilians accepted vast social inequalities. Social, moral, and ecological ills were sustained in the interest of economic growth; indeed, adherents to the faith proposed that only more growth could resolve such ills. Economic growth became the indispensable ideology of the state nearly everywhere. How?
>
> This state religion had deep roots in earlier centuries, at least in imperial China and mercantilist Europe. But it succeeded fully only after the Great Depression of the 1930s. . . . After the Depression, economic rationality trumped all other concerns except security. Those who promised to deliver the holy grail became high priests.
>
> These were economists, mostly Anglo-American economists. They helped win World War II by reflating and managing the American and British economies. The international dominance of the United States after 1945 assured wide acceptance of American ideas, especially in economics, where American success was most conspicuous. Meanwhile the USSR proselytized within its geopolitical sphere, offering a version of the growth fetish administered by engineers more than by economists.

---

[1]New York: Norton, 2000, pp. 334–336.

[2]For another discussion of the place of ecological economics in recent intellectual and historical context, see Peter Hay, *Main Currents in Western Environmental Thought*, Bloomington: Indiana University Press, 2002, Chapter 8.

American economists cheerfully accepted credit for ending the Depression and managing the war economies. Between 1935 and 1970 they acquired enormous prestige and power because, or so it seemed, they could manipulate demand through minor adjustments in fiscal and monetary policy so as to minimize unemployment, avoid slumps, and assure perpetual economic growth. They infiltrated the corridors of power and the groves of academe, provided expert advice at home and abroad, trained legions of acolytes from around the world, wrote columns for popular magazines—they seized every chance to spread the gospel. Their priesthood tolerated many sects, but agreed on fundamentals. Their ideas fitted so well with social and political conditions of the time that in many societies they locked in as orthodoxy. All this mattered because economists thought, wrote, and prescribed as if nature did not.

This was peculiar. Earlier economists, most notably the Reverend Thomas Malthus (1766–1834) and W.S. Jevons (1835–1882), tried hard to take nature into account. But with industrialization, urbanization, and the rise of the service sector, economic theory by 1935 to 1960 crystallized as a bloodless abstraction in which nature figured, if at all, as a storehouse of resources waiting to be used. Nature did not evolve, nor did it twitch and adjust when tweaked. Economics, once the dismal science, became the jolly science. One American economist in 1984 cheerfully forecast 7 billion years of economic growth—only the extinction of the sun could cloud the horizon. Nobel Prize winners could claim, without risk to their reputations, that "the world can, in effect, get along without natural resources." These were extreme statements, but essentially canonical views. If Judeo-Christian monotheism took nature out of religion, Anglo-American economists (after about 1880) took nature out of economics.

The growth fetish, while on balance quite useful in a world with empty land, shoals of undisturbed fish, vast forests, and a robust ozone shield, helped create a more crowded and stressed one. Despite the disappearance of ecological buffers and mounting real costs, ideological lock-in reigned in both capitalist and communist circles. No reputable sect among economists could account for depreciating natural assets. The true heretics, economists who challenged the fundamental goal of growth and sought to recognize value in ecosystem services, remained outside the pale to the end of the century. Economic thought did not adjust to the changed conditions it helped to create; thereby it continued to legitimate, and indeed indirectly to cause, massive and rapid ecological change. The overarching priority of economic growth was easily the most important idea of the twentieth century.

From about 1880 to 1970 the intellectual world was aligned so as to deny the massive environmental changes afoot. While economists ignored nature, ecologists pretended humankind did not exist. Rather than sully their science with the uncertainties of human affairs, they sought out pristine patches in which to monitor energy flows and population dynamics. Consequently they had no political, economic—or ecological—impact.

Is McNeill correct in his assessment that "the overarching priority of economic growth was easily the most important idea of the twentieth century"? It's hard to imagine a more important one. There are still very few who question the priority of economic growth.[3] Yet many students are turned off by economics for the reasons also given by McNeill, namely the economists' total abstraction from nature and their extreme devotion to economic growth as the *summum bonum*. While this aversion is understandable, it would be very sad if the only students who studied economics were those who didn't realize the fundamental limits of the discipline or those who, realizing that something was wrong, didn't have the energy or courage to try to reform it.

Professor McNeill specifically meant ecological economics in his reference to the "true heretics" who remained outside the pale to the end of the century. The purpose of our textbook is to try to change that deplorable situation—to help the next generation of economists take proper account of nature and nature's limits. Achieving this objective will require a fusion of insights and methodologies from numerous disciplines to create a transdisciplinary approach to economics. Such an approach is necessary if we hope to understand nature's limits and create policies that allow our economy to develop within those limits. However, to achieve anything more than random outcomes, we must direct available means toward specific ends. McNeill convincingly argues that ever-greater material consumption provided by never-ending economic growth is the agreed-upon end for the majority of modern society. This emphasis on an impossible and probably undesirable end is arguably a more serious shortcoming to traditional economics than a limited understanding of means.

■ THE CALL FOR CHANGE

As this is written, there are news reports of a group of economics students in French and British universities who are rebelling against what they are being taught. They have formed a Society for Post-Autistic Economics. Their implicit diagnosis is apt, since autism, like conventional economics, is characterized by "abnormal subjectivity; an acceptance of fantasy rather than reality." Ecological economics seeks to ground economic thinking in the dual realities and constraints of our biophysical and moral environments. Current "canonical assumptions"[4] of insatiable wants and infinite resources, leading to growth forever, are simply not founded in reality.

---

[3]For an interesting political history of how growth came to dominate U.S. politics in the postwar era, see R. M. Collins, *More: The Politics of Growth in Postwar America*, New York: Oxford University Press, 2000.

[4]*Canonical* literally means "according to religious law" and commonly means according to accepted usage.

Their dire consequences are evident. And that truly is something new under the sun.

In the early days of ecological economics, it was hoped that the ecologist would take over the economist's territory and redeem the failures of an economics that neglected nature. While ecologists made many important contributions, one is forced to accept McNeill's assessment that their influence has basically been disappointing—most have been unwilling "to sully their science with the uncertainties of human affairs." The reasons that many ecologists appear to have had difficulty in dealing with policy will be the subject of speculation in Chapter 2. While ecology may not share the same inadequacies as economics, studying ecosystems as if they were isolated from human affairs on a planet of six billion humans also suggests an inclination to accept fantasy over reality. Ecological economics, therefore, is not simply bringing the light of ecology into the darkness of economics. Both disciplines need fundamental reform if their marriage is to work.

Nor is autistic tunnel vision limited to economics and ecology. Most universities these days educate students within the narrow confines of traditional disciplines. Rather than training students to examine a problem and apply whatever tools are necessary to address it, universities typically train students in a set of discipline-specific tools that they are then expected to apply to all problems. The difficulty is that the most pressing problems we face today arise from the interaction between two highly complex systems: the human system and the ecological system that sustains it. Such problems are far too complex to be addressed from the perspective of a single discipline, and efforts to do so must either ignore those aspects of the problem outside the discipline or apply inappropriate tools to address them. "Abnormal subjectivity" (autism) is the inevitable result of education in a single subject. Applying insights from one discipline to another can serve to dispel the fantasies to which each alone is prone. For example, how could an economist conversant with ecology or physics espouse infinite growth on a finite planet? Effective problem-solving research must produce a mutually intelligible language for communication across disciplines. Otherwise, each discipline shall remain isolated in its own autistic world, unable to understand the world around it, much less to resolve the problems that afflict it.

■ A TRANSDISCIPLINARY SCIENCE

Ecological economists must go well beyond the fusion of ecology and economics alone. The complex problems of today require a correspondingly complex synthesis of insights and tools from the social sciences, natural sciences, and humanities. We frequently see research in which teams of researchers trained in different disciplines separately tackle a single problem and then strive to combine their results. This is known as multidisciplinary research, but the result is much like the blind men who examine an elephant, each describing the elephant according to the single body part they touch. The difference is that the blind men can readily pool their information, while different academic disciplines lack even a common language with which their practitioners can communicate. Interdisciplinary research, in which researchers from different disciplines work together from the start to jointly tackle a problem, allowing them to reduce the language barrier as they go, is a step in the right direction. But while universities have disciplines, the real world has problems. Ecological economics seeks to promote truly transdisciplinary research in which practitioners accept that disciplinary boundaries are academic constructs irrelevant outside of the university and allow the problem being studied to determine the appropriate set of tools, rather than vice versa.

Just as effective problem solving requires the insights and tool sets of a variety of disciplines, defining the goals toward which we should strive would benefit from open discussion of the value sets of different ideologies. Unfortunately, the two dominant ideologies of the twentieth century seem to lack sufficient diversity within their value sets to stimulate this discussion. Specifically, the former USSR and the West, though differing in important ways, shared a fundamental commitment to economic growth as the first priority. The Marxist's deterministic ideology of dialectical materialism refused any appeals to morality or justice. The "new socialist man" would emerge only under objective conditions of overwhelming material abundance, which in turn required maximum economic growth. Bourgeois selfishness would disappear only with the disappearance of scarcity itself. In the U.S. and the West generally, the bulk of society did not reject appeals to justice and morality, but we did come to believe that our moral resources were very scarce relative to our natural resources and technological powers. Our strategy was to grow first, in the hope that a bigger pie would be easier to divide than a smaller one.[5] But in practice, McNeill's "growth fetish" dominated both systems, and both were unmindful of the costs of growth.

Infinite growth in a finite system is an impossible goal and will eventually lead to failure. The USSR failed first because its system of central

---

[5]Though for a large and powerful group in the West, the primacy of growth and the belief in the powers of the market has turned common conceptions of morality on their head. Ayn Rand, a highly influential philosopher and author who counts Alan Greenspan (former chairman of the Federal Reserve and arguably once one of the most powerful men in the world) among her fervent admirers, argues that altruism is evil and selfishness is a virtue (*Selfishness: A New Concept of Egoism*, New York: Signet, 1964). Milton Friedman, Nobel Prize–winning economist of the Chicago school, argues in the same vein: "Few trends could so thoroughly undermine the very foundation of our free society as the acceptance by corporate officials of a social responsibility other than to make as much money for their stockholders as possible" (Chicago: University of Chicago Press, 1962, p. 135).

planning, along with its neglect of human rights, was more inefficient than the decentralized markets and greater respect for human rights in the West. The USSR was less mindful of the social and environmental costs of growth than the West, so it collapsed sooner. Because of its greater efficiency, the West can keep going for a bit longer in its impossible quest. But it, too, will collapse under the accumulating cost of growth. However, thus far the collapse of the USSR has been recognized by the West only as a validation of our superior efficiency. The possibility that efficiency only buys time and that unlimited growth must eventually fail in the West as well is something we have not yet considered. Alternatives to our misguided goal of infinite growth and limitless material consumption will be discussed throughout this text.

> **THINK ABOUT IT!**
> *Think about a problem society currently faces, one that you know something about. Does the information needed to resolve this problem fit snugly within the boundaries of one academic discipline, or are insights from several different disciplines needed to solve it? What disciplines might be involved? In your university, how much interaction exists between the departments (professors and students) of those different disciplines?*

■ AN OVERVIEW OF THE TEXT

Part I of this book is an introduction to the subject of ecological economics. Ecological economics seeks not only to explain how the world works but also to propose mechanisms and institutions for making it work better. Chapter 1 explains the basic subject matter of neoclassical and ecological economics in order to show the full scope of the new transdiscipline of ecological economics. Having defined the territory, we first establish basic agreement on the fundamental nature of the system we propose to analyze. Chapter 2 begins by describing the core (preanalytic) vision of ecological economics, that the economic system is a part or subsystem of a larger global ecosystem that sustains it. This view is contrasted with the fundamental vision of neoclassical economics, that the economic system is a self-sufficient whole entity unto itself. If we seek to make a system work better, we need to know the resources available to us—the means—and the desired outcomes—the ends. Chapter 3 focuses on the ends-and-means spectrum, an essential step for understanding a science that defines itself as a mechanism for connecting scarce means to alternative ends.

Part II focuses on the containing and sustaining Whole, the Earth and its atmosphere. In these chapters, we delve deeper into the nature of the Whole—the global ecosystem that sustains us by providing the resources that feed the economic process and the sinks where we dispose of our wastes. Chapter 4 establishes the fundamental importance of low entropy (useful, ordered matter-energy) in economic production and the inevitability of its conversion via the economic process to high-entropy, disordered, useless waste. Chapter 5 addresses the tangible forms in which low entropy manifests itself, the abiotic goods and services provided by nature, and examines their specific market-relevant characteristics. Chapter 6 does the same for biotic resources. Chapter 7 shows that many of the goods and services provided by nature were formerly superabundant, and it made little difference if an economic system dedicated to allocating scarce resources ignored them. Now, however, as we have discussed in this Introduction, these resources have become scarce, and their allocation has become critically important.

Part III begins our examination of the part of the whole in which we are most interested, the economic subsystem. We draw out the useful elements of neoclassical economic theory and integrate them into ecological economics. Microeconomics, macroeconomics, or international trade each provides sufficient material for years of study, and this text conveys no more than the essentials.

Chapters 8 and 9 introduce microeconomics, the study of mechanisms for efficiently allocating specific scarce resources among specific alternative ends. They explain the self-organizing properties of a competitive market economy, through which millions of independent decision makers, freely acting in their own self-interest, can generate the remarkable outcomes alluded to at the start of Chapter 1. These chapters also explain how neoclassical production and utility functions must be modified to address the concerns of ecological economics.

In Chapter 10, we take a step back from the traditional microeconomic analysis of allocation to examine the specific characteristics resources must have if they are to be efficiently allocated by the market mechanism. We find that few of the goods and services provided by nature exhibit all of them. Attempts to allocate resources that do not have the appropriate characteristics via the unregulated market result in inefficient, unfair, and unsustainable outcomes. Rather than individual self-interest creating an invisible hand that maximizes social well-being, market allocation of such "nonmarket" goods creates an invisible foot that can kick the common good in the pants. Careful analysis of these market-relevant characteristics of scarce resources is an essential prelude to policy formulation. Thus, Chapter 11 applies the concepts of market failures to abiotic resources, and Chapter 12 applies them to biotic ones.

In Chapter 13, we turn our attention to human behavior, with three major goals. First, we hope to clarify the desirable ends of economic activity by assessing what things and activities contribute to satisfying and fulfilling lives. Second, conventional economic models are built on the assumption that humans are insatiable, rational, and self-interested utility maximizers. Such behavior is a serious obstacle to developing the cooperative mechanisms needed to address the market failures described in Chapters 10–12. We look at the empirical evidence regarding these assumptions and find them lacking. Finally, we assess the empirical evidence regarding cooperative behavior. We conclude that cooperation is an integral part of human behavior that is somewhat suppressed by market economies but that can be effectively elicited by a variety of different institutions. This is a fortuitous conclusion, given the evidence that cooperative behavior is necessary to solve the most serious problems we currently face.

In Part IV, we turn to macroeconomics. As we stated earlier, ecological economics views the economy as a part of a larger finite system. This means that the traditional goal of macroeconomic policy—unlimited economic growth in the physical dimension—is impossible. Thus, in ecological economics, optimal scale replaces growth as a goal, followed by fair distribution and efficient allocation, in that order. Scale and distribution are basically macroeconomic issues. Therefore, in addition to the fiscal and monetary policy tools that dominate the discussion in traditional texts, we will introduce policies that can help the economy reach an optimal scale.[6] Chapter 14 focuses on the basic macroeconomic concepts of GNP and welfare. It starts by examining economic accounting, or the measurement of desirable ends ranging from gross national product to human needs assessment. Chapter 15 discusses the role of money in our economy. Chapter 16 focuses on the issue of distribution within and between generations, and Chapter 17 briefly develops the basic macroeconomic model of how saving and investing behavior combines with the supply and demand of money to determine the interest rate and level of national income. We then relate the macroeconomic model to policy levers designed to achieve the ecological economics goals of sustainable scale and just distribution.

Part V addresses international trade. In Chapters 18 and 19, we discuss how different economies interact and the troublesome issue of global economic integration. We consider especially the consequences of global integration for policy making. Chapter 20 looks at financial issues, with an emphasis on speculation and financial crises, and examines the implications of globalization for macroeconomic policy.

Part VI focuses on policy. Chapter 21 presents the general design prin-

---

[6]Optimal scale is the point where the marginal benefits of additional growth are just equal to the marginal costs of the reduction in ecosystem function that this growth imposes. As we will show in the text, numerous factors can affect optimal scale.

ciples of policy. Chapter 22 reviews a number of specific policy options that primarily affect scale, Chapter 23 reviews policies that primarily affect distribution, and Chapter 24 reviews policies that primarily affect allocation.

Our concluding chapter, Looking Ahead, once again reflects on the ethical assumptions of ecological economics. We call for a return to the beginnings of economics as a moral philosophy explicitly directed toward raising the quality of life of this and future generations.

In summary, neoclassical microeconomic theory arose primarily as an effort to explain the market economy. Macroeconomics arose in response to the failure of microeconomic theory to explain and respond to recessions and depressions. Ecological economics is emerging in response to the failures of microeconomics and macroeconomics to address unsustainable scale and inequitable distribution. Ecological economics takes a more inclusive, and activist, position. We describe the nature of scarce resources and the ends for which they should be used and proactively prescribe appropriate institutions for their efficient allocation in a social context of just distribution and sustainable scale. We have the basic allocative institution of the market—it needs improvement, but at least it exists. We have no institution for limiting scale, and our institutions for governing distribution (antitrust, progressive taxation) have been allowed to atrophy. We know that building institutions is a political task and that "politics is the art of the possible." That is a wise conservative counsel. Yet that dictum also prohibits attempting true physical impossibilities in a vain effort to avoid apparent political "impossibilities." When faced with the unhappy dilemma of choosing between a physical and a political impossibility, it is better to attempt the politically "impossible."

PART I

An Introduction to Ecological Economics

CHAPTER

1

Why Study Economics?

■ WHAT IS ECONOMICS?

Economics is the study of the **allocation** of limited, or scarce, resources among alternative, competing ends.[1] We can choose, for example, to allocate steel to plowshares or SUVs. These products in turn are apportioned to different individuals—Somalian farmers or Hollywood stars, for example. Of course, as a society we don't consciously choose to allocate steel to a particular number of plows or SUVs. But we do have collective desires, the sum of the individual choices that each of us makes to buy one thing or another. Really, economics is about what we desire and what we're willing to give up to get it.

In fact, three critical questions guide economic inquiry, and there is a clear order in which they should be asked:

1. What ends do we desire?
2. What limited, or scarce, resources do we need to attain these ends?
3. What ends get priority, and to what extent should we allocate resources to them?

This last question cannot be answered without deep reflection on the answers to the first two questions. Only after we have answered all these questions can we decide which are the best mechanisms for allocating those resources.

Traditionally, economists have said that the answer to the first question is "utility" or human welfare.[2] Welfare depends on what people want, which they reveal through market transactions—by what goods and services

Allocation is the process of apportioning resources to the production of different goods and services. *Neoclassical economics* focuses on the market as the mechanism of allocation. *Ecological economics* recognizes that the market is only one possible mechanism for allocation.

---

[1]We will re-examine this definition later in the text, because not all resources are scarce in an economic sense. For example, no matter how much you use information, there is just as much left for someone else, or more if your use leads to improvements. Since we live in "the information age" this is an important point. Many of nature's services are similarly not depleted through use.

they buy and sell. Naturally, this only reveals preferences for market goods and implicitly assumes that nonmarket goods contribute little to welfare. Humans are assumed to be insatiable,[3] so welfare is increased through the ever-greater provision of goods and services, as measured by their market value. Thus, unending economic growth is typically considered an adequate, measurable proxy for the desirable end.

This view is fundamental to the main school of economics today, known as **neoclassical economics (NCE)**. Since neoclassical economists assume that markets reveal most desired ends and that most scarce resources are market goods, they devote most of their attention to the mechanism for allocating resources to alternative ends, which is, of course, the market. The reason the market is considered the appropriate mechanism is that under certain restrictive assumptions it is efficient, and efficiency is considered a value-free, objective criterion of "the good." **Efficient allocation** is shorthand for **Pareto efficient allocation**, a situation in which no other allocation of resources would make at least one person better off without making someone else worse off (the name Pareto is for economist Vilfredo Pareto). Efficiency is so important in neoclassical economics that it is sometimes taken to be an end in itself.[4]

Efficient allocation is shorthand for *Pareto efficient allocation,* a situation in which no other allocation of resources would make at least one person better off without making someone else worse off.

But we should bear in mind that if our ends were evil, then efficiency would just make things worse. After all, Hitler was rather efficient in killing Jews. Efficiency is worthwhile only if our ends are in fact good and well-ordered—a job not worth doing is not worth doing well. We will return to this in our discussion of an ends-means spectrum in Chapter 3.

Ecological economics takes a different approach than its neoclassical counterpart. In **ecological economics**, efficient allocation is important but far from being an end in itself. Take the example of a ship. To load a ship efficiently is to make sure that the weight on both sides of the keel is the same, and the load is distributed from front to back so that the ship floats evenly in the water. While it is extremely important to load the cargo efficiently, it is even more important to make sure that not too much cargo is placed on the ship. It is of little comfort if an overloaded ship founders efficiently! Who is entitled to place their cargo on the ship is also important; we wouldn't want the passengers in first class to hog all the cargo space so that those in steerage lack adequate food and clothing for their voyage.

Ecological economists look at the Earth as a ship and gross material production of the economy as the cargo. The seaworthiness of the ship is

---

[2]Many neoclassical economists actually argue that economics is a positive science (i.e., based on value neutral propositions and analysis). Since desired ends are normative (based on values), they would therefore lie outside the domain of economic analysis.

[3]Insatiability means that we can never have enough of all goods, even if we can get enough of any one good at a given time.

[4]D. Bromley, The Ideology of Efficiency: Searching for a Theory of Policy Analysis, *Journal of Environmental Economics and Management* 19: 86–107 (1990).

determined by its ecological health, the abundance of its provisions, and its design. Ecological economists recognize that we are navigating unknown seas and no one can predict the weather for the voyage, so we don't know exactly how heavy a load is safe. But too heavy a load will cause the ship to sink.

Neoclassical economists focus solely on allocating the cargo efficiently. **Environmental economics**, a subset of neoclassical economics, recognizes that welfare also depends to a large extent on ecosystem services and suffers from pollution but is still devoted to efficiency. As markets rarely exist in ecosystem services or pollution, environmental economists use a variety of techniques to assign market values to them so that they, too, may be incorporated into the market model. Ecological economists insist on remaining within the weight limits (or in nautical terms, respecting the Plimsoll line[5]) determined by the ship design and the worst conditions it is likely to encounter and making sure that all passengers have sufficient resources for a comfortable voyage. Once those two issues have been safely resolved, the hold is efficiently loaded.

Substantial evidence exists that the cargo hold is already too full for a safe voyage, or at least nearing capacity, and many passengers have not been allowed to load the basic necessities for the voyage. Certainly we seem to have too many greenhouse gases in the hold, too many toxic compounds. To make room for an ever-growing cargo, we have ripped out components of the ship we deem unimportant. But we live on a very complicated ship, and we know very little about its design and the impact of our choices on its structural integrity. How many forests and wetlands are needed to keep it afloat? What species are crucial rivets, whose loss will compromise the ship's seaworthiness? Ecological economics addresses these issues. It also assumes that our goal is not simply to load the ship to the limit but to maintain areas of the ship for our comfort and enjoyment, to revel in the exquisite beauty of its craftsmanship, and to maintain it in excellent condition for future generations.

So why study economics? If we do not, we will probably end up serving less important ends first and running out of resources while more important ends remain unmet. We are also likely to overload and swamp the ship unless we have studied the seas in which it will be sailing, as well as the ship's own design and functioning.

---

[5]In 1875, Samuel Plimsoll supported Britain's Merchant Shipping Act, requiring that a load-limit line be painted on the hull of every cargo ship using British ports. If the waterline exceeded the Plimsoll line, the ship was overloaded and prohibited from entering or exiting the port. Because of England's seafaring dominance, the practice was adopted worldwide. Yet shipowners who profited from overloading their ships fiercely resisted the measure. They could buy insurance at rates that made it profitable to occasionally risk losing an overloaded ship. The Plimsoll line has saved the lives of many sailors.

■ The Purpose of This Textbook

This textbook is designed to introduce ecological economics as a necessary evolution of conventional economic thought (neoclassical economics) that has dominated academia for over a century. Our text will critique not only neoclassical economic theory but also the pro-growth market economy that in many people's minds has come to be virtually synonymous with American democracy. Ecological economists do not call for an end to markets. Markets are necessary. What must be questioned is the prevailing belief that markets reveal all our desires, that they are the ideal system not only for allocating all resources efficiently but also for distributing resources justly among people, and that markets automatically limit the overall macroeconomy[6] to a physical scale that is sustainable within the biosphere.

Part of our goal is to explain markets and show what they do well. Another part of our goal is to show why the unregulated market system is inadequate for allocating most of the goods and services provided by nature. This portion of the text should not be controversial—most of the basic arguments actually come from neoclassical economics, and it is only by drawing attention to their full implications that we depart from orthodoxy.

More contentious (and more important) is the call by ecological economics for an end to growth. We define **growth** as an increase in **throughput**, which is the flow of natural resources from the environment, through the economy, and back to the environment as waste. It is a quantitative increase in the physical dimensions of the economy or of the waste stream produced by the economy. This kind of growth, of course, cannot continue indefinitely, as the Earth and its resources are not infinite. While growth must end, this in no way implies an end to **development**, which we define as qualitative change, realization of potential, evolution toward an improved but not larger structure or system—an increase in the quality of goods and services (where quality is measured by the ability to increase human well-being) provided by a given throughput. Most of you have ceased growing physically yet are probably studying this text in an effort to further develop your potential as humans. We expect human society to continue developing, and indeed argue that only by ending growth will we be able to continue developing for the indefinite future. Fortunately, many desirable ends require few physical resources.

The idea of "sustainable development," to be discussed later, is development without growth—that is, qualitative improvement in the ability to satisfy wants (needs and desires) without a quantitative increase in through-

Growth is a quantitative increase in size or an increase in throughput.

Throughput is the flow of raw materials and energy from the global ecosystem, through the economy, and back to the global ecosystem as waste.

Development is the increase in quality of goods and services, as defined by their ability to increase human well-being, provided by a given throughput.

---

[6]Microeconomics focuses primarily on how resources are allocated toward the production and consumption of different goods and services. Macroeconomics traditionally focuses primarily on economic growth (i.e., the size of the economy), employment, and inflation.

put beyond environmental carrying capacity. Carrying capacity is the population of humans that can be sustained by a given ecosystem at a given level of consumption, with a given technology. Limits to growth do not necessarily imply limits to development.

Conventional neoclassical economists might define economic growth as the increase in an economy's production of goods and services, typically measured by their market value, that is, an increase in gross national product (GNP). However, an economy can develop without growing, grow without developing, or do both at the same time. GNP lumps together quantitative growth with qualitative development—two very different things that follow very different laws—and is thus not a very useful measure.

In spite of the distinction between growth and development, calling for an end to growth requires an almost revolutionary change in social perceptions of the good (our ends and their ranking), a theme that will recur throughout this text. As we are all aware, the transition from adolescence to maturity is a difficult time for individuals and will be for society as well.

The market economy is an amazing institution. Market forces are justly credited with contributing to an unprecedented and astonishingly rapid increase in consumer goods over the past three centuries. Poor people in affluent countries today have many luxuries that kings of Europe could not have dreamed of in centuries past, and we have achieved this through a system that relies on free choice. In the market in its pure form, individuals are free to purchase and produce any market good they choose, and there is no controlling authority apart from the free will of individual humans. Of course, the pure form exists only in textbooks, but competitive markets do show impressive powers of self-regulation. Arguments for modifying such an admittedly impressive system must be persuasive indeed. However, a brief detour into the history of markets and economics suggests that such modifications occur all the time.

■ COEVOLUTIONARY ECONOMICS[7]

As Karl Polanyi showed in his classic *The Great Transformation*,[8] the economic system is embedded as a component of human culture, and like our culture, it is in a constant state of evolution. In fact, our ability to adapt to changing environmental circumstances through cultural evolution is

---

[7]Many of the basic ideas here come from the work of Richard Norgaard, including R. Norgaard, Coevolutionary Development Potential, *Land Economics* 60: 160–173 (1984) and R. Norgaard, Sustainable Development: A Coevolutionary View, *Futures*: 606–620 (1988).

[8]K. Polanyi. *The Great Transformation: The Political and Economic Origins of Our Time*. Boston: Beacon Press (2001).

something that most clearly distinguishes humans from other animals. Economic, social, and political systems, as well as technological advances, are examples of cultural adaptations. All these systems have adapted in response to changes in the environment, and these adaptations in turn provoke environmental change, to which we must again adapt in a co-evolutionary process. Examples of some of the major coevolutionary adaptations and their implications for future change will help illustrate this concept.

From Hunter-Gatherer to Industrialist

For more than 90% of human history, humans thrived as small bands of nomadic hunter-gatherers. Anthropology and archaeology together provide us with a reasonable understanding of the hunter-gatherer economy. Rather than the "nasty, brutish and short" life that many imagine, early people met their basic needs by working only a few hours a day, and resources were sufficient to provide for both young and old who contributed little to gathering food. A recent study of the !Kung, who live in a very arid, marginal environment, found that 10% of the population was over 60, which compares favorably with populations in many industrialized countries.[9]

Small bands of hunter-gatherers would deplete local resources and then move on to places where resources were more abundant, allowing the resource base in the previous encampment to recover. Mobility was essential to survival, and accumulating goods reduced mobility. Numerous chronicles by anthropologists attest that hunter-gatherers show very little concern for material goods, readily discarding their possessions, confident in their ability to make new ones as needed.[10] Property rights to land made no sense in a nomadic society, and prior to domestication some 10,000 years ago, property rights to animal herds were virtually impossible. Food was also shared regardless of who provided it, perhaps partly because of technological limits. Some food simply cannot be harvested in discrete bundles, and if hunters bring home a large game animal, un-

---

[9]R. Lee, "What Hunters Do for a Living." In J. Gowdy, ed. *Limited Wants, Unlimited Means: A Reader on Hunter-Gatherer Economics and the Environment*. Washington, DC, Island Press, 1998.

[10]M. Sahlins, "The Original Affluent Society." In J. Gowdy, op. cit.

[11]Recent anecdotal evidence supports this relationship between storage technology and property rights. In an indigenous community in Alaska, the government provided freezers for food storage, and the impact was dramatic. Where successful hunters previously shared their game with the community, freezers (probably contemporaneous with the breakdown of other social structures) enabled hunters to store their game for their own leisurely consumption. Older, younger, or weaker members of the community were left without a source of subsistence.

[12]Lee, op. cit.

shared food would simply rot or attract dangerous predators.[11] Studies of the !Kung and other tribes found that both young and old were generally exempt from food gathering, and even many mature men and women simply chose not to participate in this activity very often yet were given equal shares of the harvest.[12]

If private property and wealth accumulation were impractical and absent from human society for most of human existence, it is hard to argue that these are inherent characteristics of human nature rather than cultural artifacts.

Gradually hunter-gatherer societies developed the technology to store large quantities of food for months on end, an essential precursor to agriculture. Agriculture ended the nomadic lifestyle for many early peoples. People began to settle in towns or small communities, which led to greater population concentrations than had previously been possible.[13] The technologies of storage and agriculture changed the nature of property rights. Certainly agriculture itself made some form of property rights to land essential. Surplus production allowed greater division of labor and specialization, which in turn led to ever-greater production, fostering extensive trade and eventually the development of money. Greater populations, the need to protect increasing riches against other groups, and the need to defend property rights within the community meant more need for government, and ruling classes developed.[14] Ruling classes and the needs of the state clearly had to be supported through the productive capacity of others, which inevitably led to some sort of tax system and concentrations of wealth in the upper echelons of the hierarchy.

The chain of evolutionary events did not end there, of course. Higher populations and agriculture would have disrupted local ecosystems, eventually decreasing their capacity to produce food and materials independently of agriculture. This only increased the demands society would place on agriculture. These demands, accompanied by a more rapid exchange of ideas in denser communities, stimulated new technologies, such as large-scale irrigation.[15] Irrigation over time led to increased soil salinity, eventually reducing the capacity of the ecosystem to sustain such high population levels without further agricultural innovations or migration.

---

[13]J. Diamond, *Guns, Germs, and Steel: The Fates of Human Societies*, New York: Random House, 1997.

[14]Many political philosophers argue that the primary purpose of government is to protect private property. In the words of John Locke, "Government has no other end but the preservation of property," from "An Essay Concerning the True Original, Extent, and End of Civil Government."

[15]Diamond, op. cit.

The Industrial Revolution

Ever-greater surplus production, accompanied by better ships, allowed trade on an expanding scale. Traders exchanged not only goods but also ideas, further speeding up the rate of technological progress. Among the crucial technological leaps was the ability to extract and use fossil fuels and other nonrenewable mineral resources. It is no coincidence that the market economy and fossil fuel economy emerged at essentially the same time.[16] Trade also allowed specialization to take place across regions, not only across individuals within a society. Technological advance, fossil energy, and global markets laid the groundwork for the Industrial Revolution.

The Industrial Revolution had profound impacts on the economy, society, and the global ecosystem. For the first time, human society became largely dependent on fossil fuels and other nonrenewable resources (partially in response to the depletion of forests as fuel). Fossil fuels freed us from dependence on the fixed flow of energy from the sun, but it also allowed the replacement of both human and animal labor by chemical energy. This increased energy allowed us ever-greater access to other raw materials as well, both biological and mineral. New technologies and vast amounts of fossil energy allowed unprecedented production of consumer goods. The need for new markets for these mass-produced consumer goods and new sources of raw material played a role in colonialism and the pursuit of empire. The market economy evolved as an efficient way of allocating such goods, and stimulating the production of even more.

THINK ABOUT IT!

There are an estimated 25,000 person hours of labor in a barrel of oil, and humanity uses on the order of 85 million barrels per day. How much of the surge in economic production since the eighteenth century do you think was due to the magic of the market and how much to the magic of fossil fuels?

International trade exploded, linking countries together as never before. A greater ability to meet basic needs, and advances in hygiene and medical science, resulted in dramatic increases in population, whose needs were met through greater energy use and more rapid depletion of resources. Growing populations quickly settled the last remaining frontiers,[17] removing the overflow valve that had allowed populations to relo-

---

[16]Coal became a viable source of energy only after the commercialization of the Newcomen steam engine in 1712, which was used to pump water from the mines. James Watt improved on Newcomen's design and produced his first commercial engine in 1776, the same year that Adam Smith's *Wealth of Nations* appeared.

[17]Many of these 'frontiers' were of course already inhabited, and simply seized by better armed colonizers.

cate as local resources ran out. Per-capita consumption soared, and with it the waste output that now threatens to degrade our ecosystems.

■ THE ERA OF ECOLOGICAL CONSTRAINTS

As we stated earlier, economics is the science of the allocation of scarce resources among alternative ends. The success of the Industrial Revolution dramatically reduced the scarcity of consumer goods for much of the world's population. The accompanying economic growth, however, now threatens the former abundance of the goods and services produced by nature upon which we ultimately depend. These have become the newly scarce resources,[18] and we must redesign our economic system to address that reality. Unfortunately, our ability to increase consumption while depleting our resource base has led people to believe that humans and the economy that sustains us have transcended nature. In the current system, the greatest claims to wealth have seemingly nothing to do with natural resources but rather are acquired through financial transactions on computers that physically do nothing more than move electrons. While knowledge and information are important, ultimately wealth requires physical resources. A recipe is no substitute for a meal, even though a good recipe may improve the meal.

Though the current economic system has been around for a remarkably short time in relation to past systems, it has wrought far greater environmental changes. These changes have redefined the notion of scarce resources, and they demand correspondingly dramatic changes in economic theory and in our economic system. Change in our economic system is inevitable. The only question is whether it will occur as a chaotic response to unforeseen disruptions in the global life support system or as a carefully planned transition toward a system that operates within the physical limits imposed by a finite planet and the spiritual limits expressed in our moral and ethical values. The answer depends largely on how fast we act, and the burning question is: How much time do we have?

The Rate of Change

For the vast majority of human history, technological, social, and environmental changes occurred at a glacial pace. The agricultural revolution was really not a revolution but a case of evolution. For example, it probably took several thousand years to create corn from the ancestral stock of teosinthe.[19] People generally saw no evidence of change from one gener-

---

[18]See R. Hueting, *The New Scarcity and Economic Growth: More Welfare Through Less Production?*, Amsterdam: North Holland, 1980.

[19]Diamond, op. cit.

ation to the next, and human culture could evolve at a correspondingly slow pace to adapt to the changes that did occur. Only with the Industrial Revolution did change really begin to accelerate to the extent that we could notice it from one generation to the next. And much of what the Industrial Revolution did was to increase the extraction of nonrenewable resources, thereby increasing human material consumption. As a result of this subsidy from nature, the general perception was that the future would always get better, and all that was needed was more of the same. Our response has been to use up this finite subsidy at ever greater rates, so that now, for the first time in human history, we can dramatically change the Earth's systems on a human time scale (a truly new thing under the sun). In fact, it threatens to alter the ability of the Earth to support life. While cultures have continually and slowly evolved in adapting to new technologies and new constraints, the unprecedented rate of change in technology and ecological degradation means we no longer have the luxury of biding our time. Most likely we will have to change our cultural institutions and values in response, particularly the economic institutions and values that have led to this state of affairs. Since there is certainly some limit to how fast we can adapt culturally, we need also to consider seriously how to slow down the rate of change that is forcing the adaptations. It is worth remembering that not all change is desirable and that even desirable change can be too fast.

Scale is the physical size of the economic subsystem relative to the ecosystem that contains and sustains it.

The Difficulty of Achieving Desirable Change

It would be foolish to underestimate the difficulty of finding the right balance between limiting and adapting to change. Currently, our economic system is focused primarily on the microeconomic issue of efficient allocation. Applied economics also focuses on the macroeconomic issue of maximizing growth. Ecological economics, however, focuses primarily on the larger macroeconomic issue of how big is too big. This is the question of **scale**. How large, in its physical dimensions, should the economic system be relative to the ecosystem that sustains it? As soon as we ask this question, we imply that there is an optimal scale (and many believe we have already surpassed it) and hence a need to end growth. If we accept a need to end growth, we must also accept a need to address the distribution issue much more seriously.

The Link Between Sustainable Scale and Just Distribution

Distribution is the apportionment of resources among different individuals. Why does ending growth require us to focus on distribution?

First, it seems pretty likely that the negative impacts of our excessive resource use will be worse for future generations than for our own. Thus, concern with scale involves a concern for future generations, or intergenerational distribution. Yet some 1.2 billion people alive today live in abject poverty, while many others have so much wealth they scarcely know what to do with it. It would be a peculiar set of ethical beliefs that would have us care about generations not yet born while ignoring the plight of the miserable today.

> **THINK ABOUT IT!**
> *Why might excessive resource use have greater impacts on future generations than on the current one? Look back at the definition of Pareto efficient allocation. If the current generation is the de facto owner of all resources, could it be Pareto efficient for the current generation to consume fewer resources so that future generations are better off?*

Second, as long as the economy is growing, we can always offer to the poor the future prospect of a slice of a larger pie. We do not need to redistribute now, some argue, because concentrated capital feeds the capitalist system, and if the poor remain patient, their misery will soon be relieved. This is certainly a much more politically palatable option than redistribution, but as soon as we call for an end to growth, this option is gone. We certainly can't ask today's poor to sacrifice their hopes for a better future so that unborn generations will enjoy necessities of which they can only dream—especially when a reluctance to redistribute wealth today would suggest that the future generations for whom the poor are asked to sacrifice are likely to be someone else's children. Thus, distribution is of central importance to ecological economics.[20]

Neoclassical economics is concerned almost solely with efficient allocation. Ecological economics also considers efficient allocation important, but it is secondary to the issues of scale and distribution. As we will see, in fact an efficient allocation cannot even be theoretically determined without a prior resolution to the distribution and scale questions. Typically that resolution is to take the existing distribution and scale as given.

Fortunately, as McNeill reminds us, it is only since the Depression that the growth fetish has taken control of economics. And as readers of this book will learn, if they don't know it already, there is a lot in economics that is true and useful—that is independent of the growth ideology and that we could hardly do without. Indeed, as we shall show, the basic economic tools of optimization themselves provide the best means for arguing against the preoccupation with growth.

Why study economics, and in particular ecological economics? As we

---

[20]Chapter 1 in *Ecological Economics: A Workbook for Problem-Based Learning* that accompanies this text discusses scale, distribution, and allocation as desirable ends of economic activity, and Exercise 1.2 asks you to apply these concepts.

noted at the beginning of this chapter, economics is about what we want and what we have to give up to get it. Growth is one more thing we may want and like anything else, we have to give up something to get it. Ecological economists always ask if the extra growth is worth the extra sacrifice it entails. Neoclassical economists tend to forget this question or to believe that the answer is always affirmative.

BIG IDEAS to remember

- Ends and means
- Pareto efficient allocation
- Allocation, distribution, scale

- Growth versus development
- Throughput
- Coevolutionary economics

CHAPTER

2

The Fundamental Vision

■ THE WHOLE AND THE PART

Ecological economics shares many concepts with conventional neoclassical economics. For example, both take as basic the concept of **opportunity cost**, defined as the best alternative that has to be sacrificed when you choose to do something. But ecological economics has a fundamentally different starting point—a different vision at its core of the way the world really is. To put it starkly, conventional economics sees the economy, the entire macroeconomy, as the whole. To the extent that nature and the environment are considered at all, they are thought of as parts or sectors of the macroeconomy—forests, fisheries, grasslands, mines, wells, ecotourist sites, and so on. Ecological economics, by contrast, envisions the macroeconomy as part of a larger enveloping and sustaining Whole—namely, the Earth, its atmosphere, and its ecosystems. The economy is seen as an open subsystem of that larger "Earthsystem." That larger system is finite, nongrowing, and materially closed, although open to solar energy.

It is important to understand the distinctions between open, closed, and isolated systems. An **open system** takes in and gives out both matter and energy. The economy is such a system. A **closed system** imports and exports energy only; matter circulates within the system but does not flow through it. The Earth closely approximates a closed system. An **isolated system** is one in which neither matter nor energy enters or exits. It is hard to think of an example of an isolated system, except perhaps the universe as a whole. We say the Earth is approximately a closed system because it does not exchange significant amounts of matter with outer space—an occasional meteor comes in, an occasional rocket never returns, and we have a moon rock in a stained glass window in the National Cathedral. Maybe material exchanges will be greater someday, but so far they are negligible.

Opportunity cost is the best alternative given up when a choice is made. For example, if a farmer cuts down a forest to expand his cropland, and if the consequently lost wildlife habitat, water purification, and climate regulation would have been the next best "use" of the land, then the value of wildlife habitat, water purification, and climate regulation is the opportunity cost of the expanded cropland.

However, we do have a significant flow-through or throughput of energy in the form of incoming sunlight and exiting radiant heat. That through-put, like the ecosystem, is also finite and nongrowing. For the Earth, the basic rule is: Energy flows through, material cycles within.

Back to the problem of the whole and the part. Why is it so important? Because if the economy is the whole, then it can expand without limit. It does not displace anything and therefore incurs no opportunity cost—nothing is given up as a result of physical expansion of the macroeconomy into unoccupied space. But if the macroeconomy is a part, then its physical growth encroaches on other parts of the finite and nongrowing whole, exacting a sacrifice of something—an opportunity cost, as economists would call it. In this case, if we choose to expand the economy, the most important natural space or function sacrificed as a result of that expansion is the opportunity cost. The point is that growth has a cost. It is not free, as it would be if we were expanding into a void. The Earth-ecosystem is not a void; it is our sustaining, life-supporting envelope. It is therefore quite conceivable that at some point the further growth of the macro-economy could cost us more than it is worth. Such growth is known as **uneconomic growth**. This leads to another insight that is fundamental to ecological economics and distinguishes it from conventional economics: Growth can be uneconomic as well as economic. There is an optimal scale of the macroeconomy relative to the ecosystem.[1] How do we know we have not already reached or passed it?

■ Optimal Scale

The idea of **optimal scale** is not strange to standard economists. It is the very basis of microeconomics. As we increase any activity, be it producing shoes or eating ice cream, we also increase both the costs and the benefits of the activity. For reasons we will investigate later, it is generally the case that after some point, costs rise faster than benefits. Therefore, at some point the extra benefits of growth in the activity will not be worth the extra costs. In economist jargon, when the **marginal costs** (extra costs) equal the marginal benefits, then the activity has reached its optimal scale.[2] If we grow beyond the optimum, then costs will go up by more

---

[1]Beyond optimal scale, physical expansion becomes uneconomic growth, even if we mislead-ingly still call it "economic" growth. We use the word "economic" in two senses: (1) of or per-taining to the economy and (2) yielding net benefits above costs. If the entity we call "the economy" physically grows, then we call that economic growth in sense 1. But growth in sense 1 may be economic or uneconomic in sense 2. Our linguistic habit of using sense 1 often leads us to prejudge the issue in terms of sense 2.

[2]"Marginal" means the last unit, in this case, the last unit of something obtained, produced, or consumed. Marginal cost (benefit) is the cost (benefit) of a very small increase in some activity.

than benefits. Subsequently, growth will make us poorer rather than richer. The basic rule of microeconomics, that optimal scale is reached when marginal cost equals marginal benefit (MC = MB), has aptly been called the "when to stop rule"—that is, when to stop growing. In macroeconomics, curiously, there is no "when to stop rule," nor any concept of the optimal scale of the macroeconomy. The default rule is "grow forever." Indeed, why not grow forever if there is no opportunity cost of growth? And how can there be an opportunity cost to growth of the macroeconomy if it is the whole?

Even if one adopts the basic vision of ecological economics and considers the economy as a subsystem of the ecosystem, there still would be no need to stop growing as long as the subsystem is very small relative to the larger ecosystem. In this "empty-world vision," the environment is not scarce, and the opportunity cost to expansion of the economy is insignificant. But continued growth of the physical economy into a finite and nongrowing ecosystem will eventually lead to the "full-world economy" in which the opportunity cost of growth is significant. We are already in such a full-world economy, according to ecological economists.

This basic ecological economics vision is depicted in Figure 2.1. As growth moves us from the empty world to the full world, the welfare from economic services increases while the welfare from ecological services diminishes. For example, as we cut trees to make tables, we add the economic service of the table (holding our plates so we won't have to eat off the floor) and lose the ecological service of the tree in the forest (photosynthesis, securing soil against erosion, providing wildlife habitat, etc.). Traditionally, economists have defined capital as produced means of production, where "produced" implies "produced by humans." Ecological economists have broadened the definition of capital to include the means of production provided by nature. We define capital as a stock that yields a flow of goods and services into the future. Stocks of manmade capital include our bodies and minds, the artifacts we create, and our social structures. **Natural capital** is a stock that yields a flow of natural services and tangible natural resources. This includes solar energy, land, minerals and fossil fuels, water, living organisms, and the services provided by the interactions of all of these elements in ecological systems.

We have two general sources of welfare: services of manmade capital (dark gray stuff) and services of natural capital (light gray stuff), as represented by the thick arrows pointing to "Welfare" in Figure 2.1. Welfare is placed outside the circle because it is a psychic, not a physical, magnitude (an experience, not a thing). Within the circle, magnitudes are physical. If we object to having a nonphysical magnitude in our basic picture of the economy on the grounds that it is metaphysical and unscientific, then we will have to content ourselves with the view that the economic system is

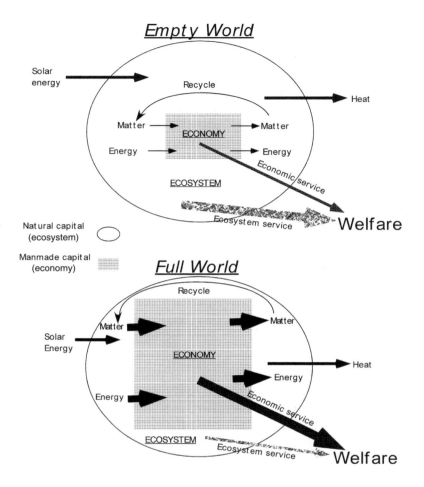

Figure 2.1 • From empty world to full world.

just an idiotic machine for turning resources into waste for no reason. The ultimate *physical* output of the economic process is degraded matter and energy—waste. Neglecting the biophysical basis of economics gives a false picture. But neglecting the psychic basis gives a meaningless picture. Without the concept of welfare or enjoyment of life, the conversion of material resources first into goods (production) and then into waste (consumption) must be seen as an end in itself—a pointless one. Both conventional and ecological economics accept the psychic basis of welfare, but they differ on the extent to which manmade and natural capital contribute to it.

■ DIMINISHING MARGINAL RETURNS AND UNECONOMIC GROWTH

As the economy grows, natural capital is physically transformed into manmade capital. More manmade capital results in a greater flow of services from that source. Reduced natural capital results in a smaller flow of services from that source. Moreover, as growth of the economy continues, the services from the economy grow at a decreasing rate. As rational beings, we satisfy our most pressing wants first, hence the law of diminishing marginal utility (to which we will return). As the economy encroaches more and more on the ecosystem, we must give up some ecosystem services. As rational beings, we presumably will sequence our encroachments so that we sacrifice the least important ecosystem services first. This is the best case, the goal. In actuality we fall short of it because we do not understand very well how the ecosystem works and have only recently begun to think of ecosystem services as scarce. But the consequence of such rational sequencing is a version of the law of increasing marginal cost (to which we will return): for each further unit of economic expansion beyond some threshold, we must give up a more important ecosystem service. Marginal costs increase while marginal benefits decrease. At some point increasing marginal costs will equal declining marginal benefits.

| Box 2-1 | MARGINAL UTILITY VS. MARGINAL COST |

- **Marginal utility:** The marginal utility of something is the additional benefit or satisfaction you derive from obtaining an additional unit of that thing. The **law of diminishing marginal utility** states that the more one has of something, the less satisfaction an additional unit provides. For example, the first slice of pizza on an empty stomach offers considerable satisfaction, but each additional slice provides less satisfaction than the previous one.
- **Marginal cost:** Marginal cost is the additional cost of producing one more unit. The **law of increasing marginal cost** is similar to that of diminishing marginal utility. For each additional ton of wheat harvested, you have to make use of inferior land and workers (you used the best first). Also, once you've used all the land for wheat, adding more labor, fertilizer, and so on is the only way to increase the wheat harvest. But with fixed land, we will have diminishing returns to the variable factors (labor, fertilizer)—more and more laborers and fertilizer will be needed for each additional ton of harvest. Diminishing returns is a further reason for increasing marginal costs. Neoclassical economics is constantly comparing increasing marginal costs with declining marginal benefits, looking for their point of intersection that

defines the optimal scale of each microeconomic activity. It does not apply this logic to the macroeconomy or recognize that it has an **optimal scale.** Ecological economics insists that the logic of optimal scale is relevant to the entire macroeconomy, as well as to its parts.

This first step in analyzing the core or preanalytic vision of ecological economics can be expressed graphically (Figure 2.2). The basic logic goes back to William Stanley Jevons (1871) and his analysis of labor supply in terms of balancing the marginal utility of wages with the marginal disutility of labor to the worker. Put another way, Jevons asked: When does the effort of working begin to exceed the value of the wage to the worker? Ecological economists ask: When does the cost to all of us of displacing the Earth's ecosystems begin to exceed the value of the extra wealth produced? In Figure 2.2, the marginal utility (MU) curve reflects the dimin-

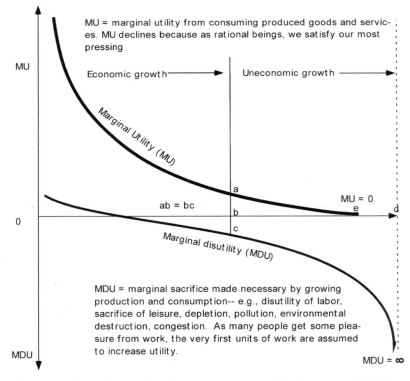

Figure 2.2 • Limits to growth of the macroeconomy. Point b = economic limit or optimal scale, where marginal utility (MU) = marginal disutility (MDU) (maximum net positive utility); e = futility limit, where MU = 0 (consumer satiation); d = catastrophe limit, where MDU = infinity. At point d, we have gone beyond sustainable scale.

ishing marginal utility of additions to the stock of manmade capital. The marginal disutility (MDU) curve reflects the increasing marginal cost of growth (sacrificed natural capital services and disutility of labor), as more natural capital is transformed into manmade capital. The optimal scale of the macroeconomy (economic limit to growth) is at point b, where MU = MDU, or where ab = bc, and net positive utility is a maximum (the area under the MU curve minus the area above the MDU curve).

Two further limits are noted: point e, where MU = 0 and further growth is futile even with zero cost; and point d, where an ecological catastrophe is provoked, driving MDU to infinity. For example, excessive CO_2 emissions from the fossil fuels used to power economic growth might destabilize the climate, reduce agricultural productivity, and cause billions of deaths. These outer limits need not occur in the order depicted. We could have an ecological disaster before reaching satiation. The diagram shows that growth out to point b is literally economic growth (benefiting us more than it costs), while growth beyond point b is literally uneconomic growth (costing us more than it benefits us). Beyond point b, GNP, "that which seems to be wealth," does indeed become "a gilded index of far-reaching ruin," as John Ruskin predicted over a century ago.[3] The nice thing about point b, the economic limit, is that it occurs first, allowing us to maximize net benefits while stopping us from destroying the capacity of the Earth to support life.

The concepts of optimal scale and uneconomic growth have a universal logic; they apply to the macroeconomy just as much as to microeconomic units.[4] How did we come to forget this in macroeconomics? How did we come to ignore the existence of the MDU curve and the issue of optimal scale of the macroeconomy? We suggest two possibilities. One is the "empty-world vision" that recognizes the concept of uneconomic growth but claims that we are not yet at that point; neoclassical economists tend to think that MU is still very large and MDU is still negligible. In this case we can look at the factual evidence to resolve the difference, as will be done later.

The other possibility for explaining the total neglect of the costs of growth is a paradigm difference: The economy is simply not seen as a subsystem of the ecosystem but rather the reverse—the ecosystem is a subsystem of the economy (Figure 2.3). Here we are discussing different

---

[3] J. Ruskin, *Unto This Last* (1862), in Lloyd J. Hubenka, ed., *Four Essays on the First Principles of Political Economy*, Lincoln: University of Nebraska Press, 1967.

[4] It is a mistake to think that microeconomics is about little things and macroeconomics is about big things. Microeconomics means the economics of the part, macroeconomics means the economics of the whole or aggregate. Parts can be big, aggregates can be small. Although MB = MC is a rule of microeconomic analysis, we can apply it to something big, the economic subsystem, as long as the big thing is a part, not the whole.

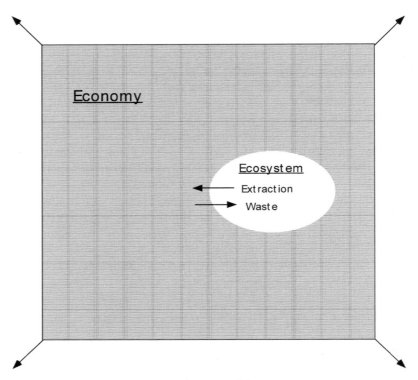

Figure 2.3 • The ecosystem as a subsystem of the economy.

conceptual worlds, and no empirical measurements will resolve the difference.

In the vision of Figure 2.3, the ecosystem is merely the extractive and waste disposal sector of the economy. Even if these services become scarce, growth can still continue forever because technology allows us to "grow around" the natural sector by substituting of manmade for natural capital, following the dictates of market prices. Nature is, in this view, nothing but a supplier of various indestructible building blocks, which are substitutable and superabundant. The only limit to growth, in this view, is technology, and since we can always develop new technologies, there is no limit to economic growth. The very notion of "uneconomic growth" makes no sense in that paradigm. Since the economy is the whole, the growth of the economy is not at the expense of anything else; there is no opportunity cost to growth. On the contrary, growth enlarges the total to be shared by the different sectors.[5] Growth does not increase

---

[5]A note of caution: The dark gray stuff in Figure 2.1 is in physical dimensions. The dark gray stuff in Figure 2.3 is probably thought of by neoclassical economists as GNP; it is in units of value and therefore not strictly physical. But value is price times quantity, and the latter has an irreducible physical component. Indeed it is mainly changes in that physical component that economists seek to measure in calculating *real* GNP—i.e., changes in GNP not due to changes in prices.

the scarcity of anything; rather, it diminishes the scarcity of everything. How can one possibly oppose growth?

■ A PARADIGM SHIFT

Where conventional economics espouses growth forever, ecological economics envisions a steady-state economy at optimal scale. Each is logical within its own preanalytic vision, and each is absurd from the viewpoint of the other. The difference could not be more basic, more elementary, or more irreconcilable.

In other words, ecological economics calls for a "paradigm shift" in the sense of philosopher Thomas Kuhn,[6] or what we have been calling, following economist Joseph Schumpeter,[7] a change in preanalytic vision. We need to pause to consider more precisely just what these concepts mean. Schumpeter observes that "analytic effort is of necessity preceded by a preanalytic cognitive act that supplies the raw material for the analytic effort" (p. 41). Schumpeter calls this preanalytic cognitive act "Vision." One might say that vision is the pattern or shape of the reality in question that the right hemisphere of the brain abstracts from experience and then sends to the left hemisphere for analysis. Whatever is omitted from the preanalytic vision cannot be recaptured by subsequent analysis. Correcting the vision requires a new preanalytic cognitive act, not further analysis of the old vision. Schumpeter notes that changes in vision "may reenter the history of every established science each time somebody teaches us to see things in a light of which the source is not to be found in the facts, methods, and results of the preexisting state of the science." (p. 41). It is this last point that is most emphasized by Kuhn (who was apparently unaware of Schumpeter's discussion).

Kuhn distinguished between "normal science," the day-to-day solving of puzzles within the established rules of the existing preanalytic vision, or "paradigm" as he called it, and "revolutionary science," the overthrow of the old paradigm by a new one. It is the common acceptance by scientists of the reigning paradigm that makes their work cumulative and that separates the community of serious scientists from quacks and charlatans. Scientists are right to resist scientific revolutions. Most puzzles or anomalies, after all, do eventually get solved, one way or another, within the existing paradigm. And it is unfortunate when people who are too lazy to master the existing scientific paradigm seek a shortcut to fame by summarily declaring a "paradigm shift" of which they are the leader. Nevertheless, as Kuhn demonstrates, paradigm shifts, both large and small, are

---

[6]T. Kuhn, *The Structure of Scientific Revolutions*, Chicago: University of Chicago Press, 1962.

[7]J. Schumpeter, *History of Economic Analysis*, New York: Oxford University Press, 1954.

undeniable episodes in the history of science—the shift from the Ptolemaic (Earth-centered) to the Copernican (sun-centered) view in astronomy and Newton's notions of absolute space and time versus Einstein's relativity of space and time are only the most famous. As Kuhn demonstrates, there does come a time when sensible loyalty to the existing paradigm becomes stubborn adherence to intellectual vested interests.

Paradigm shifts are obscured by textbooks whose pedagogical organization is, for good reason, logical rather than historical.[8] Physics students would certainly be unhappy if, after learning in the first three chapters all about the ether and its finely grained particles, they were suddenly told in Chapter 4 to forget all that stuff about the ether because we just had a Newtonian paradigm shift and now accept action at a distance unmediated by fine particles (gravity)!

Thirty years ago, a course in the history of economic thought was required in all graduate economics curricula. Today such a course is usually not even available as an elective. This is perhaps a measure of the (over)confidence economists have in the existing paradigm. Why study the errors of the past when we now know the truth? Consequently, the several changes in preanalytic vision in the history of economic thought are unknown to students and to many of their professors.

A change in vision from seeing the economy as the whole to seeing it as a part of the relevant Whole—the ecosystem—constitutes a major paradigm shift in economics. In subsequent chapters, we will consider more specific consequences of this shift.

The Circular Flow and the Linear Throughput

Differing preanalytic visions lead to a few basic analytical differences as well, although many tools of analysis remain the same between standard and ecological economics, as we'll discuss later.

Given that standard economics has a preanalytic vision of the economy as the whole, what is its first analytic step in studying this whole? It is depicted in Figure 2.4, the familiar circular flow diagram with which all basic economics texts begin. In this view, the economy has two parts: the production unit (firms) and the consuming unit (households). Firms produce and supply goods and services to households; households demand goods and services from firms. Firm supply and household demand meet in the goods market (lower loop), and prices are determined there by the interaction of supply and demand.

---

[8]Textbooks are designed to initiate the student into the reigning paradigm as efficiently as possible. Chapter 2 builds on Chapter 1, Chapter 3 builds on Chapters 1 and 2, etc. Efficient pedagogy is logical and cumulative. But the history of science is not so tidy. In history there are times when we have to throw out earlier chapters and start over. This textbook is not immune to this danger, although we have tried to be sensitive to it.

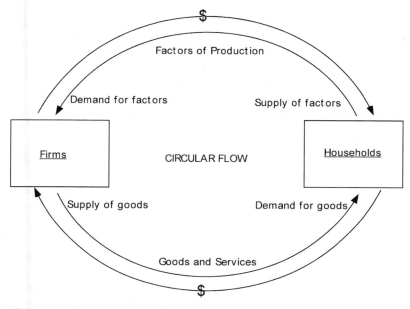

Figure 2.4 • The circular flow of the economy.

At the same time, firms demand factors of production from the households, and households supply factors to the firms (upper loop). Prices of factors (land, labor, capital) are determined by supply and demand in the factor market. These factor prices, multiplied by the amount of each factor owned by a household, determine the income of the household. The sum of all these factor incomes of all the households is National Income. Likewise, the sum of all goods and services produced by firms for households, multiplied by the price at which each is sold in the goods market, is equal to National Product. By accounting convention, National Product must equal National Income. This is so because profit, the value of total production minus the value of total factor costs, is counted as part of National Income.

THINK ABOUT IT!
Would the equality still hold if profits were negative? Explain.

The upper and lower loops are thus equal, and in combination they form the circular flow of exchange value. This is a very important vision. It unifies most of economics. It shows the fundamental relationship between production and consumption. It is the basis of microeconomics, which studies how the supply-and-demand plans of firms and households emerge from their goals of maximizing profits (firms) and maximizing utility (households). It shows how supply and demand interact under different market structures to determine prices and how price changes lead to changes in the allocation of factors to produce a different mix of goods and services. In addition, the circular flow diagram also provides the basis for macroeconomics—it shows how the aggregate behavior of firms and households determines both National Income and National Product.

■ Say's Law: Supply Creates Its Own Demand

The equality of National Income and National Product, as mentioned, guarantees that there is always enough purchasing power in the hands of households in the aggregate to purchase the aggregate production of firms. Of course, if some firms produce things households do not want, the prices of those things will fall, and if they fall below what it cost to produce them, those firms will make losses and go out of business. The circular flow does not guarantee that all firms will sell whatever they produce at a profit. But it does guarantee that such a result is not impossible because of an overall glut of production in excess of overall income. This comforting feature of the economy is known as **Say's Law**: supply creates its own demand. For a long time, economists believed Say's Law ruled out any possibility of long-term and substantial unemployment, as occurred during the Great Depression. However, the experience of the Depression led John Maynard Keynes to reconsider Say's Law and the comforting conclusion of the circular flow vision.

There may indeed always be enough income generated by production to purchase what is produced. But there is no guarantee that all the income will be spent, or spent in the current time period, or spent on goods and services, or spent in the national market. In other words, there are leakages out of the circular flow. There are also corresponding injections into the circular flow. But there is no guarantee that the leakages and injections will balance each other.

■ Leakages and Injections

What are these leakages and injections? One leakage from the expenditure stream is savings. People refrain from spending now in order to be able to spend later. The corresponding injection is investment. Investment results in expenditure now but increased production only in the future. Thus, the circular flow can be restored if saving equals investment. This recycling of savings into investment is accomplished through financial markets and interest rates. In Figure 2.5, the shaded rectangles represent the financial institutions that collect savings and lend to investors.

A second leakage from the circular flow is payment of taxes. The corresponding injection is government expenditure. The rectangle represents the institutions of public finance. Public finance policies can balance the taxes and government spending, or intentionally unbalance them to compensate for imbalances in savings and investment. For example, if saving exceeds investment, the government might avoid a recession by allowing government expenditures to exceed taxes by the same amount.

The third leakage from the national circular flow is expenditure on imports. The corresponding injection is expenditure by foreigners for our exports. International finance and foreign exchange rates are mechanisms for balancing exports and imports. Again the corresponding institutions are represented by the rectangle. The circular flow is restored if the sum of leakages equals the sum of injections, that is, if savings plus taxes plus imports equals investment plus government spending plus exports. If the sum of leakages is greater than the sum of injections, unemployment or deflation tends to result. If the sum of injections is greater than the sum of leakages, we tend to have either expansion or inflation.

Leakages and injections are shown in Figure 2.5, an expanded circular

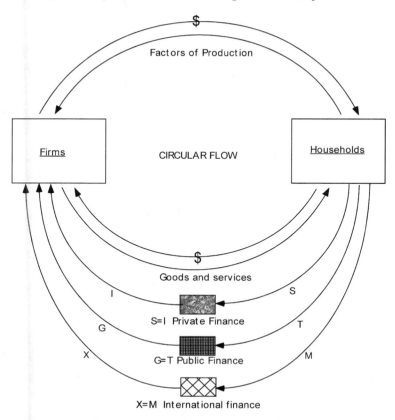

Figure 2.5 • The circular flow with leakages and injections. S = savings, I = investment, G = government expenditure, T = taxes, X = exports, M = imports.

flow diagram. For simplicity we have assumed that households are net savers, net taxpayers, and net importers, while firms are net investors, net recipients of government expenditure, and net exporters.

The circular flow diagram not only unites micro- and macroeconomics but also shows the basis for monetary, fiscal, and exchange rate policy in the service of maintaining the circular flow so as to avoid unemployment and inflation. With so much to its credit, how could one possibly find fault with the circular flow vision?

There is no denying the usefulness of the circular flow model for analyzing the flow of exchange value. However, it has glaring difficulty as a description of a real economy. Notice that the economy is viewed as an isolated system. Nothing enters from outside the system; nothing exits the system to the outside. But what about all the leakages and injections just discussed? They are just expansions of the isolated system that admittedly make the concept more useful, but they do not change the fact that nothing enters from outside and nothing exits to the outside. The whole idea of analyzing leakages and injections is to be able to reconnect them and close the system again. Why is the isolated system a problem? Because an isolated system has no outside, no environment. This is certainly consistent with the view that the economy is the whole. But a consequence is that there is no place from which anything can come or to which it might go. If our preanalytic vision is that the economy is the whole, then we cannot possibly analyze any relation of the economy to its environment. The whole has no environment.

What is it that is really flowing around and around in a circle in the circular flow vision? Is it really physical goods and services, and physical laborers and land and resources? No. It is only abstract exchange value, the purchasing power represented by these physical things.[9] The "soul" embodied in goods by the firms is abstract exchange value. When goods arrive at the households, the "soul" of exchange value jumps out of its embodiment in goods and takes on the body of factors for its return trip to the firms, whereupon it jumps out of the body of factors and reincorporates itself once again into goods, and so on. But what happens to all the discarded bodies of goods and factors as the soul of exchange value transmigrates from firms to households and back *ad infinitum*? Does the

---

[9]We are careful to say "abstract exchange value" rather than "money" because not even money in the sense of currency can circulate as an isolated system. Money wears out and has to be replaced by new money. The physical wear and tear of hand-to-hand circulation means that even money has to have a throughput to maintain its circulation. Because fractional money circulates more rapidly than notes of higher denomination, we usually adopt metal coins rather than paper to withstand the higher velocity of circulation of small denominations. For this reason, as inflation has eroded its value, the U.S. Treasury has periodically attempted to issue the dollar in coin form, though without much success.

system generate wastes? Does the system need *new* inputs of matter and energy? If not, then the system is a perpetual motion machine, a contradiction to the Second Law of Thermodynamics (about which more later). If it is not to be a perpetual motion machine (a perfect recycler of matter and energy), then wastes must go somewhere and new resources must come from somewhere outside the system. Since there is no such thing as perpetual motion, the economic system cannot be the whole. It must be a subsystem of a larger system, the Earth-ecosystem.

The circular flow model is in many ways enlightening, but like all abstractions, it illuminates only what it has abstracted out of reality and leaves in darkness all that has been abstracted from. What has been abstracted from, left behind, in the circular flow model is the linear throughput of matter-energy by which the economy lives off its environment. Linear throughput is the flow of raw materials and energy from the global ecosystem's sources of low entropy (mines, wells, fisheries, croplands), through the economy, and back to the global ecosystem's sinks for high-entropy wastes (atmosphere, oceans, dumps). The circular flow vision is analogous to a biologist describing an animal only in terms of its circulatory system, without ever mentioning its digestive tract. Surely the circulatory system is important, but unless the animal also has a digestive tract that connects it to its environment at both ends, it will soon die of starvation or constipation. Animals live from a metabolic flow—an entropic throughput from and back to their environment. The **law of entropy** states that energy and matter in the universe move inexorably toward a less ordered (less useful) state. An entropic flow is simply a flow in which matter and energy become less useful; for example, an animal eats food and secretes waste and cannot ingest its own waste products. The same is true for economies. Biologists, in studying the circulatory system, have not forgotten the digestive tract. Economists, in focusing on the circular flow of exchange value, have entirely ignored the metabolic throughput. This is because economists have assumed that the economy is the whole, while biologists have never imagined that an animal was the whole or was a perpetual motion machine.

■ LINEAR THROUGHPUT AND THERMODYNAMICS

The linear throughput is in physical units and is strictly subject to the laws of conservation of mass and energy and the law of entropy. The circular flow is in units of abstract exchange value and is not subject to any obvious physical limits. The circular flow can nominally grow forever by virtue of inflation, but we set this case aside to ask if the real economic value in the sense of satisfying wants, of qualitative development, can grow forever.

The Fallacy of Misplaced Concreteness

Obviously, a model that abstracts from the environment and considers the economy in isolation from it cannot shed any light on the relation of the economy to the environment. This kind of mistake was given a name by philosopher and mathematician Alfred North Whitehead. He called it the **fallacy of misplaced concreteness**. By that he meant the error of mistaking the map for the territory, the error of treating an abstract model, made with the purpose of understanding one aspect of reality, as if it were adequate for understanding everything, or entirely different things, things that had been abstracted from in making the model. Whitehead was no enemy of abstract thought. He emphasized that we cannot think without abstraction. All the more important, therefore, to be aware of the limits of our abstractions. The power of abstract thought comes at a cost. The fallacy of misplaced concreteness is to forget that cost.

Let's take a closer look at what standard economists have abstracted from in the circular flow model—namely, the throughput, the metabolic flow from raw material inputs to waste outputs. The throughput is in physical units. Consequently, the laws of physics apply strictly to it.

By the **First Law of Thermodynamics**, the conservation of matter and energy, we know that throughput is subject to a balance equation: Input

> The *First Law of Thermodynamics* states that neither matter nor energy can be created or destroyed.

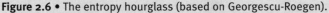

BOX 2-2 | THE LAWS OF THERMODYNAMICS

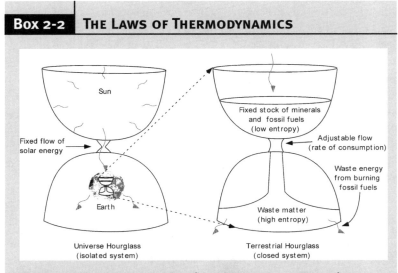

Figure 2.6 • The entropy hourglass (based on Georgescu-Roegen).

The hourglass on the left is an isolated system; no sand enters, no sand exits. Also, within the hourglass there is neither creation nor destruction of sand; the amount of sand in the hourglass is constant. This, of course, is the analog of the First Law of Thermodynamics, the conservation of matter and energy. Finally, there is a continual running down of sand in the top chamber and an accumulation of sand in the bottom chamber. Sand in the bottom chamber has used up its potential to fall and thereby do work. It is high-entropy or unavailable matter-energy. Sand in the top chamber still has potential to fall; it is low-entropy or available matter-energy (still useful). This is the analogy of the Second Law of Thermodynamics: Entropy, or "used-up-ness," increases in an isolated system. The hourglass analogy is particularly apt because entropy is "time's arrow" in the physical world—that is, we can tell earlier from later by whether or not entropy has increased. However, unlike a real hourglass, the entropy hourglass cannot be turned upside down!

With a bit of license, we can extend the basic analogy by considering the sand in the upper chamber to be the stock of low-entropy fossil fuel on Earth, depicted in the right-hand figure. Fossil energy is used at a rate determined by the constricted middle of the hourglass, but unlike a normal hourglass, humans alter the width (i.e., they change the rate of consumption of fossil fuels). Once consumed, the sand falls to the bottom of the chamber, where it accumulates as waste and interferes with terrestrial life processes.

To represent solar energy, the top of the hourglass on the left would be vast (from the human perspective), as would the bottom; solar energy, too, ends as waste heat, but it is not confined to the Earth. It does not disappear, but it radiates into outer space, and unlike waste matter does not accumulate on Earth. The constricted middle, however, would be quite small, and humans would be unable to adjust it. The solar source of low entropy is stock-abundant but flow-limited. In other words, there is a lot of it, but we get only a little at a time. The terrestrial source is stock-limited but flow-abundant, until the stock runs out. The asymmetry is important. With industrialization we have come to depend more and more on the least abundant source of low entropy. However convenient in the short run, this will be uneconomic in the long run.

equals output plus accumulation. If there is accumulation, the economic subsystem is growing. In steady-state equilibrium, growth and accumulation would be zero, and input flow would equal output flow. In other words, all raw material inputs eventually become waste outputs. The throughput has two ends: depletion of environmental sources and pollution of environmental sinks. Ignoring throughput is the same as ignoring depletion and pollution. Unlike exchange value, the flow of throughput is not circular; it is a one-way flow from low-entropy sources to high-entropy sinks. This is a consequence of the **Second Law of Thermodynamics**, the entropy law. We can recycle materials, but never 100%; recycling is a circular eddy in the overall one-way flow of the river. Energy, by the entropy law, is not recyclable at all. More precisely, it is recyclable, but

The *Second Law of Thermodynamics* states that entropy never decreases in an isolated system. Although matter and energy are constant in quantity (First Law), they change in quality. The measure of quality is entropy, and basically it is a physical measure of the degree of "used-up-ness" or randomization of the structure or capacity of matter or energy to be useful to us. Entropy increases in an isolated system. We assume the universe is an isolated system, so the Second Law says that the natural, default tendency of the universe is "shuffling" rather than "sorting." In everyday terms, left to themselves, things tend to get mixed up and scattered. Sorting does not occur by itself.

it always takes more energy to do the recycling than the amount that can be recycled. Thus, recycling energy is not physically impossible but always economically a loser—regardless of the price of energy. No animal can directly recycle its own waste products as its own food. If it could, it would be a perpetual motion machine. In strict analogy, no economy can function by directly reusing only its own waste products as raw materials.

The circular flow diagram gives the false impression that the economy is capable of direct reuse. Some very good textbook writers have explicitly affirmed this false impression. For example, Heilbroner and Thurow,[10] in a standard economics text, tell us that "the flow of output is circular, self-renewing, self-feeding." In other words, the economy is a perpetual motion machine. To drive the point home, the first study question at the end of the chapter is, "Explain how the circularity of the economic process means that the outputs of the system are returned as fresh inputs." It would have been reasonable to ask how dollars spent reappear as dollars earned in the circular flow of exchange value and how purchasing power is regenerated in the act of production. But explaining how outputs are returned as inputs, indeed *fresh* inputs, requires the student to discover the secret of perpetual motion! Of course, the authors do not really believe in perpetual motion; they were trying to get across to the student the importance of replenishment—how the economic process reproduces itself and keeps going for another round. Certainly this is an important idea to stress, but the key to understanding it is precisely that replenishment must come from *outside* the economic system. This is a point conventional economists tend to neglect, and it leads to the mistaking of the part for the whole. If the economy is the whole, it has no outside; it is an isolated system.

The error in the text cited is fundamental but not unique. It is representative of most standard texts. Heilbroner and Thurow have the virtue of clear expression—a virtue that makes it easier to spot errors. Other texts leave the student with the same erroneous impression but without forthrightly stating the implication in words that cause us to think again. Nor is the error confined to standard economists. Karl Marx's models of simple and expanded reproduction are also isolated circular flow models. Marx, with his theory that labor was the source of all value, was even more eager than standard economists to deny any important role to nature in the functioning of the economy and creation of value. For Marx, the idea that nature embodied scarcity was an abomination. All poverty was the result of unjust social relations, or class exploitation, not the "niggardliness of nature." Thomas Malthus had argued that overpopulation relative to

---

[10]R. Heilbroner and L. Thurow, *The Economic Problem*, New York: Prentice-Hall, 1981, pp. 127, 135.

natural capacities was also an independent cause of poverty and that social revolution could not eliminate poverty. Marx felt that Malthus' ideas
were a threat to his and treated him with contempt and vituperation. Political debates between neo-Marxists and neo-Malthusians continue to this
day.[11]

The Importance of Throughput

Let's turn from the theoretical reasons for the importance of throughput to
an empirical look at its size and composition in modern economies. The
following paragraph is from a book about the dependence of the economy
on the environment:

> *Researchers have calculated that industry moves, mines, extracts, shovels,*
> *burns, wastes, pumps, and disposes of 4 million pounds of material in order to*
> *provide one average middle-class American family's needs for a year. In 1990,*
> *the average American's economic and personal activities mobilized a flow of*
> *roughly 123 dry-weight pounds of material per day—equivalent to a quarter*
> *of a billion semitrailer loads per year. This amounts to 47 pounds of fuel, 46*
> *of construction materials, 15 of farmland, 6 of forest products, 6 of industrial*
> *minerals, and 3 of metals of which 90% is iron and steel. Net of 6 pounds of*
> *recycled materials, that Average American's daily activities emitted 130*
> *pounds of gaseous material into the air, created 45 pounds of material arti-*
> *facts, generated 13 pounds of concentrated wastes, and dissipated 3.5 pounds*
> *of nongaseous wastes into the environment in such scattered forms as pesti-*
> *cides, fertilizers, and crumbs of material rubbed off tires. In addition, the per-*
> *son's daily activities required the consumption of about 2,000 pounds of water*
> *that after use is sufficiently contaminated that it cannot be reintroduced into*
> *marine or riparian systems, and produced 370 pounds of rock, tailings, over-*
> *burden, and toxic water as a result of extracting oil, gas, coal, and minerals.*
> *. . . In sum, Americans waste or cause to be wasted nearly 1 million pounds*
> *of materials per person per year.[12]*

That's a lot of throughput to abstract from—to leave out of our model!
It all ends up as waste, but necessary waste to support our population at
our standard of consumption, with our present technology. Better technologies, as well as a better ordering of our priorities, can reduce the
throughput without lowering the quality of life. However, by how much,
and by what policies, are big issues in ecological economics.

In 1997, a coalition including the World Resources Institute (WRI,
U.S.), the Wuppertal Institute (Germany), the Netherlands Ministry for

---

[11]See H. Daly, A Marxian-Malthusian View of Poverty and Exploitation, *Population Studies*,
May 1971.

[12]P. Hawken, A. Lovins, and H. Lovins, *Natural Capitalism*, Boston: Little, Brown, 1999, pp.
51–52.

Housing, Spatial Planning and the Environment, and the National Institute for Environmental Studies (Japan) attempted to measure throughput in each of their industrial countries for the period 1975–1993.[13] Their basic finding was that total material requirements (per-capita annual flows) for each of the four countries did not change much between 1975 and 1993. The range was 45–85 metric tons of natural resources per person per year, with the U.S. at the high end, Japan at the low end, and Germany and the Netherlands in between. Over the period, the U.S. flows declined slightly and those of the other countries rose slightly. Most of the decline in U.S. requirements was accounted for by better soil erosion control, not better industrial efficiency. The roughly constant total resource requirements over time are the product of a declining resource requirement per dollar of gross domestic product (GDP) with a growing number of dollars of GDP, in all four countries. We have become more *efficient* but not more *frugal*. It is as if we developed cars that got twice as many miles per gallon and then drove twice as many miles, thereby burning the same number of gallons.

> **THINK ABOUT IT!**
> *Which goal do you think should come first, efficiency or frugality? We will come back to this, but maybe you can answer it already.*

While it is important to have empirical information on the physical size, composition, and change over time in the throughput, we also must have some basis for judging the environmental costs of these flows. How large are they relative to the capacity of the ecosystem to absorb and regenerate them? Exactly what opportunity costs do these flows inflict on us? On other species? Partial answers are given by the World Wildlife Fund (WWF):[14]

> *While the state of the Earth's natural ecosystems has declined by about 33 per cent over the last 30 years ("Living Planet Index"), the ecological pressure of humanity on the Earth has increased by about 50 per cent over the same period ("World Ecological Footprint"), and exceeds the biosphere's regeneration rate.*

In terms of our "economy as subsystem" diagram (Figure 2.1), this means that the capacity of natural capital (light gray stuff) to supply life-support services has declined by about 33% and that the demand generated by manmade capital (dark gray stuff) for life-support services, provided by the light gray stuff, has increased by about 50%—and this has occurred over the past 30 years. There are two blades to this scissors:

---

[13]*Resource Flows: The Material Basis of Industrial Economies*, Washington, DC: WRI, 1997.

[14]World Wildlife Fund, UNEP, *Living Planet Report 2000*, Gland, Switzerland: WWF International, 2000, p. 1.

increasing demand for carrying capacity (ecological footprint) and decreasing supply of carrying capacity (living planet index). Both blades of the scissors are being squeezed by the same hand—namely, growth. The "ecological footprint" is the number of hectares of productive land or sea needed to support one average person at the world average consumption level. The study estimates that as of 1997 the ecological footprint of the Earth's total population was at least 30% higher than the Earth's biological reproductive capacity. This deficit is made up by consuming or drawing down natural capital, thus "borrowing from" or perhaps "robbing" the future. Scholars may have statistical arguments over the best measures of carrying capacity demanded and supplied, but the basic qualitative conclusion of unsustainable trends is hard to deny.

BIG IDEAS to remember

- Whole and part
- Open, closed, and isolated systems
- Optimal scale
- Full world versus empty world
- Diminishing marginal utility
- Increasing marginal costs
- Paradigm and preanalytic vision
- Circular flow
- Linear throughput
- Say's Law
- Leakages and injections
- Laws of thermodynamics
- Fallacy of misplaced concreteness
- Entropy hourglass
- Measures of throughput volume

CHAPTER

3

Ends, Means, and Policy

■ ENDS AND MEANS: A PRACTICAL DUALISM

Ecological economics has at least as much in common with standard economics as it has differences. One important common feature is the basic definition of economics as the study of the allocation of scarce means among competing ends (though we will explain in later chapters why focusing on scarce resources is necessary but not sufficient). There are disagreements about what is scarce and what is not, what are appropriate mechanisms for allocating different resources (means), and how we rank competing ends in order of importance—but there is no dispute that using means efficiently in the service of ends is the subject matter of economics. Using means in the service of ends implies policy. Alternatively, policy implies knowledge of ends and means. Economics, especially ecological economics, is inescapably about policy, although the rarefied levels of abstraction sometimes reached by economists may lead us to think otherwise.

THINK ABOUT IT!

Economic anthropologist Karl Polanyi states, "The substantive meaning of economics derives from man's dependence for his living upon nature and his fellows. It refers to the interchange with his natural and social environment, in so far as this results in supplying him with the means of material want satisfaction."

Does this contradict in any way the definition we offer? In what important ways does it add to our definition?

If economics is the study of the allocation of scarce means in the service of competing ends, we have to think rather deeply about the nature of ends and means. Also, policy presupposes knowledge of two kinds: of possibility and purpose and of means and ends. Possibility reflects how the world works. In addition to keeping us from wasting time and money on impossibilities, this kind of knowledge gives us information about tradeoffs between real alternatives. Purpose reflects desirability, our ranking of ends, our criteria for distinguishing better from worse states of the world. It does not help much to know how the world works if we cannot distinguish better from worse states of the world. Nor is it useful to pursue a better state of the world that happens to be impossible. Without both kinds of knowledge, policy discussion is meaningless.[1]

To relate this to economic policy, we need to consider two questions. First, in the realm of possibility, the question is: What are the means at our disposal? Of what does our ultimate means consist? By "ultimate means" we mean a common denominator of possibility or usefulness that we can only use up and not produce, for which we are totally dependent on the natural environment. Second, what ultimately is the end or highest purpose in whose service we should employ these means? These are very large questions, and we cannot answer them completely, especially the latter. But it is essential to raise the questions. There are some things, however, that we say by way of partial answers, and it is important to say them.

Means

Ultimate means, the common denominator of all usefulness, consist of low-entropy matter-energy.[2] Low-entropy matter-energy is the physical coordinate of usefulness, the basic necessity that humans must use up but cannot create and for which the human economy is totally dependent on nature to supply. Entropy is the qualitative physical difference that distinguishes useful resources from an equal quantity of useless waste. We do not use up matter and energy per se (First Law of Thermodynamics), but we do irrevocably use up the quality of usefulness as we transform matter and energy to achieve our purposes (Second Law of Thermodynamics). All technological transformations entail a before and after, a gradient or metabolic flow from concentrated source to dispersed sink, from high to low temperature.[3] The capacity for entropic transformations of matter-energy to be useful is therefore reduced both by the emptying of finite sources and by the filling up of finite sinks. If there were no entropic gradient between source and sink, the environment would be incapable of

---

[1]Chapters 1 of *Ecological Economics: A Workbook for Problem-Based Learning* that accompanies this text asks you to consider the desirable ends for solutions to specific problems, and Chapter 2 asks you to think about the means available to achieve those ends.

[2]By "matter-energy" we mean just matter and energy, but with the recognition that they are convertible according to Einstein's famous formula, $E = mc^2$.

[3]For a scholarly development of this theme, see N. Georgescu-Roegen, *The Entropy Law and the Economic Process*, Cambridge, MA: Harvard University Press, 1971.

serving our purposes or even sustaining our lives. Technical knowledge helps us use low entropy more efficiently; it does not enable us to eliminate or reverse the direction of the metabolic flow.

Matter can of course be recycled from sink back to source by using more energy (and more material implements) to carry out the recycling. Energy can be recycled only by expending more energy to carry out the recycling than the amount recycled, so it is never economic to recycle energy—regardless of price. Recycling also requires material implements for collection, concentration, and transportation. The machines used to collect, concentrate, and transport will themselves wear out through a process of **entropic dissipation**, the gradual erosion and dispersion of their material components into the environment in a one-way flow of low-entropy usefulness to high-entropy waste. Any recycling process must be efficient enough to replace the material lost to this process. Nature's bio-geochemical cycles powered by the sun can recycle matter to a high degree—some think 100%. But this only underlines our dependence on nature's services, since in the human economy we have no source equivalent to the sun, and our finite sinks fill up because we are incapable of anything near 100% material recycling.

Information: The Ultimate Resource? There is a strong tendency to deny our dependence on nature to achieve our purposes. Among the more explicit denials is that from George Gilder:[4]

> Gone is the view of a thermodynamic world economy, dominated by "natural resources" being turned to entropy and waste by human extraction and use. . . . The key fact of knowledge is that it is anti-entropic: it accumulates and compounds as it is used. . . . Conquering the microcosm, the mind transcends every entropic trap and overthrows matter itself.

According to *The Economist*, George Gilder is "America's foremost technology prophet" whose recommendation can cause the share price of a company to increase by 50 percent the next day.[5] If Gilder is really that influential, it simply proves that stock prices are often based on erroneous information and irrational expectations. To cast further doubt on Gilder's Gnostic[6] prophecy, one need only recall the aphorisms of Nobel chemist

---

[4]G. Gilder, *Microcosm: The Quantum Revolution in Economics and Technology*, New York: Simon & Schuster, 1989, p. 378. Similar views are expressed by the late Julian Simon in *The Ultimate Resource*, Princeton, NJ: Princeton University Press, 1981. Recently Peter Huber has continued the tradition in *Hard Green: Saving the Environment from the Environmentalists*, New York: Basic Books, 2000.

[5]March 25, 2000, p. 73.

[6]Gnosticism was an early Christian heresy teaching that salvation was available only to those who through esoteric spiritual knowledge could transcend matter and who believed that Christ was noncorporeal, hence any view that denigrates the material world and sees knowledge as an escape therefrom.

Frederick Soddy, "No phosphorus, no thought,"[7] and of Loren Eisley, "The human mind . . . burns by the power of a leaf." As Kenneth Boulding— one of the pioneers of ecological economics—pointed out, knowledge has to be imprinted on physical structures in the form of improbable arrangements of matter before it is effective in the economy. And low entropy is the quality of matter-energy that increases its capacity to receive and retain the improbable imprint of human knowledge. For example, to receive the imprint, a typical computer microelectronics plant producing 5000 wafers per day generates some 5 million liters of organic and aqueous solvent waste (i.e., high entropy) per year,[8] in addition to raw materials and energy used. With regard to retaining the imprint, recent estimates suggest that the information technology (IT) economy in the U.S. currently consumes 13% of the electricity we use as a nation, and this level is increasing rapidly.[9]

Furthermore, as important as knowledge is, it is misleading to say it grows by compounding accumulation. New dollars from compound interest paid into a bank account are not offset by any decline in old dollars, that is, the principal. Yet new knowledge often renders old knowledge obsolete, as we saw in our discussion of scientific revolutions and paradigm shifts. Do scientific theories of phlogiston[10] and the ether[11] still count as knowledge? And when knowledge becomes obsolete, the artifacts that embody that knowledge become obsolete as well. Again the IT economy is the best example. According to the US EPA, Americans purchased some 65 million computers and monitors loaded with toxic materials in 2007 and stored or disposed of 72 million. This is just part of the 1322 tons of toxin-laden computer products that reached their end of life that year.[12] For every three computers that enter the market, two become obsolete. The corollary to Moore's law—that computer speed will double every 18 months while prices will fall—is that brand-new IT devices are never far

---

[7]Phosphorus is an essential component of chlorophyll, which is needed for photosynthesis, which in turn is needed for life, which is needed for thought.

[8]J.M. Desimone, Practical Approaches to Green Solvents, *Science* 297 (5582) (2002).

[9]M.P. Mills, "Kyoto and the Internet: The Energy Implications of the Digital Economy," Testimony of Mark P. Mills, Science Advisor, The Greening Earth Society, Senior Fellow, The Competitive Enterprise Institute, President, Mills McCarthy & Associates, before the Subcommittee on National Economic Growth, Natural Resources, and Regulatory Affairs, U.S. House of Representatives, 2000. Globally, Internet servers doubled their energy use between 2000 and 2005 (Koomey, Jonathan. 2007. *Estimating Total Power Consumption by Servers in the U.S. and the World.* Oakland, CA: Analytics Press).

[10]Phlogiston is a hypothetical substance formerly thought to be a volatile constituent of all combustible material, released as flame in combustion.

[11]Ether is an all-pervading, infinitely elastic, massless medium once thought to fill the upper regions of space, as air fills the lower regions. Its hypothetical existence avoided confronting the mystery of "action at a distance," later recognized in the concept of gravity.

[12]*Electronics Waste Management in the United States: Approach One.* Final Report, US EPA, July 2008.

from becoming electronic waste. This is hardly antientropic. Physicists will not be surprised, because they have never found anything that is antientropic.

As E. J. Mishan noted, technological knowledge often unrolls the carpet of increased choice before us by the foot while simultaneously rolling it up behind us by the yard.[13] Yes, knowledge develops and improves, but it does not grow exponentially like money compounding in the bank. Furthermore, new knowledge need not always reveal new possibilities for growth; it can also bring serious harm and reveal new limitations. The new knowledge of the fire-resisting properties of asbestos increased its usefulness; subsequent new knowledge of its carcinogenic properties reduced its usefulness. New knowledge can cut both ways. Finally, and most obviously, knowledge has to be actively learned and taught every generation—it cannot be passively bequeathed like an accumulating stock portfolio. When society invests little in the transfer of knowledge to the next generation, some of it is lost, and its distribution often becomes more concentrated, contributing to the growing inequality in the distribution of income, as well as to a general dumbing-down of the future.

It is a gross prejudice to think that the future will always know more than the past. Every new generation is born totally ignorant, and just as we are always only one failed harvest away from starvation, we are also always only one failed generational transfer of knowledge away from darkest ignorance. Although it is true that today many people know many things that no one knew in the past, it is also true that large segments of the present generation are more ignorant than were large segments of past generations. The level of policy in a democracy cannot rise above the average level of understanding of the population. In a democracy, the distribution of knowledge is as important as the distribution of wealth.

Waste as a Resource? The common view among economists and many others is that waste is just a resource we have not yet learned to use, that nature supplies only the indestructible building blocks of elemental atoms, and that all the rest either is or can be done by humans. What counts to economists is value added by human labor and capital—that to which value is added is thought to be totally passive stuff, not even worthy of the name *natural resources*, as evidenced by Gilder's putting the term in quotation marks. Natural processes, in this view, do not add value to the elemental building blocks—and even if they did, manmade capital is thought to substitute for such natural services.

The brute facts remain, however, that we can only get so much energy from a lump of coal, we cannot burn the same lump twice, and the resulting ashes and heat scattered into nature's sinks really are polluting

---

[13] E. J. Mishan, *The Costs of Economic Growth*, New York: Praeger, 1967.

wastes and not just matter-energy of equally useful potential, if only we knew how to use it. Eroded topsoil washed to the sea and chlorofluorocarbons in the ozone layer are also polluting wastes on a human time scale, not just "resources out of place." No one denies the enormous importance of knowledge.[14] But this denigration of the importance of the physical world, and exclusive emphasis on knowledge as our ultimate resource, seems to be a modern version of Gnosticism. It appears to be religiously motivated by a denial of our creaturehood as part of the material world, by the belief that we have, or soon will have, transcended the world of material creation and entered an unlimited realm of esoteric knowledge, albeit technical now rather than spiritual. Thus, even in the discussion of means we are pushed out of the purely biophysical realm to consider alternative religious philosophies, including most prominently the revival of the ancient Christian heresy of Gnosticism.

Ends

We argued earlier that there is such a thing as ultimate means and that it is low-entropy matter-energy. Is there such a thing as an **ultimate end**, and if so, what is it? Following Aristotle, we think there are good reasons to believe that there must be an ultimate end, but it is far more difficult to say just what it is. In fact we will argue that, while we must be dogmatic about the existence of the ultimate end, we must be very humble and tolerant about our hazy and differing perceptions of what it looks like.

In an age of pluralism, the first objection to the idea of ultimate end is that it is singular. Do we not have many ultimate ends? Clearly we have many ends, but just as clearly they conflict and we must choose between them. We rank ends. We prioritize. In setting priorities, in ranking things, something—only one thing—has to go in first place. That is our practical approximation to the ultimate end. What goes in second place is determined by how close it came to first place, and so on. Ethics is the problem of ranking plural ends or values. The ranking criterion, the holder of first place, is the ultimate end (or its operational approximation), which grounds our understanding of objective value—better and worse as real states of the world, not just subjective opinions.

We do not claim that the ethical ranking of plural ends is necessarily done abstractly, *a priori*. Often the struggle with concrete problems and policy dilemmas forces decisions, and the discipline of the concrete decision helps us implicitly rank ends whose ordering would have been too obscure in the abstract. Sometimes we have regrets and discover that our

---

[14]For interesting discussions of the limitations of knowledge, see P. R. Ehrlich et al., Knowledge and the Environment, *Ecological Economics* 30:267–284 (1999), and M. H. Huesemann, Can Pollution Problems Be Effectively Solved by Environmental Science and Technology? An Analysis of Critical Limitations, *Ecological Economics* 37:271–287 (2001).

ranking really was not in accordance with a subsequently improved understanding of the ultimate end. Like scientific theories, desirable ends should also be subjected to empirical testing and falsification.

Neoclassical economists reduce value to the level of individual tastes or preferences, about which it is senseless to argue. But this apparent tolerance has some nasty consequences. Our point is that we must have a dogmatic belief in objective value, an objective hierarchy of ends ordered with reference to some concept of the ultimate end, however dimly we may perceive the latter. This sounds rather absolutist and intolerant to modern devotees of pluralism, but a little reflection will show that it is the very basis for tolerance. If A and B disagree regarding the hierarchy of values, and they believe that objective value does not exist, then there is nothing for either of them to appeal to in an effort to persuade the other. It is simply A's subjective values versus B's. B can vigorously assert her preferences and try to intimidate A into going along, but A will soon get wise to that. They are left to resort to physical combat or deception or manipulation, with no possibility of truly reasoning together in search of a clearer shared vision of objective value because, by assumption, the latter does not exist. Each knows his own subjective preferences better than the other, so no "values clarification" is needed. If the source of value is in one's own subjective preferences, then one does not really care about the other's preferences, except as they may serve as means to satisfying one's own. Any talk of tolerance becomes a sham, a mere strategy of manipulation, with no real openness to persuasion.[15]

Of course, we must also be wary of dogmatic belief in a too explicitly defined ultimate end, such as those offered by many fundamentalist religions.[16] In this case, again, there is no possibility of truly reasoning together to clarify a shared perception, because any questioning of revealed truth is heresy.

■ THE PRESUPPOSITIONS OF POLICY

Ecological economics is committed to policy relevance. It is not just a logical game for autistic academicians. Because of our commitment to policy, we must ask: What are the necessary presuppositions for policy to make sense, to be worth discussing? We see two.

First, we must believe that there are real alternatives to choose from. If there are no alternatives, if everything is determined, then it hardly makes

---

[15]For a fuller exposition of this argument, see C. S. Lewis, *The Abolition of Man*, New York: Macmillan, 1947.

[16]Many economists seem to view "efficiency" or growth as the ultimate end and border on religiosity in their convictions. See Robert Nelson, *Economics as Religion*, University Park: Pennsylvania State University Press, 2002.

sense to discuss policy—what will be, will be. If there are no options, then there is no responsibility, no need to think.

Second, even if there are real alternatives, policy dialogue would still make no sense unless there were a real criterion of value to use for choosing between the alternatives. Unless we can distinguish better from worse states of the world, it makes no sense to try to achieve one state of the world rather than another. If there is no value criterion, then there is no responsibility, no need to think.

In sum, serious policy must presuppose (1) nondeterminism—that the world is not totally determined, that there is an element of freedom that offers us real alternatives; and (2) nonnihilism—that there is a real criterion of value to guide our choices, however vaguely we may perceive it.

The fact that many people engaged in discussing and making policy reject one or both of these presuppositions is, in A. N. Whitehead's term, "the lurking inconsistency," a contradiction at the basis of the modern worldview that enfeebles thought and renders action halfhearted. If we even halfway believe that purpose is an illusion foisted on us by our genes to somehow make us more efficient at procreation,[17] or that one state of the world is as good as another, then it is hard to get serious about real issues. And ecological economics must be serious about real issues. As Whitehead noted, "Scientists animated by the purpose of proving that they are purposeless constitute an interesting subject for study."[18]

■ Determinism and Relativism

The preceding section may seem pretty obvious and consistent with common sense. What is the point of stating the obvious? The point is that many members of the intelligentsia deny nondeterminism or nonnihilism, yet they want to engage in a policy dialogue. It is not just that we disagree about exactly what our alternatives are in a particular instance or about what our value criterion implies for a concrete case—that's part of the reasonable policy dialogue. The point is that determinists who deny the effective existence of alternatives, and nihilists or relativists who deny the existence of a value criterion beyond the level of subjective personal tastes, have no logical basis for engaging in policy dialogue—and yet they do! We cordially and respectfully invite them to remember and reflect deeply upon their option of remaining silent—at least about policy.[19]

---

[17]As asserted in E. O. Wilson's *Consilience* (New York: Knopf, 1998) and R. Dawkins' *The Blind Watchmaker* (New York: Norton, 1996).

[18]A. Whitehead, *The Function of Reason*, Princeton, NJ: Princeton University Press, 1929, p. 12.

One may well agree with the logic of our position—that policy rules out determinism and nihilism—but argue that there are so few real determinists and nihilists around that in effect we are kicking at an open door or attacking a straw man. We hope this is true. However, one leading biologist, Paul R. Ehrlich, who has contributed much to ecological economics, recently wrote a book with this stated purpose:[20] "to give an evolutionist's antidote to the extreme hereditary determinism that infests much of the current discussion of human behavior—the idea that we are somehow simply captives of tiny, self-copying entities called genes" (p. x). In other words, Ehrlich felt that the influence of the hard-line determinists is sufficiently toxic to require a 500-page antidote, even if a rather mild and general one.

A stronger and more specific antidote was thought necessary by Wendell Berry, who took particular aim at the influential writings of Edward O. Wilson, especially his recent book *Consilience*. Berry deserves to be quoted at length:[21]

> *A theoretical materialism as strictly principled as Mr. Wilson's is inescapably deterministic. We and our works and acts, he holds, are determined by our genes, which are determined by the laws of biology, which are determined ultimately by the laws of physics. He sees that this directly contradicts the idea of free will, which even as a scientist he seems unwilling to give up, and which as a conservationist he cannot afford to give up. He deals with this dilemma oddly and inconsistently.*
>
> *First, he says that we have, and need, "the illusion of free will," which, he says further, is "biologically adaptive." I have read his sentences several times, hoping to find that I have misunderstood them, but I am afraid that I understand them. He is saying that there is an evolutionary advantage in illusion. The proposition that our ancestors survived because they were foolish enough to believe an illusion is certainly optimistic, but it does not seem very probable. And what are we to think of a materialism that can be used to validate an illusion? Mr. Wilson nevertheless insists upon his point; in another place he speaks of "self-deception" as granting to our species the "adaptive edge." Later, in discussing the need for conservation, Mr. Wilson affirms the Enlightenment belief that we can "choose wisely." How a wise choice can be made on the basis of an illusory freedom of the will is impossible to conceive, and Mr. Wilson wisely chooses not to try to conceive it. (p. 26)*

---

[19]In the sciences, these are hard-line neo-Darwinists and sociobiologists; in the humanities, postmodern deconstructionists; in theology, hard-line Calvinist believers in predestination; and in the social sciences, some evolutionary psychologists, and so-called value-free economists who reduce value to subjective individual tastes, any one of which is as good as another.

[20]P. R. Ehrlich, *Human Natures: Genes, Cultures, and the Human Prospect*, Washington, DC: Island Press, 2000.

[21]W. Berry, *Life Is a Miracle: An Essay Against Modern Superstition*, Washington, DC: Counterpoint Press, 2000.

We have learned from personal conversation with Wilson that he considers the question of how one squares scientific determinism with purposeful policy to be "the mother of all questions." Mutual humility in the face of mystery and paradox is more easily expressed, and understood, in friendly conversation over wine and dinner than in dry academic print. No one can, in practice, live by the creeds of determinism or nihilism. In this sense, no one takes these creeds seriously, not even the advocates themselves. So we tend to discount any effect on policy of these doctrines. However, many open-minded citizens halfway suspect that the learned scholars who publicly proclaim these views might know something that they do not. Maybe I really am just a robot controlled by my selfish genes; maybe purpose really is just an epiphenomenal illusion; maybe *better* and *worse* really are just meaningless terms for lending undue authority to subjective personal preferences or to class-based, gender-based, or race-based interests. The fact that determinist or nihilist views cannot consistently be lived out in practice by individuals does not mean that their existence, lurking in the back of the collective mind, is not capable of disabling policy.

In the Introduction, we referred briefly to the difficulty some ecologists have in dealing with policy, the messy world of human affairs. To the extent that the ecologist, like some biologists, is a determinist, policy of any kind would be silly. Such an ecologist would necessarily be more laissez-faire than the most extreme free market economist. Hence our view that ecological economics is not simply a matter of bringing the light of ecology to dispel the darkness of economics. There is that to be sure, but there is also some darkness within ecology that economists do not need to import.

Perhaps we should take some cues from modern physics, just as traditional economics takes cues from nineteenth-century mechanical physics. Quantum indeterminacy and chaos theory have upset the "scientific" foundations of determinacy. And many of our greatest modern physicists, those who have best come to understand the physical matter underlying the scientific materialism paradigm, increasingly question its ability to provide any ultimate truths. For example, Einstein points out that scientific knowledge "of what is does not open the door directly to what should be." He goes on to ask, "What should be the goal of our human aspirations? The ultimate goal itself and the longing to reach it must come from another source."[22] In Schrodinger's words, "The scientific picture of the real world around me is very deficient. It gives a lot of factual information, puts all our experience in a magnificently consistent order, but it is ghastly silent about

---

[22]A. Einstein, *Ideas and Opinions*, New York: Crown, 1954. Quoted in T. Maxwell, Integral Spirituality, Deep Science, and Ecological Awareness, Zygon: *Journal of Religion and Science* 38(2): 257–276 (2003).

all and sundry that is really near to our heart, that really matters to us—we do not belong to the material world the science constructs for us."[23]

Policy students, including economists, implicitly assume that the world offers more than one possibility to choose from and that some choices really are better than others. This is also true, of course, for ecological economists, who, while continuing to take biology and ecology very seriously, must not fall into the metaphysical traps of determinism or nihilism that seem to have ensnared some in those disciplines.

To be sure, not every conceivable alternative is a real alternative. Many things really are impossible. But the number of viable possibilities permitted by physical law and past history is seldom reduced to only one. Through our choices, value and purpose lure the physical world in one direction rather than the other. Purpose is independently causative in the world.

Box 3-1   DETERMINISM IN THE HISTORY OF PHILOSOPHY

Materialism, determinism, and mechanism are closely related metaphysical doctrines about the basic nature of reality. If you study the history of philosophy, you will see that they go back to Epicurus, Democritus, and Lucretius, over 2000 years ago, and these doctrines are still very much with us today. It would be arrogant for two economists to think that they can resolve this ancient puzzle but also naïve to think that we can sidestep it, since economics is unavoidably about choice. If choice is an illusion, what does that say about economics?

Because humans are part of reality, it follows that if matter in motion is all there is to reality, then that is all there is to humans as well. Since the motions of matter are determined by mechanical law, it follows that the same laws ultimately determine human action. This **determinism** rules out free will—it means that our purposes are not independently causative in the world. Only mechanical motion of matter is causative. Purposes, intentions, values, choices are all dreams or subjective hallucinations. They are effects, not causes.

Nihilism, the rejection of all moral values, is the ethical consequence of the materialist, determinist cosmology. Things are what they are, and you can do nothing about it because your will and purpose have no power to change things. You can have no responsibility for what cannot be otherwise. For Epicurus this was a great relief—much better than worrying about the gods' anger and retribution, about responsibility and guilt and punishment. Relax, don't worry, do your best to enjoy life. Nothing can really hurt you, because when you are dead, that's the end and you can no longer suffer. This view is still very much alive in the

[23]E. Schrodinger, *My View of the World*, Cambridge: Cambridge University Press, 1964. Quoted in Maxwell, op. cit.

modern secular world, although it has a long history of conflict with Christianity, Judaism, and Islam, as well as other philosophies that reject materialism as an adequate view of reality. They insist that good and evil are as real in our experience as matter and that humans have at least some capacity for choice between them. To ignore our direct experience of good, evil, and freedom is considered antiempirical and against the deeper spirit of science.

It is not our intent to convert you either to or from Epicurianism, Christianity, or any other position. Maybe you do not yet have any position on this question. But logic does have its demands, and no doctrine is exempt from them. Even the early materialists recognized the contradictions involved in a doctrine that ruled out freedom, novelty, and choice. Epicurus tried to restore a modicum of freedom in an ad hoc manner by introducing the notion of the "clinamen"—the idea that atoms swerved from their determined motions for unexplained reasons and that this was the source of novelty, and perhaps some degree of freedom. Our advice is to be skeptical of any easy answer to a problem that has been around for 2500 years and also to be humble in the face of any logical contradictions that you cannot resolve.

■ The Ends-Means Spectrum

Ultimate means and the ultimate ends are two extremes of an **ends-means spectrum** in the middle of which economic value is determined. In everyday life, it is our mid-range ends and means that interact, not their ultimate origins in the realms of the spirit or the electron. We will discuss this intermediate, mid-spectrum interaction in our consideration of the function of markets and relative prices (see Chapter 8). But for now it is useful to think of the entire ends-means spectrum depicted in Figure 3.1. The economic choices that exist in the mid-range of the spectrum are not illusory. They are not totally determined by material causes from below, nor are they rendered meaningless by an absence of final cause from above or the presence of a predestining final cause. As we will discuss later, prices, relative values, are determined by supply and demand. But supply reflects alternative conditions of relative possibility, of the reality of ultimate means, while demand reflects independent conditions of relative desirability, rooted in notions of better and worse, of ethical choices based on some perception of the ultimate end.

In its largest sense, humanity's ultimate economic problem is to use ultimate means efficiently and wisely in the service of the ultimate end. Stated in this way, the problem is overwhelming in its inclusiveness. Therefore, it's not hard to understand why in practice it has been broken up into a series of sub-problems, each dealt with by a different discipline, as indicated on the right side of the ends-means spectrum.

At the top of the spectrum, we have the ultimate end, studied by religion and philosophy. It is that which is intrinsically good and does not derive its goodness from any instrumental relation to some other or higher good. It is the highest good, to which all other good is instrumental and derivative. Needless to say, it is not well defined. As noted earlier, there are unacceptable consequences from denying its existence, but the dimness of our vision of the ultimate end is part of the human condition and requires a great deal of mutual tolerance. The error of treating as ultimate that which is not is, in theological terms, idolatry.

At the bottom of the spectrum is ultimate means, the useful stuff of the world—low-entropy matter-energy, which we can only use up and cannot create or replenish, and whose net production cannot possibly be the end result of any human activity. The ultimate end is much harder to define than ultimate means. Our current approximation to the ultimate end, unfortunately, seems to be economic growth, and part of the critique of economic growth is that our devotion to it has become idolatrous, worshipping a false god, so to speak, because it is not really ultimate. But it is

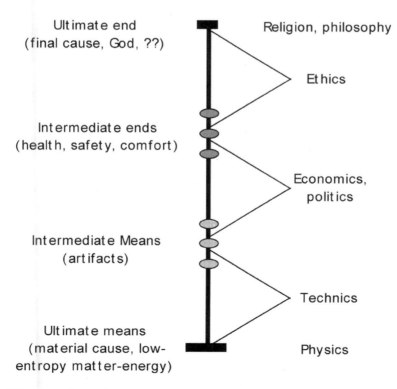

Figure 3.1 • The ends-means spectrum.

not easy to formulate a central organizing principle of society that does not border on idolatry.

To reiterate, since we are forced by scarcity to choose which of our many intermediate ends will be satisfied and which will be sacrificed, we must rank our intermediate ends. Ranking means establishing priority. Priority means that something goes in first place. That holder of first place is our operational estimate of the ultimate end. It provides the ordering criterion for ranking other intermediate ends. Second place goes to whatever is nearest to or best serves first place, and so on. This ranking of intermediate ends relative to our vision of the ultimate end is the problem of ethics. Economists traditionally take the solution to the ethical problem as given and start their analysis with a given ranking of intermediate ends, or with the assumption that one person's ranking is as good as another's, so that ethics is indistinguishable from personal subjective tastes.

At the bottom of the spectrum, physics studies ultimate means, and technics studies the problem of turning ultimate means into artifacts specifically designed to satisfy each of our intermediate ends. Economists also habitually assume the technical problem to have been solved; that is, technology is taken as given. Thus, the remaining segment of the spectrum is the middle one of allocating *given* intermediate means to the service of a *given* hierarchy of intermediate ends. This is the significant and important economic problem, or rather political economic problem, quite distinct from the ethical and technical problems.

The middle-range nature of the problem of political economy is significant. It means that, from the perspective of the entire spectrum, economics is, in a sense, both too materialistic and not materialistic enough. In abstracting from the ethical and religious problem it is too materialistic, and in abstracting from the technical and biophysical problem it is not materialistic enough. Economic value has both physical and moral roots. Neither can be ignored. Yet many thinkers are attracted to a monistic philosophy that focuses only on the biophysical or only on the psychic root of value. Ecological economics adopts a kind of practical dualism. Dualism is not as simple as monism, and it entails the mysterious problem of how the material and the spiritual interact. That is indeed a large and enduring mystery. But on the positive side, dualism is more radically empirical than either monism, refusing to deny or ride roughshod over inconvenient facts just to avoid confronting a mystery.[24]

---

[24]Honesty requires facing up to mystery. Although we suspect that mystery is an enduring part of the human condition and not just another word for future knowledge not yet discovered, we nevertheless respect the scientific and philosophical quest to solve mysteries, including the mystery inherent in the dualism we advocate as a practical working philosophy.

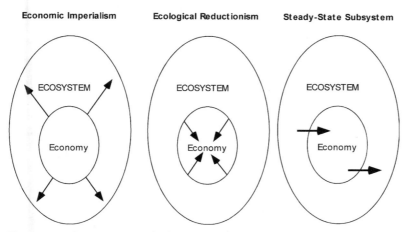

Figure 3.2 • Three strategies for integrating ecology and economics.

■ THREE STRATEGIES FOR INTEGRATING ECOLOGY AND ECONOMICS

Previous attempts to integrate economics and ecology have been based on one of three strategies: (1) economic imperialism, (2) ecological reductionism, or (3) steady-state subsystem. Each strategy may be thought of as beginning with the picture of the economy as a subsystem of the ecosystem. The differences concern the way each treats the boundary between the economy and the rest of the ecosystem (Figure 3.2).

Economic Imperialism

Economic imperialism seeks to expand the boundary of the economic subsystem until it encompasses the entire ecosystem. The goal is one system, the macroeconomy as the whole. This is to be accomplished by complete internalization of all external costs and benefits into prices. Price, of course, is the ratio (e.g., dollars per gallon) at which something is exchanged for money (or for some other commodity) by individuals in the market. The aspects of the environment not customarily traded in markets can be treated as if they were by imputation of "shadow prices"—the economist's best estimate of what the price of the function or thing would be if it were traded in a competitive market. Everything in the ecosystem is theoretically rendered comparable in terms of its ability to help or hinder individuals in satisfying their wants. Implicitly, the single end pursued is ever-greater levels of consumption, and the only intermediate means to effectively achieve this end is growth in market goods. Economic imperialism is basically the neoclassical approach.

Subjective individual preferences, however whimsical, uninstructed, or ill-considered, are taken as the source of all value. Since subjective wants are thought to be infinite in the aggregate, as well as sovereign, there is a tendency for the scale of activities devoted to satisfying them to expand. The expansion is considered legitimate as long as all costs are internalized. But most of the costs of growth we have experienced have come as surprises. We could not have internalized them if we could not first imagine and foresee them. Furthermore, even after some external costs have become visible to all (e.g., greenhouse warming), internalization has been very slow and partial. As long as the evolutionary fitness of the environment to support life is unperceived by economists, it is likely to be destroyed in the imperialistic quest to make every molecule in creation, including every strand of DNA, pay its way according to the pecuniary rules of present value maximization.

Furthermore, this imperialism sacrifices the main virtue of free-market economists, namely their antipathy to the arrogance of central planners. Putting a price tag on everything in the ecosystem requires information and calculating abilities far beyond anything attempted by Gosplan in the old Soviet Union.[25] As an example, let's take a look at what calculations would be needed to accurately quantify and internalize the costs associated with global warming. Currently we are incapable of accounting for even carbon dioxide flows, the most basic piece of the puzzle. How much carbon is being absorbed by oceans or terrestrial ecosystems? How will it affect these ecosystems? Will global warming lead to positive feedback loops, such as a release of methane from a thawing arctic tundra and increased atmospheric water vapor from more rapid evaporation from the oceans (both potent greenhouse gases), or negative feedback loops via increased sequestration of carbon by forests? How will temperature changes affect global weather patterns over the next century? (And how certain can we be about such estimates, when we cannot even accurately predict the weather next week?) What changes would have occurred even in the absence of global warming? What technologies will evolve to cope with these problems, and how much will changing our rate of greenhouse gas emissions affect the rate of technological advance? Finally, how will these factors affect the economy? Bear in mind that while a meteorologist cannot accurately predict the weather in a week, she can at least stick her head out the window and say, "It's raining." Economists, on the other hand, at the time of this writing are in the midst of a heated debate over whether or not the economy is in a recession *right now*.

---

[25]Gosplan is the Russian acronym for the State Planning Committee, which centrally developed 5-year and annual plans for the Soviet economy, at all levels from individual enterprises to the national level.

These calculations, a small fraction of those that would be needed to estimate the costs of global warming, are clearly beyond the capabilities of modern science and quite probably beyond the capacity of the human mind. And calculating all the costs at the time they occur is the straightforward part. How do we determine the present value of costs to future generations? The currently favored approach, intertemporal discounting (to which we return in Chapter 10), gives less value to future costs and benefits than those that occur today, and the discount rate we choose in this calculation is likely to be as important as any other of the variables mentioned earlier. But discounting in this case implies that future generations have no inalienable right to a stable climate, economic growth will continue throughout the discount period, and economic growth is a satisfactory substitute. Yet the discount rate we choose for internalizing costs will itself affect the rate of growth.

THINK ABOUT IT!

The Stern Review on the economics of climate change concludes that society should spend about 1% of global GNP to reduce the risk of climate catastrophe.[26] Economist William Nordhaus uses a higher discount rate in a similar study and concludes that 1% of GNP annually greatly exceeds the benefits of avoiding catastrophe. In 2007, per-capita global GNP grew at about 3%. In other words, Nordhaus argues that accepting our living standards from four months ago is too high a price to pay to avert catastrophe. Do you think Nordhaus appropriately discounts future impacts? Do you think those impacts should be discounted at all?[27]

The global warming example brings up another serious problem with economic imperialism: the assumption that the most efficient mechanism for allocating almost any means among any ends is the market. In fact, markets are incapable of allocating goods that cannot be owned and inefficient at allocating goods for which use does not lead to depletion (either or both of which are properties of the bulk of ecosystem services). Even if we could put an appropriate charge on greenhouse gas emissions to internalize their costs, who would receive the charge? It would seem only fair that it would go to those who bear the costs. Would it even be a market transaction if when we purchased something, we did not pay the person who bore the costs of production? However, global warming is likely to affect the entire population of the planet for countless generations into the future. This would imply that not only would we need to calculate all

[26]N. Stern, *Stern Review: The Economics of Climate Change*. HM Treasury, London, 2006.

[27]W. Nordhaus, "Critical Assumptions in the Stern Review on Climate Change," *Science*. 317(13)(2007):201–202.

the costs, we would need to do so for *each individual*. Strangely enough, as we will discuss later, some neoclassical economists argue that those who bear the costs of externalities should not receive the payments,[28] but in this case, how could we say that the result resembles a market solution? A major goal of this text will be to explain exactly why many goods and services are not amenable to market solutions, independently of whether or not we are able to internalize all costs.

Let's play the role of the stereotypical economist and assume away all these problems. There is, then, no doubt that once the scale of the economy has grown to the point that formerly free goods become scarce, it is better that these goods should have a positive price reflecting their scarcity than to continue to be priced at zero. But there remains the prior question: Are we better off at the new scale with formerly free goods correctly priced or at the old scale with free goods also correctly priced at zero? In both cases, the prices are right. This is the suppressed question of optimal scale, and it is not answered by market prices.

Ecological Reductionism

Ecological reductionism begins with the true insight that humans are not exempt from the laws of nature. It then proceeds to the false inference that human action is totally explainable by, reducible to, the laws of nature. It seeks to explain whatever happens within the economic subsystem by exactly the same naturalistic principles that it applies to the rest of the ecosystem. It shrinks the economic subsystem to nothing, erasing its boundary. Taken to the extreme, in this view energy flows, embodied energy costs, and relative prices in markets are all explained by a mechanistic system that has no room for purpose or will. This may be a sensible vision from which to study some natural systems. But if one adopts it for studying the human economy, one is stuck from the beginning with the important policy implication that policy makes no difference. We encounter again all the problems of determinism and nihilism already discussed.

Economic imperialism and ecological reductionism have in common that they are monistic visions, albeit rather opposite monisms. It is the monistic quest for a single substance or principle by which to explain all value that leads to excessive reductionism on both sides. Certainly one should strive for the most reduced or parsimonious explanation possible without ignoring the facts. But respect for the basic empirical facts of chance and necessity on one hand and self-conscious purpose and will on the other hand should lead us to a kind of practical dualism or polarity

---

[28]E. T. Verhoef, "Externalities." In J. C. J. M. van den Bergh, ed., *Handbook of Environmental and Resource Economics*, Northampton, MA: Edward Elgar, 1999.

reflected in the ends-means spectrum. After all, the fact that our being should consist of two fundamental elements offers no greater inherent improbability than that it should rest on one only. How these two fundamental elements of our being interact is a mystery—precisely the mystery that the monists of both kinds are seeking to avoid. But economists are too much in the middle of the spectrum to adopt either monistic "solution." Economists are better off denying the tidy-mindedness of either monism than denying the untidy and mysterious facts.

The Steady-State Subsystem

The remaining strategy is the **steady-state subsystem**, the one adopted here. It does not attempt to eliminate the subsystem boundary, either by expanding it to coincide with the whole system or by reducing it to nothing. Rather, it affirms the fundamental necessity of the boundary and the importance of drawing it in the right place. It says that the scale of the human subsystem defined by the boundary has an optimum and that the throughput by which the ecosystem maintains and replenishes the economic subsystem must be ecologically sustainable. Once we have drawn this boundary in the appropriate place, we must further subdivide the economic subsystem into regions where the market is the most effective means of allocating resources and regions where it is inappropriate. These regions are determined by inherent characteristics of different goods and services, to be discussed at length in this text.

Box 3-2   THE STEADY-STATE ECONOMY

The idea of a steady-state economy comes from classical economics and was most developed by John Stuart Mill (1857), who referred to it as the "stationary state." The main idea was that population and the capital stock were not growing. The constancy of these two physical stocks defined the scale of the economic subsystem. Birth rates would be equal to death rates and production rates equal to depreciation rates, so that both the stock of people (population) and the stock of artifacts (physical capital) would be constant—not static, but in a state of dynamic equilibrium. Most classical economists dreaded the stationary state as the end of progress, but not Mill:[a]

> It is scarcely necessary to remark that a stationary condition of capital and population implies no stationary state of human improvement. There would be as much scope as ever for all kinds of mental culture, and moral and social progress; as much room for improving the Art of Living and much more likelihood of its being improved, when minds cease to be engrossed by the art of getting on.

Mill thought we would pay more attention to getting better once we ceased to be so preoccupied with getting bigger. He also recognized that growth could become uneconomic:

> If the earth must lose that great portion of its pleasantness which it owes to things that the unlimited increase of wealth and population would extirpate from it, for the mere purpose of enabling it to support a larger, but not a happier or better population, I sincerely hope, for the sake of posterity, that they will be content to be stationary, long before necessity compels them to it.

In physical terms, populations of both human bodies and things are what physicists call "dissipative structures," things that fall apart, die, and decay if left to themselves. People die, goods wear out. To keep a population of dissipative structures constant requires births equal to deaths and production equal to depreciation—in other words, input equal to output equal to throughput, a concept with which you are now familiar. But births can equal deaths at low rates or at high rates. Either one will keep the population constant. Which do we want? If we want a long life expectancy for individuals, we must choose low birth rate equal to low death rate. For an equilibrium population with birth equal to death rates at 40 per thousand per year, the average age at death must be 25 years. If we want people to live to be 67 rather than 25, we will have to lower birth and death rates to 15 per thousand per year. Can you explain why? Can you apply the same logic to lifetime or durability of the stock of goods?

To summarize: The main idea of a steady-state economy is to maintain constant stocks of wealth and people at levels that are sufficient for a long and good life. The throughput by which these stocks are maintained should be low rather than high, and always within the regenerative and absorptive capacities of the ecosystem. The system is therefore sustainable—it can continue for a long time. The path of progress in the steady state is no longer to get bigger but to get better. This concept was a part of classical economics but unfortunately was largely abandoned by NCE. More precisely, the terms *stationary* and *steady state* were redefined to refer not to constant population and capital stock but to their proportional growth—a constant ratio between ever-growing stocks of people and things!

---

[a]J. S. Mill, Principles of Political Economy, Book IV, *Chapter VI (1848)*. Online: http://www.econlib.org/library/Mill/mlPbl.html.

BIG IDEAS to remember

- Practical dualism versus monisms
- Presuppositions of policy
- Ultimate means
- Ultimate end
- Information and knowledge

- Determinism and materialism
- Ends-means spectrum
- Economic imperialism
- Ecological reductionism
- Steady-state subsystem

Conclusions to Part I

In Part I, we have defined economics as the science of the allocation of scarce resources among alternative desirable ends. Ecological economics distinguishes itself from mainstream economics in its preanalytic vision of the economic system as a subsystem of the sustaining and containing global ecosystem. Economic growth is not an end in itself, continuous physical economic growth is not possible, and eventually the costs imposed by growth on the sustaining system become greater than the benefits of that growth. Economic systems change the environment, then adapt in response to those changes in a coevolutionary process. Economic growth has increased the scarcity of ecological goods and services relative to manmade goods and services, and the economic system must adapt to that fact. The way in which the economy evolves is not predetermined; it can be influenced by policy in ways that are better or worse.

In the ecological economy, there is an ultimate end, although it is hard to perceive and not universally agreed upon; the ultimate means is low-entropy matter-energy, and the market is a very useful but by no means sufficient institution for allocating means to the service of ends. Although we shrink from trying to define the ultimate end, as a basis for future discussion in this text we suggest a working definition of the penultimate end for the ecological economy: the maintenance of ecological life-support systems far from the edge of collapse (which requires an end to material growth of the economy) and healthy, satisfied human populations free to work together in the pursuit and clarification of a still vague ultimate end—for a long, long time.

PART II

The Containing and Sustaining Ecosystem: The Whole

CHAPTER

4

The Nature of Resources and the Resources of Nature

The economic system is a subsystem of the global ecosystem, and one of the major goals of ecological economics is to determine when the benefits of continued growth in the economic subsystem are outweighed by the increasing opportunity costs of encroaching on the sustaining ecosystem. Achieving this goal demands a clear understanding of how the global ecosystem sustains the economy and how economic growth affects the sustaining ecosystem. In addition to determining when economic growth becomes uneconomic, ecological economists must provide the policies necessary to keep the economy within its optimal size range. Currently, the dominant tool for determining economic optimality is the market. However, markets function effectively only with goods and services that have certain specific attributes, and they really do not function at all with goods that cannot be exclusively owned. Developing effective policies requires a clear understanding of the specific attributes of goods and services that the economic system must allocate between alternative ends. In this chapter, we introduce you to several concepts that will be useful for understanding scarce resources. These include the difference between stock-flow and fund-service resources and the concepts of rivalness and excludability. We will also consider further the laws of the thermodynamics.

Chapters 5 and 6 will apply these concepts to the abiotic and biotic scarce resources on which our economy depends. Chapters 9 and 10 will explain why these concepts are so important to policy analysis.

◼ A FINITE PLANET

With the exception of inconsequential bits of material arriving from space, there is only so much water, so much land, and so much atmosphere to our planet. We have finite supplies of soils, minerals, and fossil fuels. Even if we argue that natural processes make more soil and fossil fuels, the rate at which they do so is not only finite, it is exceedingly slow from a human perspective. Fortunately, we are blessed with a steady influx of solar energy that will undoubtedly continue long past the extinction of the human race,[1] but the rate at which this energy arrives is also fixed and finite. Of course, for this energy to be useful, it must be captured, and at present virtually all of that capture is performed by a finite stock of photosynthesizing organisms. In other words, it appears that we live on a finite planet. Why waste words on such an obvious fact?

Continued economic growth is the explicit goal of most economists and policy makers. Many economists even argue that economic growth is not only compatible with a clean environment, it is a prerequisite for achieving one. A clean environment is a luxury good, the story goes. People who are struggling simply to feed themselves cannot be concerned with pollution. The fact that throughout the Third World, poverty forces people to actually live and work in garbage dumps, finding food to eat, clothes to wear, and goods and materials to recycle speaks for itself—survival takes precedence over environment. And work in a factory, no matter how much it pollutes, must be better than life in a dump. Only in rich nations can we afford the luxury of clean water and clean air. This would explain the fact that water quality and air quality in the United States have improved since the 1970s (we will return to this apparent paradox later), and even forest cover is expanding in many areas. The best way to clean up the planet and preserve its remaining ecosystems, it is often argued, is through economic growth.

THINK ABOUT IT!

Do you think the environment in wealthy countries has improved over the past 20 years? Have the global environmental impacts of wealthy countries diminished over the past 20 years? Where do most of the things you buy come from? Do you think their production has negative impacts on the environment?

In contrast to this scenario, the laws of physics tell us that we cannot create something from nothing. Economic production therefore requires

---

[1]The average life span of a mammalian species is only one million years, while the sun is expected to last for several billion years. See R. Foley, "Pattern and Process in Hominid Evolution." In J. Bintliff, ed. *Structure and Contingency*. Leicester, England: Leicester University Press, 1999.

raw material inputs, and the finite supply of those inputs limits the size of the economy. The economic system cannot grow indefinitely, no matter how much we can substitute a new resource for an exhausted one. For example, human populations cannot continue growing forever. A simple calculation shows that even at a continuous 1% rate of growth, the human population would have a mass greater than the entire planet in just over 3000 years.[2] Similarly, we could not conceivably continue to increase the physical mass of artifacts we own and consume over the next 1000 years at the same rate as we have during the past 50. But population growth rates are already slowing. A recent U.N. report estimates that the worldwide population will stabilize at about 9 billion people by the year 2300,[3] though many ecologists believe planetary ecosystems would have difficulty sustaining even half that number.[4] Some also argue that we can produce more using less, so the physical mass of artifacts need not increase. It is true that we can now produce 12 aluminum cans from the same material it once took to produce one, but we still use more aluminum than ever before, and aluminum can only be rolled so thin. Still others assert that economic value is not a measure of a physical quantity, and therefore it is not at all obvious that the production of economic value has physical limits.

It is true that economic value is not a physical quantity. Economic production is really all about creating welfare, quality of life, utility, or whatever else we choose to call this psychic flux of satisfaction. Does it really matter, then, that we live on a finite planet? Certainly it matters in terms of economic production. Economic production, as it is typically understood, is the transformation of raw materials supplied by the ecosystem into something of value to humans. Transformation requires energy, and it inevitably generates waste. Even the service sector requires physical inputs to sustain those who provide the service. We have finite supplies of

---

[2]Some would say that this type of calculation is really just a straw man argument and that no one argues that human population growth will continue indefinitely. However, University of Maryland Professor Julian Simon once claimed that human populations could continue growing at the same rate for the next 7 million years with existing technologies (J. Simon, *The Ultimate Resource*, 2nd ed. Princeton, NJ: Princeton University Press, 1996). A steady 1% growth rate of the current population for that long would leave us with more people than the estimated number of atoms in the universe. Simon is widely and favorably cited by the *New York Times* "science" writer John Tierney and in a recent influential book by Bjorn Lomborg, *The Skeptical Environmentalist*, Cambridge: Cambridge University Press, 2001.

[3]J. Chamie, United Nations Population Division, Department of Economic and Social Affairs, 2004. World Population to 2300: Executive Summary. Online: www.un.org/esa/population/publications/longrange2/WorldPop2300final.pdf. The 2009 report has revised 2050 population estimates and growth rates upwards but has no projections beyond that date.

[4]For example, G. Daily, A. Ehrlich, and P. Ehrlich, Optimum Human Population Size, *Population and Environment* 15(6) (1994), argue that 5.5 billion people clearly exceeds the planet's carrying capacity and suggest optimal populations of 1.5 to 2 billion.

energy, finite supplies of raw materials, finite absorption capacities for our wastes, and poorly understood but finite capacities for ecosystems to provide a host of goods and services essential for our survival. And evidence suggests that we are reaching the limits with respect to these resources, as we will describe in greater detail below. With continued growth in production, the economic subsystem must eventually overwhelm the capacity of the global ecosystem to sustain it.

All of this does not mean economic value cannot continue to grow indefinitely. Indeed, we believe that perhaps it can if we define economic value in terms of the psychic flux of human satisfaction, and we learn to attain this satisfaction through nonmaterial means. Ecological economics does not call for an end to economic development, merely to physical growth, while mainstream economists' definitions of economic progress confusingly conflate the two. The problem is that the existing market economy is ill-suited to providing nonmaterial satisfaction. Even if one accepts some variant of the NCEs' assertion that infinite economic "growth" is possible by redefining growth as the ever-greater provision of psychic satisfaction (what we call economic development), the conventional economic paradigm is probably an inadequate guide for achieving this goal— but we'll come back to that later. Our point for now is that constant growth in physical throughput is impossible. Once we understand this, the question becomes how to decide when economic production becomes uneconomic, particularly if this has already happened. Before we address this last question, however, we need to look more closely at the assertion stated above—that infinite growth is impossible in a closed system. The branch of science most relevant to this issue, and indeed most relevant to the economic problem, is thermodynamics.

> **THINK ABOUT IT!**
> *Of all the activities and objects that give you satisfaction, which ones consume the fewest resources and produce the least waste? Which consume the most resources and produce the most waste? Which of these are produced by the market economy?*

■ THE LAWS OF THERMODYNAMICS

A Brief History of Thermodynamics

With the advent of the Industrial Revolution and the machine age at the end of the eighteenth century, scientists became intrigued by the idea of a perpetual motion machine—a machine fueled by the very same heat it generated while it worked. In 1824, French scientist Sadi Carnot, while trying to calculate the greatest amount of work that could be done by a given amount of heat, realized that a heat engine (e.g., a steam engine) could perform work only by taking heat from one reservoir and transferring it to another at a lower temperature. In fact, the performance of work in general required a temperature differential between two reservoirs, and all else being equal, the greater the differential, the more work that could be performed. However, even with a temperature differential, it was impossible to convert heat or any kind of energy directly into work with 100% efficiency. It turned out that this was related to the obvious fact that heat would naturally flow from a hotter item to a colder one, and not vice versa. While heat could be made to flow from a colder object to a hotter one, the amount of work needed to make this happen was greater than the amount of energy latent in the increased temperature of the hotter object.[5] To the dismay of industrialists, physical laws did not allow a perpetual motion machine.

Within the course of the next few decades, some other important facts were established. Robert Mayer and Herman Helmholtz showed that energy cannot be created or destroyed, and James Joule performed experiments demonstrating that energy and work are equivalent. Rudolf Clausius recognized that there were two related principles at work here, which came to be called the First and Second Laws of Thermodynamics. The First Law established that energy could not be created or destroyed, and the Second Law established that energy moved inevitably toward greater homogeneity. Because work requires a temperature differential, homogeneity means that energy becomes increasingly unavailable to perform work. In the words of Georgescu-Roegen, "all kinds of energy are gradually transformed into heat, and heat becomes so dissipated in the end that mankind can no longer use it."[6] Clausius coined the term *entropy* for the Second Law, derived from the Greek word for transformation, in recognition of the fact that entropy was a one-way street of irreversible change, a continual increase in disorder in the universe. While the First Law of Thermodynamics relates to quantity, the Second Law relates to quality.

A dictionary definition of **entropy** is "a measure of the unavailable energy in a thermodynamic system." "Unavailable" means unavailable to do work. Unavailable energy is also known as bound energy, and available energy as free energy. For example, gasoline carries a form of free energy: It can be burned in an internal combustion engine to generate work. Work can be transformed into free energy in a different form (e.g., it can carry a car to the top of a big hill, where it has the potential energy to coast back down) or into heat, which diffuses into the surrounding environment. The energy in the gasoline transformed into heat has not disappeared but has

---

[5]L. P. Wheeler, *Josiah Willard Gibbs: The History of a Great Mind*, New Haven, CT: Archon Books, 1999.

[6]N. Georgescu-Roegen, 1976. *Energy and Economic Myths: Institutional and Analytic Economic Essays*, New York: Pergamon Press, p. 8.

instead become bound energy, unavailable to perform work. In the well-cited example used by Georgescu-Roegen, the ocean contains enormous amounts of energy, but that energy is not available to run a ship.[7] It is bound energy, because there is no reservoir of a lower temperature to which the energy within the ocean can be transferred, and Carnot showed that such a temperature differential was essential to perform work.

THINK ABOUT IT!
Would you invest in a revolutionary new automobile designed to capture its own exhaust and burn it again?

Does matter, as well as energy, obey the laws of thermodynamics? Einstein's famous $E = mc^2$ established the equivalence between matter and energy and thus the fact that the First Law applies to matter as well as energy. Georgescu-Roegen argued that the entropy law also applies to matter and proposed that this be recognized as the fourth law of thermodynamics.[8] Although physicists dispute the idea of a formal "fourth law," there is no dispute about matter being subject to entropy in the sense of a natural tendency to disorder. When a cube of sugar is dropped into a cup of water it gradually dissolves, losing its order. Nor will that order spontaneously reappear. This is equally obvious for mixing liquids and gases, or more generally for any substance that is soluble in another. It is less obvious for materials in environments in which they are not soluble. However, friction, erosion, and chemical breakdown inexorably lead to the breakdown and diffusion of even the hardest metals over sufficient time, resulting in increased disorder.

It is important to recognize that the laws of thermodynamics were developed more from experimental evidence than from theory, and the mechanism behind entropy is still not completely understood.[9] When the laws of thermodynamics were first proposed, mechanical physics was the dominant paradigm in science. In a mechanical system, every action has an equal and opposite reaction and is thus inherently reversible. One theoretical explanation of entropy comes from efforts to harmonize the irreversibility inherent to entropy with the reversibility that characterizes mechanical physics. This has resulted in the field of statistical mechanics, best explained by referring to the example of the sugar cube used above. When in a cube on a shelf, sugar molecules are not free to disperse—there is only one state space available to them. When placed into a container of

---

[7]Ibid., p. 6.

[8]The third law of thermodynamics states that the entropy of any pure, perfect crystalline element or compound at absolute zero (0 K) is equal to zero. This is not particularly relevant to economics.

[9]R. Beard and G. Lozada, *Economics, Entropy and the Environment: The Extraordinary Economics of Nicholas Georgescu-Roegen*, Cheltenham, England: Edward Elgar, 2000.

water, in which sugar is soluble, sugar molecules are free to move. Suddenly, there are almost countless possible arrangements the sugar molecules can take within that container. Each arrangement may have an equal probability, but only one of those arrangements is that of the cube. Thus, the probability of the cube remaining intact is almost immeasurably small. According to this statistical version of thermodynamics, or statistical mechanics, a sugar cube dissolved in water could spontaneously reassemble, and a cold pot of water could spontaneously come to a boil; it is simply not very likely. But unlikely events are quasi-certain to happen if we wait long enough, and indeed they might happen tomorrow with the same (low) probability as for the day after a billion years from now. So the fact that we have never observed a cold pot of water spontaneously come to a boil, or even less significant instances of spontaneous increases in low entropy, remains an empirical difficulty for statistical mechanics.

Statistical mechanics is a far from universally accepted explanation of entropy, and while it does seem to allow for reversibility, which is compatible with mechanical physics, it also depends entirely on random motion, which is incompatible. If the defenders of statistical mechanics believe that it reconciles entropy theory with mechanical physics, they must also believe that if every atom in the universe happened to be traveling in the opposite direction to which it now moves, then heat would move from colder objects to warmer ones, and order would spontaneously appear.[10] If the statistical view of entropy is correct, the gradual dispersion of material via physical and chemical erosion may not be entropy per se, because the physical and chemical erosion of matter is fundamentally different from the dissipation of heat. Regardless of the explanation, however, the end result and the practical implications are the same: Both matter and energy move irreversibly toward less-ordered states, and lower-entropy states can be restored only by converting low entropy to high entropy elsewhere in the system—and the increase in entropy elsewhere will be greater than the local decrease in entropy that it made possible.

Entropy and Life

If all matter-energy moves toward greater disorder, how, then, do we explain life? Is life not a form of spontaneous order that emerged from the chaotic maelstrom that was our early planet? Has not the continued evolution of life on Earth led to highly complex and ordered life forms? And don't ecosystems exhibit yet another level of complexity and order that arises from the mutual interactions of the organisms of which they are composed? These facts in no way contradict entropy, but to understand

---

[10]N. Georgescu-Roegen, *The Entropy Law and the Economic Process*, Cambridge, MA: Harvard University Press, 1971.

why this is so, we remember the distinction made in Chapter 2 between isolated, closed, and open systems. Isolated systems are those in which neither matter nor energy can enter or leave. The universe is such a system. The Earth, in contrast, is a materially closed system, in which radiant energy can enter and leave, but for all practical purposes matter does not. The Earth is continually bathed in the low entropy of solar radiation that has allowed the complexity and order of life to emerge and increase. Any living thing on our planet is an open system, capable of absorbing and emitting both matter and energy.[11] A biological or ecological system is capable of maintaining its low entropy only by drawing on even greater amounts of low entropy from the system in which it exists and returning high entropy back into the system. Erwin Schrodinger has described life as a system in steady-state thermodynamic disequilibrium that maintains its constant distance from equilibrium (death) by feeding on low entropy from its environment—that is, by exchanging high-entropy outputs for low-entropy inputs.[12] This exchange results in a net increase in entropy. Hence, life on our planet needs a constant flow of low-entropy inputs from the sun simply to maintain itself.

Entropy and Economics

What, then, are the implications of the entropy law for the science of economics? The goal of the early neoclassical economists was to establish economics as a science, and in the words of William Stanley Jevons, "it is clear that economics, if it is to be a science at all, must be a mathematical science."[13] The basic argument was that economics focused on quantities of goods, services, and money and therefore was amenable to quantitative (i.e., mathematical) analysis. Such analysis enabled economists to build logically consistent theories from fundamental axioms. These theories could then be applied to problems in the real world. In the words of Leon Walras, "from real type concepts, [the physico-mathematical] sciences abstract ideal-type concepts which they define, and then on the basis of these definitions they construct *a priori* the whole framework of their theorems and proofs. After that they go back to experience not to confirm but

---

[11]H.E. Daly and J. Cobb, *For the Common Good: Redirecting the Economy Towards Community, the Environment, and a Sustainable Future*, Boston: Beacon Press, 1989, p. 253. Such open systems are often called dissipative structures. The nonequilibrium thermodynamics of dissipative structures is a field of thermodynamics under development by Nobel laureate physicist Ilya Prigogine and his collaborators. See his *The End of Certainty*, New York: Free Press, 1996.

[12]E. Schrodinger, *What Is Life?* Cambridge, England: Cambridge University Press, 1944.

[13]W. S. Jevons, quoted in R. Heilbroner. 1996. *Teachings from the Worldly Philosophy*, New York: Norton, p. 210.

to apply their conclusions."[14] Mechanical physics was the best-developed and most successful application of this approach in the sciences at the time the original neoclassicals were writing and thus was explicitly accepted as a model to emulate.[15]

In mechanical physics, all processes were considered reversible. For example, if one struck a billiard ball, an equal and opposite strike would return it exactly to its initial position. In contrast, the Second Law of Thermodynamics established the existence of irreversible processes as a fundamental law of physics. Entropy meant that in any isolated system, energy and matter would move toward a thermodynamic equilibrium in which they were equally diffused throughout the closed space. This implies an absence of temperature differentials and an inability to perform work. Quality, or order, was more important than quantity, and net quality changed in one direction only. The universe as a whole is an isolated system and thus must be inevitably progressing toward a "heat death" in which all energy is evenly dispersed.

This notion was radical in the early nineteenth century and had profound implications for science as well as philosophy. If the laws of mechanical physics were universal, then the universe was governed by the same principles as a pool table. Not only was there no such thing as irreversible change, but if one could determine the position and velocity of every atom in the universe, one would know the past and could predict the future. Though this implies no free will, no alternatives, and no sense in worrying about policy, it was during the nineteenth century the reigning worldview among scientists in the West and still holds considerable sway today. In the world of mechanical physics, the circular flow vision of economics discussed in Chapter 2 makes sense, as one can continually return to the same starting point. In a world where entropy reigns, it cannot.

Indeed, if we accept the laws of thermodynamics,[16] the entire nature of the economic system is entropic. The First Law of Thermodynamics tells us that we cannot make something from nothing and hence that all human production must ultimately be based on resources provided by nature. These resources are transformed through the production process into something of use to humans, and transformation requires work. Only low entropy or free energy can provide work. The First Law also ensures

---

[14]L. Walras, quoted in Heilbronner, ibid., p. 225.

[15]Alfred Marshall, perhaps the most famous of the founding fathers of NCE, argued that in the future, the complex science of biology would provide a better model for economics, but in the meantime he relied extensively on the methodologies of physics. Heilbronner, ibid.

[16]Though in truth, physical laws, such as gravity, function the same whether we accept them or not!

that any waste generated by the economy cannot simply disappear but must be accounted for as an integral part of the production process. And the entropy law tells us that inevitably whatever resources we transform into something useful must disintegrate, decay, fall apart, or dissipate into something useless, returning in the form of waste to the sustaining system that generated the resource. The economy is thus an ordered system for transforming low-entropy raw materials and energy into high-entropy waste and unavailable energy, providing humans with a "psychic flux" of satisfaction in the process. Most importantly, the order in our economic system, its ability to produce and provide us with satisfaction, can be maintained only by a steady stream of low-entropy matter-energy, and this high-quality, useful matter-energy is only a fraction of the gross mass of matter-energy of which the Earth is composed.

> **THINK ABOUT IT!**
> *Many people have proposed putting our toxic waste output onto rockets and shooting it into space. Based on your knowledge of thermodynamics, do you think this is a feasible solution for the pollution problem? Why or why not?*

While we stress the fundamental importance of entropy to the economic process, we do *not* advocate an "entropy theory of value" similar to the classical economists' "labor theory of value." Value has psychic roots in want satisfaction, as well as physical roots in entropy. To propose an "entropy theory of value" would be to focus on the supply side only and neglect demand. And even on the supply side, entropy does not reflect many qualitative differences in materials that are economically important (e.g., hardness, strength, ductility, conductivity). On the other hand, any theory of value that ignores entropy is dangerously deficient.

■ STOCK-FLOW RESOURCES AND FUND-SERVICE RESOURCES

We now turn our attention to an important distinction between different types of scarce resources too often neglected by conventional economists: that of stock-flow and fund-service. Conventional economics uses the phrase "factors of production." Factors of production are the inputs into a production process necessary to create any output. For example, when you make a pizza, you need a cook, a kitchen with an oven, and the raw ingredients. If you think about it carefully, however, you will clearly see that the cook and kitchen are different in some fundamental ways from the raw ingredients. The cook and kitchen are approximately the same after making the pizza as before, though just a bit more worn out. The raw ingredients, however, are used up, transformed first into the pizza itself, then rapidly thereafter into waste. The cook and kitchen are not physically embodied in the pizza, but the raw ingredients are. Thousands of years ago, Aristotle discussed this important distinction and divided causation (factors) into *material cause*, that which is transformed, and *efficient cause*, that which causes the transformation without itself being transformed in the process. Raw ingredients are the material cause, and the cook and kitchen are the efficient cause.

Other differences between these factors of production also exist. If we have enough raw ingredients to make 1000 pizzas, those ingredients could be used to make 1000 pizzas in one night, or one pizza a night for 1000 nights (assuming the ingredients were frozen and wouldn't spoil, and we had enough cooks and kitchens). The economy can use the existing stock of raw materials at virtually any rate, and time is not a factor. The productivity of raw ingredients is simply measured as the physical number of pizzas into which they can be transformed. In addition, as the ingredients for a pizza are produced over time, those ingredients can be used when they are produced, or stockpiled for future use. In contrast, while a cook or a kitchen may be capable of producing many thousands of pizzas over the course of their lifetimes, they can produce no more than a few pizzas in any given evening, even if limitless ingredients are available. The productivity of cooks and kitchens is measured as a number of pizzas per hour. However, this productivity cannot be stockpiled. For example, if we rest a cook for 6 nights, his capacity to produce a week's worth of pizzas cannot be used up all on the seventh night.

Georgescu-Roegen used the terms "stock" and "fund" to distinguish between these fundamentally different types of resources. A stock-flow resource is materially transformed into what it produces. A stock can provide a flow of material, and the flow can be of virtually any magnitude; that is, the stock can be used at almost any rate desired. Time does not enter into the equation, so the appropriate unit for measuring the production of a stock-flow resource is the physical amount of goods or services it can produce. Further, a flow can be stockpiled for future use. Finally, stock-flow resources are used up, not worn out. A fund-service resource, in contrast, suffers wear and tear from production but does not become a part of (does not become embodied in) the thing produced. Instead, a fund provides a service at a fixed rate, and the appropriate unit for measuring the service is physical output per unit of time. The service from a fund cannot be stockpiled for future use, and fund-service resources are worn out, not used up.[17] Note that the classification of a

---

[17]Georgescu-Roegen, *The Entropy Law*, op. cit.

resource as stock-flow or fund service is a function of its use. A bicycle for sale in a bike store is a stock-flow, while the bicycle you ride every day is a fund-service.

Box 4-1	STOCK-FLOW AND FUND-SERVICE RESOURCES

In the academic literature, there are many distinct definitions for stocks, flows, funds, and services. To make it clear, we are discussing the specific definitions given here. Future references will be to stock-flow and fund-service resources.

Stock-flow resources
- Are materially transformed into what they produce (material cause).
- Can be used at virtually any rate desired (subject to the availability of fund-service resources needed for their transformation), and their productivity is measured by the number of physical units of the product into which they are transformed.
- Can be stockpiled.
- Are used up, not worn out.

Fund-service resources
- Are not materially transformed into what they produce (efficient cause).
- Can be used only at a given rate, and their productivity is measured as output per unit of time.
- Cannot be stockpiled.
- Are worn out, not used up.

The stock-flow and fund-service concepts are important when analyzing human production, and probably more so when focusing on the goods and services provided by nature. Note that "material cause" is always stock-flow in nature, and "efficient cause" is always fund-service.

THINK ABOUT IT!

*Think about a specific ecosystem—or better yet, go visit one, and take along a field notebook. Make a list of three stock-flow resources provided by (or found in) that ecosystem and three fund-service resources. (Note that you will need to be very specific about the use of each resource. For example, drinking water is a stock-flow, while water for swimming is a fund-service.) Check off the attributes of stock-flow and fund-service for each (see Box 4.1). See Chapter 2 in the **Workbook for Problem-Based Learning** that accompanies this text for more on stocks, funds, excludability and rivalness.*

■ EXCLUDABILITY AND RIVALNESS

Excludability and rivalness are also crucial concepts for economic analysis, and rivalness is in fact related to the stock-flow, fund-service distinction. Although conventional economists first introduced these concepts, they rarely receive the attention they deserve. We believe they are important enough to be described in some detail both here and in Chapter 10.

Excludability is a legal principle that when enforced allows an owner to prevent others from using his or her asset. An **excludable resource** is one whose ownership allows the owner to use it while simultaneously denying others the privilege. For example, in modern society, when I own a bicycle, I can prohibit you from using it. In the absence of social institutions enforcing ownership, nothing is excludable. However, the characteristics of some goods and services are such that it is impossible or highly impractical to make them excludable. While someone could conceivably own a streetlight on a public street, when that streetlight is turned on, there is no practical way to deny other people on the street the right to use its light. There is no conceivable way that an individual can own climate stability, or atmospheric gas regulation, or protection from UV radiation, since there is no feasible institution or technology that could allow one person to deny all others access. When no institution or technology exists that makes a good or service excludable, it is known as a **nonexcludable resource**.

Rivalness is an inherent characteristic of certain resources whereby consumption or use by one person reduces the amount available for everyone else. A **rival resource** is one whose use by one person precludes its use by another person. A pizza (a stock-flow resource) is clearly rival, because if I eat it, it is no longer available for you to eat. A bicycle (a fund-service resource that provides the service of transportation) is also rival, because if I am using it, you cannot. Although you can use it after I am done, the bicycle has worn out a bit from my use and is not the same as it was. A **nonrival resource** is one whose use by one person does not affect its use by another. If I use the light of a streetlight when riding my bike at night, it does not decrease the amount of light available for you to use. Similarly, if I use the ozone layer to protect me from skin cancer, there is just as much left for you to use for the same purpose. It is possible to deplete the ozone layer (through the emission of chlorofluorocarbons, for example), but depletion does not occur through use. Nonrival resources are not scarce in any conventional sense. They do not need to be allocated or distributed, and their use does not directly affect scale. Rivalness is a physical characteristic of a good or service and is not affected by human institutions. As we will discuss at length in Chapter 10, however, institutions can make nonrival resources such as information excludable, artificially enclosing them in the world of scarcity and allocation.

Note that all stock-flow resources are rival, and all nonrival goods are fund-service. However, some fund-service goods are rival. For example, my bicycle is a fund that provides the service of transportation, but it is rival; the ozone layer is a fund that provides the service of screening UV rays, but it is nonrival.

As you will see when we turn to allocative mechanisms in subsequent chapters, the concepts of rivalness and excludability are very important.

THINK ABOUT IT!

For the list of resources you made earlier, answer the following questions:

Is the resource rival or nonrival? In general, can you think of any stock-flow resources that are nonrival? Can you think of any fund-service resources provided by nature that are rival?

Is the resource excludable or nonexcludable? (Note that excludability may differ depending on the specific value in question.) If it is non-excludable, can you think of an institution or technology that could make it excludable? Do you think it should be made excludable? Why or why not?

Is the resource a market good or a nonmarket good?

In general, can you think of any stock-flow resources that cannot be made excludable? Can you think of fund-service resources provided by nature that can be made excludable?

Exercise 2.2 of the Workbook for Problem-Based Learning accompanying this text expands on this and previous questions we explore in "Think About It!"

■ GOODS AND SERVICES PROVIDED BY THE SUSTAINING SYSTEM

To make this discussion of entropy, fund-services, stock-flows, exclud-ability, and rivalness more concrete, and to really understand the implica-tions for economic theory and policy, we must see how these concepts apply to the specific scarce resources available to our economy: the goods and services provided by nature. We undertake this task in the next two chapters and conclude this one by simply introducing the scarce re-sources.

For our purposes, we will present eight types of goods and services provided by nature, divided for convenience into nonliving and living resources. Clearly this is an enormous abstraction from the number and complexity of resources our Earth actually does supply, but these categories illustrate why the specific characteristics of goods and services we have described are of fundamental importance to economic policy.

1. *Fossil fuels.* For practical purposes, fossil fuels are a nonrenewable source of low-entropy energy. They are also very important as material building blocks.

2. *Minerals.* The Earth provides fixed stocks of the basic elements in varying combinations and degrees of purity, which we will refer to hereafter simply as minerals. This is the raw material on which all economic activity and life itself ultimately depend. Rocks in which specific minerals are found in relatively pure form we refer to as ores. Ores in which minerals are highly concentrated are a nonrenewable source of low-entropy matter. We will refer to mineral resources and fossil fuels together as **nonrenewable resources** and the first five goods and services in this list as abiotic resources (see Chapter 5).

3. *Water.* The Earth provides a fixed stock of water, of which freshwater is only a miniscule fraction. All life on Earth depends on water, and human life depends on freshwater.

4. *Land.* The Earth provides a physical structure to support us that is capable of capturing the solar radiation and rain that falls upon it. Land as a physical structure, a substrate, or a *site* has economic properties unrelated to the productivity of its soil, and is thus distinct from land as a *source* of nutrients and minerals. To capture this distinction, we will refer to land as a physical structure and location as Ricardian land.[18] The quantity and quality of soil available on a given piece of Ricardian land will be grouped with minerals, discussed below.

5. *Solar energy.* The sustaining system provides solar energy, the ultimate source of low entropy upon which the entire system depends.

6. *Renewable resources.* Life is able to harness solar energy to organize water and basic elements into more useful structures (from the human perspective) that we can use as raw materials in the economic process. Only photosynthesizing organisms are capable of achieving this directly, and virtually all other organisms, including humans, depend on these primary producers. These biological resources are traditionally referred to as **renewable resources**, but they are renewable only if extracted more slowly than the rate at

---

[18]Ibid., p. 232.

which they reproduce. Clearly, species can be exploited to extinction, so, as we shall see, biological resources are exhaustible in a way that mineral resources are not.

7. *Ecosystem services*. Living species interact to create complex ecosystems, and these ecosystems generate **ecosystem functions**. When functions are of use to humans, we refer to them as **ecosystem services**. Many of these ecosystem services are essential to our survival.

8. *Waste absorption*. Ecosystems process waste, render it harmless to humans, and, in most cases, again make it available to renewable resource stocks as a raw material input. This is really a specific type of ecosystem service but one whose economic characteristics make it worth classifying on its own. We refer to these last three goods and services as biotic resources (see Chapter 6).

We refer to all the structures and systems that provide these goods and services as **natural capital**. In the following chapters, we will examine these resources in the light of entropy, fund-services, stock-flows, excludability, and rivalness.

BIG IDEAS to remember

- Laws of Thermodynamics
  –Conservation of matter-energy
  –The law of increasing entropy
- Stock-flow resource

- Fund-service resource
- Excludable and nonexcludable resources
- Rival and nonrival resources
- Eightfold classification of resources

Abiotic Resources

In this chapter, we closely examine the five **abiotic resources** introduced in Chapter 4: fossil fuels, minerals, water, land, and solar energy. Our goal is to explain how the laws of thermodynamics, the distinction between stock-flow and fund-service resources, and the concepts of excludability and rivalness relate to these resources, in order to better understand the role they play in the ecological-economic system. We will also assess the extent to which substitutes are available and the degree of uncertainty associated with each resource. As we will see, however, abiotic resources are fundamentally different from each other, and it is their even greater dissimilarity from biotic resources that binds them together more than their similarity to each other. Perhaps the most important distinction is that biotic resources are simultaneously stock-flow and fund-service resources that are self-renewing, but human activities can affect their capacity to renew. Abiotic resources are either nonrenewable (fossil fuels) or virtually indestructible (everything else).

The differences between abiotic resources probably deserve more emphasis than their similarities, and we'll start with a brief summary. Fossil fuels and mineral resources are frequently grouped together under the classification of nonrenewables. The laws of thermodynamics, however, force us to pay attention to an important difference: The energy in fossil fuels cannot be recycled, while mineral resources can be, at least partially. Water is one of the most difficult resources to categorize, precisely because it has so many different forms and uses. Fossil aquifers (those that are not being recharged) are in some ways similar to mineral resources—once used, the water does not return to the ground, and while it cannot be destroyed, it can become less useful when polluted by chemicals, nutrients, or salt. Rivers, lakes, and streams, in contrast, share similarities with biotic resources: They are renewable through the hydrological cycle, driven by solar energy, and they can exhibit stock-flow and fund-service properties simultaneously. However, human activity cannot affect the total stock of water to any meaningful extent, while we can and do irreversibly destroy biotic resources. Similarly, land as a physical substrate, a location (hereafter referred to as Ricardian land), cannot be produced or destroyed in significant amounts by human activity (with the exception of sea level rise induced by anthropogenic climate change), and solar energy flows are not meaningfully affected by humans at all, although we can affect the amount of solar energy that moves in and out of the atmosphere. We now examine these resources in more detail.

■ FOSSIL FUELS

Perhaps the simplest resource to analyze is **fossil fuels,** or hydrocarbons, upon which our economy so dramatically depends. It would take an estimated 25,000 hours of human labor to generate the energy found in a barrel of oil. The fact that the fossil fuel economy and market economy emerged simultaneously in England over the course of the eighteenth century is almost certainly no coincidence. The magic of fossil fuels may be more important than the magic of the market in generating today's living standards.[1] In 1995, crude oil supplied about 35% of marketed energy inputs into the global economy, followed by coal at 27% and natural gas at 23%. In all, 86% of the energy in the economy comes from fossil hydrocarbons.[2] In geological terms and as far as humans are concerned, fossil fuels are a fixed stock. For a variety of reasons, however, it is extremely difficult to say precisely how large that stock is.

For practical purposes, we are concerned only with recoverable supplies. But what does *recoverable* mean? Clearly, hydrocarbons are found in deposits of varying quality, depth, and accessibility, and there are different costs associated with the extraction of different deposits. In economic terms, we can define recoverable supplies as those for which total extraction costs are less than the sales revenues. However, fossil fuel prices fluctuate wildly, and recoverable supplies defined in this way show similarly chaotic variation through time. We could also define recoverable supplies in entropic terms, in which case a hydrocarbon is recoverable if there is a net energy gain from extraction; that is, it takes less than a barrel of oil to recover a barrel of oil. This measure must include all the energy costs, in-

---

[1]M. Savinar, "Life after the Oil Crash." http://www.lifeaftertheoilcrash.net/Research.html. This Web site provides full references for the calculations. D. Pimentel and M. Pimentel. *Food, Energy, and Society.* 3d edition. (Boca Raton, FL: CRC Press, 2007). Account for the entropic energy loss of converting gasoline to work in an internal combustion engine, in which case it would take only 5,000 hours of labor.

[2]Energy Information Administration, Washington, DC. International Energy Outlook, 2009.

cluding those of exploration, machinery, transportation, decommissioning, and so on. While technological change can reduce these, there is a certain irreducible limit to the energy costs of extracting fossil fuels. It takes 9.8 joules of energy to lift 1 kilogram 1 meter, and no amount of technology can change that basic fact.

As we deplete the most accessible hydrocarbon supplies first, over time it will take more and more energy to recover remaining supplies. In other words, the *energy return on investment* (EROI), which is "the ratio of gross fuel extracted to economic energy required directly and indirectly to deliver the fuel to society in a useful form," declines over time.[3] In entropic terms, the energy cost of oil and natural gas extraction in the United States increased by 40% from 1970 to the 1990s.[4] During the 1950s in the U.S., every barrel of oil invested in exploration led to the discovery of about 50 more. By 1999, the ratio was about one to five. A sustainable society probably needs an EROI of at least three to one.[5] Still, under either the economic or the entropic definition of *recoverable*, estimates of recoverable reserves change constantly. Largely this is the result of new discoveries, but it also results from dramatically different methods for calculating "proven" supplies between different companies and different countries, with frequent changes often based on political or economic motives.[6] Petroleum geologists can assign reasonable probabilities to different estimates of total stocks, however.

BOX 5-1 │ ESTIMATING OIL STOCKS

Every year, the world consumes in the neighborhood of 25 billion barrels of oil (Gbo). Yet at the end of most years, reported reserves of oil are greater than they were at the start, and there is a fairly wide range of estimates as to what those reserves actually are. The increase is possible as long as new oil discoveries are greater than oil consumed, but that has not occurred in decades. For example, in 1997 the world used about 23 Gbo and discovered 7 Gbo, yet estimated reserves *increased* by 11 Gbo. How do we explain this anomaly?

When geologists estimate the quantity of oil in any given field, they assign a probability to the estimate. For example, in the late 1990s

---

[3]C. Cleveland, R. Costanza, C. Hall, and R. Kaufmann, Energy and the US Economy: A Biophysical Perspective, *Science* 225:297 (1984).

[4]C. Cleveland and D. Stern, "Natural Resource Scarcity Indicators: An Ecological Economic Synthesis." In C. Cleveland, D. Stern, and R. Costanza, eds., *The Economics of Nature and the Nature of Economics*. Cheltenam, England: Edward Elgar, 2001.

[5]C. A. S. Hall, S. Balogh and D. J. R. Murphy, What Is the Minimum EROI That a Sustainable Society Must Have? *Energies* 2:25–47(2009).

[6]C. J. Campbell and J. H. Laherrère, The End of Cheap Oil, *Scientific American*, March 1998.

geologists estimated that the Oseberg field in Norway would supply 700 million barrels of oil with 90% certainty (known as probability 90, or P90) and 2.5 billion with 10% certainty (known as P10). Different corporations and countries generally use some number within the P10–P90 range when stating their reserves, and they are often purposefully vague about what number they use. Higher reported reserves can increase stock prices, provide greater access to credit, and for OPEC countries, increase their quotas. As oil fields are exploited, geologists can use the information acquired to make better estimates about how much they contain. Based on this information and other factors (e.g., moving from P90 to P50 estimates), countries frequently revise their reserve estimates from existing fields, often upward. In the absence of major new discoveries or technological breakthroughs in the late 1980s, six OPEC countries alone revised their estimates upward by 287 Gbo, 40% more than all the oil ever discovered in the U.S.!

When calculating global oil reserves, it makes the most sense to sum the P50 estimates across countries, but even this is no easy task. In addition, revised estimates from existing reserves are not new discoveries and should not be counted as such.[a]

---

[a]C. J. Campbell and J. H. Laherrère, The End of Cheap Oil, Scientific American, March 1998.

THINK ABOUT IT!

Economists argue that price reflects scarcity. Do you think the price of oil is a good indicator of how much oil is left in the ground? Why or why not?

Regardless of what the stocks of fossil fuels are, however, they are stocks that can be extracted as flows, and the rate of flow is determined largely by human efforts. If we had adequate infrastructure, we could theoretically extract all entropically recoverable fossil energy stocks in a single year, or we could make them last 1000 generations. How long recoverable stocks will last, therefore, is determined as much by how fast we extract them as by how much there actually is. We almost certainly will never exhaust fossil fuel stocks in physical terms, because there will always remain some stocks that are too energy-intensive or too expensive to recover. From this point of view, fossil fuel stocks are nonrenewable but not exhaustible.

As we extract fossil fuels, we will logically extract them from the most accessible and highest-quality known reserves first, where net energy gains are highest.[7] These stocks essentially offer the lowest-entropy re-

---

[7]Note that the largest and most accessible reserves are also the most likely to be discovered first.

source. Therefore, as we continue to extract fossil fuels over time, we can expect not only a *quantitative* decrease but also a *qualitative* decline in stocks. For example, the first oil to be extracted actually pooled on the surface and erupted in geysers from wells with no pumping. But as stocks diminish, it takes more and more energy to extract energy; ever-larger fractions of a barrel of oil are needed as energy inputs to retrieve a barrel of oil as output, until we have reached entropic exhaustion.

Of course, resource exhaustion is only one component of fossil fuel use. Used fuel does not disappear; it must return to the ecosystem as waste. Acid rain, global warming, carbon monoxide, heat pollution, and oil spills are unavoidably associated with the use of fossil fuels. On a small scale, some of these wastes could be readily processed by natural systems, but on the current scale, they pose serious threats. Indeed, the growing accumulation of waste products from fossil fuel use and the negative impacts these have on planetary ecosystems is probably a far more imminent threat to human welfare than depletion; the sink will be full before the source is empty.

We must reiterate here that ecosystems, via the primary producers they sustain, themselves capture solar energy, and humans make direct use of much of the energy they capture. If waste products from fossil fuel use diminish the ability of these ecosystems to capture energy, there are more energy costs to fossil fuel extraction than the direct ones discussed above. These costs are, however, several degrees of magnitude more difficult to measure—and therefore that much more likely to be ignored (Figure 5.1).

THINK ABOUT IT!

Many people are concerned by the United States' dependence on oil imports from a number of politically unstable regions and countries (e.g., the Middle East, Nigeria, Venezuela, Colombia). Proposed solutions to this problem have included increased domestic drilling and extraction, greater energy efficiency, and the development of renewable energy sources. What do you think are the pros and cons of each approach?

The basic equation here is:

net recoverable energy from oil = (initial total stock of entropically recoverable reserves) – (oil already consumed) – (energy cost of extraction) – (loss of solar energy due to induced loss of capacity to capture)

Net energy from fossil fuels must account for the damage fossil fuel use causes to the ability of the sustaining system to capture solar energy, a fund-service resource. This lost capacity is measured as energy-flow/time,

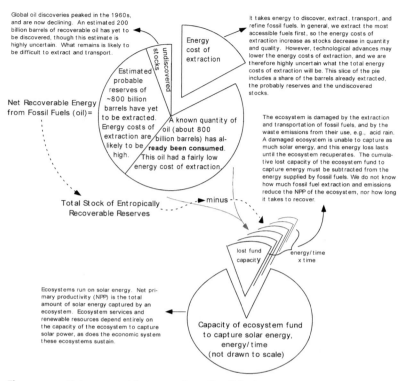

Figure 5.1 • Net recoverable energy from fossil fuels.

and we must account for the total amount of energy not captured from the time the damage occurs to the time the fund-service recovers.[8]

What points can we draw from this discussion of fossil fuels? First, once fossil fuels are used, they are gone forever—they are rival goods. While a seemingly trivial point, this has important implications for economic policy, as we will show in Chapter 11. Second, while fossil fuel stocks are finite, they are a stock-flow resource that can be extracted virtually as quickly as we wish, limited only by existing infrastructure, knowledge of stock locations, and the energy costs of extraction. We have control of the spigot and have been opening it a bit wider every decade. Eventually the reservoir must run dry. This is in stark contrast to flows of solar energy, as we pointed out in Chapter 4.

Third, our current populations and economic systems depend for survival on the use of fossil fuels. Fossil fuels not only supply 85% of our en-

---

[8]Estimates of oil already consumed, probable reserves, and oil yet to be discovered are from Campbell and Laherrère, ibid.

ergy needs, much of which is used to produce food, they also provide the raw materials for a substantial portion of our economic production, including ubiquitous plastics and, even more importantly, the fertilizers, herbicides, and pesticides that help provide food for nearly 7 billion people. At this point, we do not have the technologies available to support 7 billion people in the absence of fossil fuels.

While we may be able to substitute renewable energy for fossil fuels, it is highly uncertain that we can achieve this before the negative impacts of fossil fuel waste products force us to stop using them or the fuels themselves are depleted.

THINK ABOUT IT!
The U.S. and Canada have vast deposits of shale oil and tar sands, respectively. Both of these are fossil fuels but of fairly low quality, requiring more energy to extract and process than conventional fossil fuels and creating more associated waste. Do you think these resources present possible solutions to our energy problems? Can you dig up any information on their energy returns to investment and waste outputs?

■ MINERAL RESOURCES

Though typically grouped together with fossil fuels in economics textbooks and labeled nonrenewable resources, minerals differ in important respects from fossil fuels. Like fuels, minerals can be analyzed in terms of stocks and flows. We know the total stock is finite, and according to the First Law of Thermodynamics, this imposes a physical limit on their contribution to the material growth of the economy. Again, technology can increase the efficiency with which we extract minerals from ore, but there exists an entropic limit to efficiency. Valuable mineral deposits occur in varying degrees of purity, and, like fuels, the degree of purity can be looked at as a measure of low entropy. Highly concentrated ores are highly ordered low entropy.[9] It is much easier to extract their mineral content, and they are much more valuable. As our growing economy depletes these most valuable ores first, we must move on to ores of lower and lower purity, incurring higher and higher processing costs.

As in the case of oil, we are not exactly certain of the total stock of any particular mineral, but geologists assign reasonable probabilities to different estimates. Even the most efficient process conceivable will require

---

[9]Even if we do not accept the notion of entropy in materials, concentrated ores require much less low-entropy energy to process.

some energy to extract minerals from an ore, and the less pure the ore, the more energy will be needed. Currently, mining accounts for about 10% of global energy use.[10] However, unlike fossil fuels that cannot be burned twice, materials can be recycled (though this, too, requires energy). Therefore, we must think in terms of nonrenewable subterranean stocks as well as aboveground stocks, which accumulate as the subterranean ones are depleted. Still, we cannot avoid the laws of entropy even here, and use leads to dissipation through chemical and physical erosion; therefore, 100% recycling of any material may be impossible.

There is considerable debate over the impossibility of 100% recycling, as well as the implications. Georgescu-Roegen argues that because solar energy can provide a substitute for fossil fuels and nothing can provide a substitute for minerals, mineral depletion is actually more of a concern than fossil fuel depletion, and its inevitability means that a steady-state economy[11] is impossible (see Box 3.2). In contrast, Ayres claims that even if all elements in the Earth's crust were homogeneously distributed (the material equivalent of "heat death" mentioned above), a sufficiently efficient solar-powered extraction machine would enable us to extract these elements,[12] presumably at a rate that would provide enough raw materials to maintain the machine and still leave a material surplus. This scenario implicitly assumes that damage caused by extracting all the resources from the Earth's crust in the first place, and their consequent return to the ecosystem as waste, would not irreparably damage the Earth's ability to capture solar energy and sustain life.

Alternatively, we may be able to master the art of creating polymers from atmospheric CO_2, which could provide substitutes for many of the minerals we currently use. If such polymers were biodegradable and simply returned to the atmosphere as CO_2 we would presumably be able to achieve 100% recycling (though in this case we may not want to, at least not before atmospheric CO_2 stabilizes at preindustrial levels). Of course, none of these propositions can currently be proven empirically. Nonetheless, it appears that mineral deposits are sufficiently large, and recycling has the potential to become sufficiently efficient, that with careful use, minimizing waste and appropriate substitution where possible, we could sustain a steady-state economy for a very long time.

Figure 5.2 depicts both the accumulation of extracted minerals into

---

[10] P. Sampat, From Rio to Johannesburg: Mining Less in a Sustainable World. World Summit Policy Brief #9. Online: http://www.worldwatch.org/worldsummit/briefs/20020806.html. (World Watch).

[11] N. Georgescu-Roegen, *The Entropy Law and the Economic Process*, Cambridge, MA: Harvard University Press, 1971.

[12] R. U. Ayres, The Second Law, the Fourth Law, Recycling and Limits to Growth, *Ecological Economics* 29:473–484 (1999).

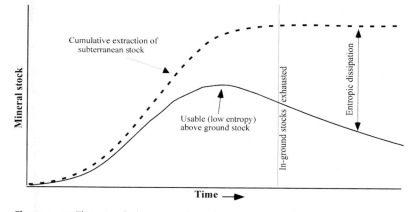

Figure 5.2 • The cumulative extraction of subterranean stock and aboveground stock of minerals over time. The distance between the two curves is a measure of entropic dissipation.

aboveground low-entropy stocks embodied in artifacts in use by (or available to) the economy (solid line) and the cumulative depletion (extraction) of subterranean stocks (dashed line) over time. We assume that initial rates of mineral extraction are low, but they increase with economic growth and greater knowledge of reserve location. Eventually, however, stocks become scarcer, the costs of extraction become greater than the benefits, and extraction ceases. The point where this occurs is labeled "In-ground stocks exhausted" on the graph, and cumulative depletion ceases. In the absence of entropy, and if 100% recycling were possible and practiced, the two lines on the graph would be identical. In the real world, some portion of aboveground stocks dissipates into waste every year. The rate of increase in the aboveground stocks is equal to net annual mineral extraction minus entropic dissipation; that is, aboveground stocks are equal to minerals currently in use plus those that can be recycled.

There are two important categories of waste. Much waste is in the form of products that have stopped working, become obsolete, or simply gone out of fashion and are discarded while still in a relatively ordered state. They are not recycled because it is either cheaper or more convenient to extract virgin mineral flows from the Earth. For our purposes, this waste returns to the subterranean stock, though with higher entropy than the ore from which it was initially extracted.[13] Eventually, as we deplete the

---

[13]Some of this material will be in a highly ordered state and have lower entropy than the same amount of mineral in the form of an ore. Georgescu-Roegen distinguishes "garbo-junk" (a bald tire useless as a tire but recyclable) from "pure waste" (the dissipated rubber particles that are not recyclable). For practical purposes, however, large stocks of ore presumably still have an overall lower entropy; otherwise waste material would be processed before the ore.

most concentrated ores, it becomes cheaper to start mining the lowest-entropy waste. For example, slag heaps near old silver mines have been mined again with newer methods. But the slag heaps resulting from the second mining will be harder to mine.

Another type of waste results from entropy in the form of mechanical or chemical erosion of the material in question. Pennies eventually wear out through use—an atom rubbed off here, another there. Other metals rust away. Hence, gross subterranean stocks are never depleted (First Law of Thermodynamics). They simply become stocks of higher and higher entropy, akin to bound energy, and are no longer of use to humans (Second Law of Thermodynamics).

We have some control over the creation of waste in the form of discarded goods, but virtually none over the effects of entropy. The entropic limit to extraction in this case occurs when the extraction process consumes more material than it can provide. As we stated earlier, some people assert that this never happens, while others assert it will happen soon enough to make a steady-state economy a pipe dream; we take the middle road.

As more minerals are brought to the surface and put to use, entropy acts on a larger stock. As more subterranean stocks are extracted, the remainder becomes more difficult to find and extract. Therefore, even before in-ground stocks are exhausted, the rate of dissipation of aboveground stocks must become greater than the net extraction of new material, and aboveground stocks begin to decline. However, even after reaching the entropic limit to extraction, we are still likely to have a large stock of material in the economy that can be reused and recycled. Over time, of course, it must gradually erode away, atom by atom. In Figure 5.2, the distance between the lines depicting cumulative extraction and aboveground stocks measures cumulative dissipation, and eventually the entire aboveground stock must succumb to entropy. This process is probably slow enough that we could achieve a steady state through material recycling for a very, very long time. However, as is the case with oil, the threat to us is probably more from the impacts of the waste itself than from the exhaustion of mineral resources. We'll put this discussion off until we get to the section on waste absorption capacity in Chapter 6.

What points can we draw from our discussion? First, mineral resources are rival goods at a given point in time. If I am using a hunk of steel in my car, it is not available for you to use. But through recycling, most of these resources could be made available for someone else to use in the future. Thus, we can think of *mineral resources as rival goods within a generation but as partially nonrival between generations*, depending on how much is wasted and how much recycled. Fossil fuels are rival both within and between generations. Second, stocks of low-entropy mineral ores are finite

A resource is nonrival between generations if the use by one generation does not leave less of the resource for future generations.

but can be extracted at virtually any rate we choose. In contrast to fossil fuels, we have control not only over the spigot of extraction but also over the drain by which extracted materials return to the ecosystem as waste. We open the spigot wider almost every year and do very little to close the drain, but if we shut the drain as much as possible (though it will always leak some), the open spigot matters less to future generations. Third, we could not sustain existing populations or levels of economic production in the absence of these minerals. While it would clearly be impossible to develop substitutes for all minerals, thus far it has been reasonably easy to develop substitutes for specific minerals as they become scarce, and it may be possible to keep this up for some time to come.

■ WATER

Earth is a water planet. Though the stock of water is finite, fully 70% of the Earth's surface is covered in water. Freshwater, however, is far less abundant, accounting for less than 3% of the total, of which less than one-third of 1% is in the form of readily exploited lakes (0.009%), rivers (0.0001%), and accessible groundwater (0.31%). Another 0.01% is found in the atmosphere, 0.31% is deep groundwater, and over 2% is in the polar ice caps and glaciers.[14] Humans are composed mostly of water, and in addition to drinking, we depend on it for agriculture, industry, hydro-electricity, transportation, recreation, and waste disposal and for sustaining the planet's ecosystem services. Water for different uses has different relevant characteristics that make generalizations difficult.

Water for drinking, irrigation, industry, and waste disposal is clearly a stock-flow resource, but a unique one. In contrast to fossil fuels and mineral deposits, many water resources are renewable as a result of the hydrologic cycle. However, for all practical purposes, many aquifers are "fossil" water, with negligible recharge rates. Many other aquifers are being mined; that is, the rate of water extraction is greater than the rate of replenishment. Even many rivers around the world, including the Colorado and the Rio Grande in North America, the Amu-Dar'ja and Syr-Dar'ja rivers that once fed the Aral Sea in Central Asia, and at times the Yellow, Hai, and Huai rivers in northern China, are so heavily used (primarily for irrigation) that they never reach the sea.

At first glance, flowing water might appear to be a fund-service resource. In any stream or river at any given time, water is flowing at a specific rate, and the proper unit of measurement is volume/time (volume per unit of time), as is the case for fund-service resources. Dams, however,

---

[14]P. Gleick, *The World's Water: The Biennial Report on Freshwater Resources*, Washington, DC: Island Press, 2002.

allow us to stockpile flowing water for later use, which is a characteristic of stock-flow resources, and water is "used up" by drinking, irrigation, industry, and waste disposal but never "wears out."

Perhaps the best way to look at flowing water is to distinguish it from the hydrologic cycle. The water itself is a stock-flow resource that is rapidly renewed by the service (provided by solar energy) of the hydrologic cycle. Hydroelectricity is produced not by water but rather by the energy transferred to water by the hydrologic cycle—it is solar energy stored in water. Solar energy is generally a fund service, but when stored in water, it can be either a stock-flow or a fund-service resource. When mechanical energy in the water is converted to electric energy by a microhydropower plant that depends on river flow, it is essentially a fund-service resource. However, damming of the river allows the energy to be stockpiled by converting mechanical energy to potential energy, which is a stock-flow resource.

When used for transportation, recreation, or sustaining all other ecosystems on the planet, water functions as a fund-service resource. Atmospheric moisture, as part of the hydrologic cycle, is essentially a fund-service resource.

Like biotic resources, water can be a stock-flow and fund-service resource simultaneously. Unlike biotic resources, however, humans cannot meaningfully affect the total stock of water on the planet. We can and do reduce the stock of usable water, and while it is possible to restore the usability of water, there are no substitutes available for its most important uses.

As one would expect from its dual nature as a stock-flow, fund-service resource, water can be rival or nonrival depending on its use; stock-flow uses are rival, and fund-service uses are nonrival. However, as flowing water is recycled through the hydrologic cycle, it is intergenerationally nonrival. Excludability varies dramatically depending on existing institutions, though for all practical purposes rainfall is nonexcludable by nature.[15]

■ RICARDIAN LAND

Ricardian land—land as a physical substrate and location, distinct from its other productive qualities—is also a fund that provides the service of a substrate capable of supporting humans and our infrastructure and of capturing solar energy and rain (Ricardian land does not include soil or the nutrients in the soil). A hectare of land may be capable of producing 1000 tons of wheat over 100 years, but one cannot produce that wheat

---

[15]The seeding of clouds with silver nitrate can produce rainfall in a specific location, but for practical purposes this is irrelevant.

from the same land in an appreciably shorter period, nor would it be possible to accumulate land's capacity as a substrate.

The services provided by land are certainly excludable, and at any given point in time, they are also rival. For example, if used for farming, land provides the service of a substrate for crops. If one farmer uses that service, no one else can in the same time period. Economists often use the term "depletable" as a synonym for "rival," but the case of land suggests that this is inappropriate.[16] Using Ricardian land does not deplete it. While rival within a generation, it is intergenerationally nonrival and absolutely nondepletable.

> **THINK ABOUT IT!**
>
> *Why do you think we distinguish between Ricardian land as a physical substrate and the more conventional definition of land that includes the soil and its mineral content? Who or what creates value in Ricardian land? What makes land in one place more valuable than a similar piece of land elsewhere? Who or what creates value in fertile topsoil?*

■ SOLAR ENERGY

The last abiotic producer of goods and services we will discuss is the sun. It bathes the Earth in 19 trillion tons of oil equivalent (toe) per year—more energy than can be found in all recoverable fossil fuel stocks—and will continue to do so for billions of years.[17] Why the fuss over the consumption of the Earth's fossil fuels?

While the flow of solar energy is vast, it reaches the Earth at a fixed rate in the form of a fine mist and hence is very difficult to capture and concentrate. Most of the sunlight that strikes the Earth is reflected back into space.[18] Over the eons, life has evolved to capture enough of this energy to maintain itself and the complex ecosystems that life creates. It would appear that the "order" of the global ecosystem over billions of years has reached a more or less stable thermodynamic disequilibrium. A better term is "meta-stable," meaning that the global ecosystem fluctuates around a steady state rather than settling into one without further varia-

---

[16]When "depletable" is used in this sense, it means that one person's *use* depletes the resource in question. Hence, the ozone layer is *nondepletable* because if I use it to protect me from skin cancer, it is still there for someone else to use. It is certainly possible to deplete the ozone layer with chemicals, but that is not a case of depletion caused by use.

[17]Unless otherwise cited, estimates of energy use and availability are from World Energy Council, *2007 Survey of Energy Resources*, London: World Energy Council, 2007. Online: http://www.worldenergy.org.

[18]N. Georgescu-Roegen, *Energy and Economic Myths: Institutional and Analytic Economic Essays*, New York: Pergamon Press, 1976.

tion.[19] Virtually all energy captured from the sun is captured by chlorophyll. In the absence of the evolution of some alternative physiological process for capturing sunlight, it would seem that our planet cannot sustain more low entropy than it currently does for any extended period. Yet through the use of fossil fuels, Americans are able to consume 40% more energy than is captured by photosynthesis by all the plants in the country. We also directly use over half of the energy captured by plants.[20]

As fossil fuels run out, we will need an alternative source of low entropy to maintain our economy at its current level of thermodynamic disequilibrium. The sun unquestionably radiates the Earth with sufficient energy to meet our needs, but how do we capture it? Global gross commercial energy consumption is about 11.3 billion toe (~14.5 TW) per year. Biomass, hydroelectricity, wind, photovoltaics, wave, and ocean thermal energy are all forms of solar energy we could potentially capture. Biomass is widely touted as a substitute for fossil fuels, but converting *all* of the net primary productivity (NPP) of the United States to liquid fuel would still not meet our liquid fuel needs. Hydroelectricity currently provides 19% of global electricity, but even fully developed it could not supply 60%. Wind currently supplies little energy (about 72,000 MW-h in 2006), but it is a promising alternative: At current installation rates, capacity is *doubling* every 3.5 years.

Photovoltaics and wave/ocean thermal technologies still play very minor roles. With all of these technologies, however, large energy investments are needed to produce the infrastructure needed to capture solar energy, and in many cases (e.g., photovoltaics), the energy returns on investment may be negligible. At the same time, human activity decreases the surface area of the planet covered in plant life and disrupts the ability of plants to capture sunlight. The net effect is likely to be an annual decrease in the amount of solar energy the Earth captures and hence a decrease in the complexity of the systems it is capable of maintaining. Figure 5.1 earlier illustrates the loss of solar energy capture that can be attributed to waste from fossil fuels.

While solar energy will bathe the Earth in more energy than humans will ever use, for practical purposes it is a fund-service resource that arrives on the Earth's surface at a fixed rate and cannot be effectively stored for later use.[21] No matter how much solar energy one nation or land-

---

[19]E. Laszlo, *Vision 2020*, New York: Gordon and Breach, 1994.

[20]D. Pimentel and M. Pimentel, *Land, Energy and Water: The Constraints Governing Ideal U.S. Population Size*, Negative Population Growth, Forum Series, 1995. Online: http://www.npg.org/forum_series/land_energy&water.htm.

[21]Solar energy can be stored in fossil fuels, in batteries, or in the form of hydrogen for later use by humans, but this energy cannot subsequently be used to power photosynthesis, the most important function of solar energy.

owner captures, there is no less left for others to capture, and it is inherently nonexcludable.[22]

■ SUMMARY POINTS

Table 5.1 summarizes some of the policy-relevant characteristics of these five abiotic resources. Why are these details important to ecological economic analysis, and what message should you take home from this chapter? The stock-flow/fund-service distinction is important with respect to scale. We have control over the rate at which we use fossil fuels, mineral resources, and water. As the economy undergoes physical growth, it must use ever-greater flows from finite stocks. Because fossil aquifers and fuels are irreversibly depleted by use, and mineral resources may be irreversibly dissipated through use, the finite stock of these resources imposes limits on total economic production over time. Limits to growth may not be apparent until the stock is seriously depleted, and once gone, it is gone forever. Funds, in contrast, provide services at a fixed rate over which we have no control (though one thing that distinguishes biotic fund services from abiotic ones is that we can damage or even destroy them). Fund-services therefore limit the size of the economy at any given time, but they do not limit total production over time.

■ Table 5.1

SELECTED POLICY-RELEVENT CHARACTERISTICS OF ABIOTIC RESOURCES

Abiotic Resource	Stock-Flow or Fund-Service	Can Be Made Excludable	Rival	Rival Between Generations	Substitutability
Fossil Fuels (nonrecyclable)	Stock-flow	Yes	Yes	Yes	Modest at margin, but possibly substitutable over time
Minerals (partially recyclable)	Stock-flow	Yes	Yes	Partially	High at margin, ultimately nonsubstitutable
Water (solar recycling)	Context-dependent	Context-dependent	Context-dependent	Stocks, yes; funds and recycled, no	Nonsubstitutable for most important uses
Ricardian Land (indestructible)	Fund-service	Yes	Yes	No	Nonsubstitutable
Solar Energy (indestructible)	Fund-service	No	No, for practical purposes	No	Nonsubstitutable

[22]Future space-based solar technologies may change this but are irrelevant at present.

Water sources are a complex mix of stock-flow and fund-service. But even the stock-flow uses of water are completely recyclable—in particular, running water is so closely linked to the fund-service of the solar-powered hydrologic cycle that it acts much like a fund-service, imposing limits on the output of the economy only at a given point in time.

Substitutability is also relevant to scale. If we can develop a substitute for a resource, then the constraints it imposes on scale are less rigid. However, developing substitutes generally relies on technology, and technology takes time to develop. In addition, truly innovative technologies are impossible to accurately predict; we could predict one only if we already knew what it would look like, in which case it would not be truly innovative.

Rivalness is relevant primarily to distribution, both within and between generations. All abiotic resources are rival except for water in some of its forms and uses, and solar energy (for practical purposes). One person's use of these rival resources means they are not available for others to use, and we must be concerned about distribution within a generation. People can use the nonrival resources of solar energy and water in its fund-service functions without leaving less for anyone else and without affecting scale, and all else being equal, we should therefore let anyone use them. When a good is nonrival between generations, we needn't worry about excessive use within a generation. When addressing distribution, we must remember that all natural resources are produced by nature, not humans, although the value of Ricardian land is generally produced by society as a whole.

Excludability is relevant primarily to allocation. The market cannot allocate nonexcludable goods, and other allocative mechanisms are needed. However, in the case of sunlight and rainfall, allocation by human institutions is simply not feasible.

BIG IDEAS to remember

- Big ideas from Chapter 4 that recur in Chapter 5
- Ricardian land
- Energy return on investment
- Recoverable reserves
- P10, P50, and P90 reserve estimates
- Net recoverable energy from fossil fuels

- Aboveground and subterranean mineral stocks
- Entropic dissipation
- Rival within versus between generations
- Garbo-junk versus pure waste
- Unique characteristics of water and solar energy

6

Biotic Resources

Biotic resources include the raw materials upon which economic production and human life depend, the ecological services that create a habitat capable of supporting human life, and the absorption capacity that keeps us from suffocating in our own waste. As nonrenewable resources are exhausted, human society will come to rely more and more on the self-renewing capacity of biotic resources. It is therefore critically important that we understand the nature of these resources.

As we turn our attention from abiotic resources to biotic ones, we must address a quantum increase in complexity, inevitably accompanied by a quantum increase in ignorance and uncertainty. One level of complexity arises from the intrinsic value we give to living systems. Abiotic resources are almost entirely considered means to various ends, where one of the foremost ends is the sustenance of life, the maintenance of biotic resources. Biotic resources not only enhance human well-being directly, they are also considered by many to be an end in their own right, especially in the case of sentient creatures. Biotic resources are also physically complex in two ways. First, the processes responsible for the sustained reproduction of individuals, populations, or species are highly complex and poorly understood. Second, individuals, populations, and species interact with other individuals, population, and species, as well as abiotic resources, to create an ecosystem. Ecosystems are extraordinarily complex and dynamic, changing over time in inherently unpredictable ways. The differences between these two types of physical complexity bear closer examination.

■ ECOSYSTEM STRUCTURE AND FUNCTION

Ecologists look at ecosystems in terms of structure and function, corresponding to the two types of physical complexity mentioned above. This distinction is very relevant to economic analysis. Conventional natural resource economics is essentially the economics of ecosystem structure. Environmental economics focuses on certain ecosystem functions. In reality, structure and function are mutually interdependent, and we need an economics that effectively integrates both. Certainly we must understand the distinctions and interactions between the two if we are to incorporate them into economic analysis.

Ecosystem structure refers to the individuals and communities of plants and animals of which an ecosystem is composed, their age and spatial distribution, and the abiotic resources discussed in Chapter 5.[1] Most ecosystems have thousands of structural elements, each exhibiting varying degrees of complexity. Scientists have learned that when enough separate elements are thrown together into a complex system, a sort of spontaneous order results. One property of such systems is their tendency to generate emergent phenomena, which can be defined as properties of the whole that could not be predicted from an understanding of the individual parts, no matter how detailed that understanding. Complex systems are also characterized by highly nonlinear behavior, which means that we cannot predict the outcomes of large interventions based on an understanding of smaller ones. For example, removing 40% of a species stock from an ecosystem may have a qualitatively different impact than removing 20%—that is, not just twice the known impact of removing 20%.

In an ecosystem, the structural elements act together to create a whole that is greater than the sum of the parts. We refer to these emergent phenomena in ecosystems as **ecosystem functions**,[2] and they include such things as energy transfer, nutrient cycling, gas regulation, climate regulation, and the water cycle. As is typical of emergent properties, ecosystem functions cannot be readily explained by even the most extensive knowledge of system components.[3] Variability, ignorance, and uncertainty play an extremely important role in the analysis of ecosystem structure and a far greater role in the analysis of ecosystem function. We have a very limited understanding of exactly how ecosystem functions emerge from the complex interactions of ecosystem structure and thus a difficult time

---

[1] It may seem strange to include such things as fossil fuels and mineral deposits as elements of ecosystem structure, but we must not forget that humans are part of the global ecosystem, and these resources affect our ability to thrive.

[2] Whether a particular element of an ecosystem is part of structure or part of function depends on perspective. Organelles are part of the structural components of a cell that enable the cell to function. Cells are structural components of an individual that enable the individual to function. In the same way, individuals are part of the structure of a population, a population is part of the structure of a local ecosystem, an ecosystem is part of the structure of a landscape, and a landscape is part of the structure of the global ecosystem.

[3] E. Odum, *Ecology: A Bridge Between Science and Society*, 3rd ed., Sunderland, MA: Sinauer, 1997.

Box 6-1   RISK, UNCERTAINTY, AND IGNORANCE

Whenever we do not know something for sure, we are uncertain, but there are different types of uncertainty. When I throw dice, I cannot say in advance what the outcome will be, but I do know the possible outcomes and their probabilities. This type of uncertainty is referred to as risk. Pure uncertainty occurs when we know the possible outcomes but cannot assign meaningful probabilities to them. Ignorance or absolute uncertainty occurs when we do not even know the range of possible outcomes.

In economics, Frank Knight pointed out that risk is a calculable or insurable cost, while pure uncertainty is not. In his view profit—the difference between revenue and calculable risk-adjusted costs—is a return for willingness to endure pure uncertainty. However, Knight was discussing the case where the entrepreneur bore the costs of failure and reaped the rewards of success. In economic decisions regarding exploitation of ecosystems, it is often the entrepreneur who reaps the rewards, while society bears the costs.[a]

Discoveries in quantum physics and chaos theory suggest that uncertainty and ignorance do not result simply from a lack of knowledge but are irreducible, inherent properties in certain systems. For example, chaos theory shows that even in a deterministic (i.e., nonrandom) system, extremely small differences in initial conditions can lead to radically different outcomes. This has been popularized as the butterfly effect, in which a butterfly flapping its wings over Japan can create a storm in North America.

Change in highly complex systems is characterized by ignorance, especially over long time spans. We cannot predict evolutionary change in organisms, ecosystems, or technologies. For example, while we can predict that computers will continue to get faster and cheaper, we cannot predict what the next big technology will be 50 years from now. Leading experts are often notoriously wrong even when predicting the future of existing technologies. Bill Gates reputedly once predicted that no one would ever need more than 540 kilobytes of computer memory.

Estimating stocks of natural resources or reproductive rates for cultivated species is basically a question of risk. Estimating reproductive rates for wild species is a question of uncertainty, since we cannot accurately predict the multitude of factors that affect these reproduction rates, but we do know the range over which reproduction is possible. Estimating ecological thresholds, conditions beyond which ecosystems may flip into alternative states, is a question of pure uncertainty, since we have limited knowledge of ecosystems and cannot predict the external conditions that affect them. Predicting the alternate state into which an ecosystem might flip when it passes an ecological threshold, and how humans will adapt, are cases of absolute ignorance involving evolutionary and technological change.

---

[a]F. H. Knight, Risk, Uncertainty, and Profit, *Boston: Houghton Mifflin, 1921; Library of Economics and Liberty, Feb. 21, 2002. Online: http://www.econlib.org/library/Knight/ knRUPl.html.*

predicting and managing the impacts of human actions on these functions. Therefore, a great deal of uncertainty attends decision making involving ecosystem functions. How we choose to treat uncertainty in economic analysis is ultimately a normative (ethical) decision, yet another source of complexity. One of the most important issues concerning any analysis of biotic resources is the degree of uncertainty involved.

Concrete examples always help clarify a concept. To illustrate the links between structure and function, and the implications of complexity, let's focus on a wet tropical forest, the terrestrial ecosystem that exhibits the greatest biodiversity of any yet studied. The forest is composed of individual plants (part of ecosystem structure). Each plant alone has little impact on climate, nutrient cycling, and habitat provision and may even be unable to reproduce. However, when we bring together hundreds of millions of plants, as in the Amazon or Congo basin, these and other ecosystem functions emerge.

The forest canopy filters out about 98% of the sunlight at ground level, dramatically reducing daytime temperatures. It traps air and insulates, increasing night temperatures under the canopy and maintaining high and constant humidity. Trees absorb the energy of tropical storms, aerate the soil to allow water absorption, and slow water flows—all of which prevent soil and nutrients from being washed out of the system. Trees create the microclimate and habitat essential to the soil fauna that help recycle nutrients, facilitating their reabsorption by the system.

On a regional scale, the water retained by forest structure is absorbed and returned to the atmosphere through evapotranspiration, increasing humidity over the forest. Greater humidity increases the frequency of rainstorms. Estimates for the Amazon forest suggest that it generates up to 50% of its own rainfall, enabling the water-dependent species there to thrive. Without the increased absorption capacity of the soil and the evapotranspiration it facilitates, rainfall would simply drain into the rivers and be flushed from the system forever.

On an even larger scale, forests absorb up to 90% of the solar energy that strikes their canopies. Much of this is released through evapotranspiration and carried high up into the atmosphere, where it is carried into the temperate zones, helping stabilize the global climate (a function provided by carbon sequestration as well).

The species and populations in the forest cannot survive without a stable climate and a steady nutrient flow. Loss of forest structure can degrade forest function to the point where the forest spontaneously declines, creating a positive feedback loop with potentially irreversible and catastrophically negative consequences. Numerous models suggest continued deforestation in the Amazon could lead to dramatic declines in rainfall, increased susceptibility to fires (such as those that occurred in 1997 in the

Amazon, Indonesia and Mexico), and spontaneous degradation of the remaining forest.[4]

In other words, ecosystem structure interacts to create ecosystem functions, and the structural elements depend on these functional attributes for their own survival. Owing to the complex nature of the whole system, as structural elements of an ecosystem are lost, in most cases we cannot say for sure to what extent ecosystem functions will be affected. Similarly, as ecosystem functions change in response to human impacts or non-anthropogenic change, we cannot say for certain what the impact will be on ecosystem structure.

> **THINK ABOUT IT!**
>
> *In Chapter 4, we asked you to make a list of stock-flow and fund-service resources provided by a local ecosystem. Which of these resources are elements of ecosystem structure? Which are elements of ecosystem function? Do you see any links between these classifications?*

Roughly speaking, conventional and natural resource economics has focused on ecosystem structure, while conventional environmental economics has focused on certain elements of ecosystem function, with a major emphasis on waste absorption capacity and the monetary valuation of other functions. In reality (as many conventional economists are fully aware), ecosystem structure and function are mutually interdependent, and conclusions based on the analysis of one dimension may not apply to the multidimensional case. With this caveat in mind, we now turn our attention to specific categories of biotic resources.

Three basic categories of biotic resources deserve attention. First are **renewable resources**, the elements of ecosystem structure that provide the raw materials for economic processes. Second are **ecosystem services**, defined as the ecosystem functions of value to humans and generated as emergent phenomena by the interacting elements of ecosystem structure. Third is **waste absorption capacity**, an ecosystem service that is sufficiently distinct from the others to warrant separate treatment.

■ RENEWABLE RESOURCES

For simplicity, we can treat biological resources as material stock-flow resources, that is, as elements of ecosystem structure. Like nonrenewable resources, biological stocks can be extracted as fast as humans desire, but

---

[4]D. Nepstad et al., Interactions Among Amazon Land Use, Forests and Climate: Prospects for a Near-Term Forest Tipping Point. *Philosophical Transactions of the Royal Society B: Biological Sciences* 363:1737–1746(2008).

they are capable of reproduction. Figure 6.1 depicts a renewable natural resource in stock-flow space: the x-axis depicts the stock, or amount of resource that exists, and the y-axis depicts the flow. The flow in this case can be the rate of reproduction (or biomass increase) that is likely for any given stock, or the rate of extraction (harvest). A 45-degree dashed line shows the theoretically maximum rate at which we can extract a given stock (i.e., we can extract the entire stock at one time; stock = flow). Actual extraction rates must lie on or below this line. We have also drawn a curve that shows the growth rate of each level of stock, which is also the sustainable yield curve. The sustainable yield is the net annual reproduction from a given stock; for every population of a resource, there is an associated average rate of population increase, and that increase represents a sustainable harvest that can be removed every year without affecting the base population.

We must caution here that there is a great deal of uncertainty concerning the position of this sustainable yield curve. Not only do we not know at precisely what rate a given population will reproduce, we are also uncertain of the exact population of any given species, though this is more true for animals than for plants, since plants sit still for the census takers while animals do not. While with careful study and census techniques we can assign reasonable probabilities to population estimates for renewable resource stocks (risk), there is greater uncertainty of a qualitatively different type concerning reproduction rates, particularly because these rates depend on a host of "external" factors such as rainfall, abundance of predator and prey species, disease, and so on. In addition, habitat destruction and degradation, pollution, climate change, and other human impacts can profoundly affect the entire curve, shifting it dramatically over time. Thus, in any given year, the actual rate of increase from a given population stock may be wildly different from the average.

As most people are at least vaguely aware, stocks of plants and animals in nature cannot grow forever. Instead, populations reach a point where they fill an available niche, and average death rates are just matched by average birth rates. Populations "stabilize" around an equilibrium, known as the **carrying capacity**. (We use the term *stabilize* loosely, because populations fluctuate in the short term depending on weather conditions, predator-prey cycles, etc. and in the longer term depending on a wide variety of factors. To paraphrase John Maynard Keynes, in the very long term, all species go extinct.) At carrying capacity, there is just enough food and habitat to maintain the existing population, and the rate of growth in biomass is zero (point K in Figure 6.1). Obviously, the rate of growth of a stock is also zero when the stock has

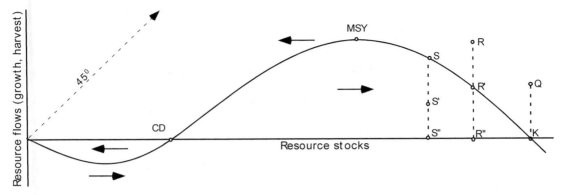

Figure 6.1 • The growth, or sustainable yield, curve.

been driven extinct (the origin in Figure 6.1). In between these two points, things get interesting.

The growth, or sustainable yield, curve for a renewable natural resource indicates the increase in stock over one time period for any given stock. The y-axis can measure growth or harvest, that is, flows from or to the existing stock depicted on the x-axis. Any harvest up to the total stock is theoretically possible. A harvest at any point on the sustainable yield curve, such as S, is just equal to the growth in the stock and hence has no net impact on stocks. Harvests above the sustainable yield curve deplete the resource, and harvests below the curve lead to an increase in the stock, as indicated by the arrows. For example, a harvest at R will reduce the stock to S″, and a harvest at S′ will allow the stock to increase to R″. MVP represents the minimum viable population, which is the level below which a species or stock cannot sustain itself even in the absence of harvest. The phrase **critical depensation** refers to the spontaneous decline of a population or ecosystem that has fallen below the minimum viable population or size.[5] K represents carrying capacity. The graph suggests extremely rapid growth rates—about 30% per time period at **maximum sustainable yield (MSY)**, which is the average maximum catch that can be removed under existing environmental conditions over an indefinite period without causing the stock to be depleted, assuming that removals and natural mortality are balanced by stable recruitment and growth. While this may be appro-

---

[5]The first edition of this textbook used the term *critical depensation level* instead of *minimum viable population*. *Critical depensation* is useful when referring to ecosystems that cannot sustain themselves when stocks fall below a certain level, since it makes little sense to talk about size of an ecosystem as a population. However, *minimum viable population*, in general, is a more intuitively obvious phrase.

priate for small, rapidly reproducing species, growth rates for many economically important species are on the order of 1% per year or less.

THINK ABOUT IT!

Prior to the arrival of humans in North America, where on the graph in Figure 6.1 would you locate American bison populations? Ten thousand years later, approximately where on the graph do you think bison harvests occurred? After the introduction of the horse, where do you think they occurred? After the introduction of the rifle? After the settling of the frontier by Europeans? What impact would the conversion of the Great Plains to agriculture have on the sustainable yield curve for bison?

What happens when we remove some fish, for instance, from a population at carrying capacity? In terms of the graph, we represent this as a harvest at point Q. As there is zero net productivity at our starting point K, harvest Q reduces the stock of fish by the quantity Q-K, and we find ourselves at stock R″. At the lower population stock, there are fewer fish competing for available food, shelter, and breeding grounds, and the remaining fish get more of each than they would under more crowded conditions. This greater resource abundance per individual leads to increased growth rates and fertility. With less competition for spawning areas, a higher percentage of eggs are laid in desirable locations, thereby increasing recruitment.[6] In addition, most species' growth is fastest in their youth (as a percentage of biomass) and slower as they age. If we harvest larger, older fish, the remaining population has a higher percentage of recruits and potentially faster net growth.

At stock R″, we could continue annual harvests forever at point R′, just equal to the annual rate of increase of the stock. Any harvest below the sustainable yield curve would be less than the annual growth rate. Flows can accumulate into stocks, and the population would increase. Any harvest above the sustainable yield curve would reduce the stock even further. The arrows above and below the curve indicate the direction of change in stock for harvests in each of these regions. For example, a harvest at point R would reduce the population to stock S″. At S″, per-capita resource abundance would be even greater than at R″, increasing net annual growth and sustainable harvest from R′ to S.

Over a certain range, lower population stocks can generate higher sustainable yields, but this obviously cannot go on forever. Eventually, the breeding population is insufficient to sustain high yields. Insufficient

---

[6]Note that a harvest at point R decreases the stock in the subsequent period by an amount equal to R-R′ (as measured on the y-axis), and a harvest at point S′ increases the next period's stock by S-S′. Thus, R-R′ and S-S′ as measured on the y-axis equal S″-R″ as measured on the x-axis.

Box 6-2 | MINIMUM VIABLE POPULATION, MAXIMUM SUSTAINABLE YIELD, AND UNCERTAINTY

Using the notion of maximum sustainable yield to help communicate ideas in ecological economics is a far cry from using the concept as a tool in resource management. In reality, the MSY will vary dramatically from year to year in response to climatic cycles such as the El Niño Southern Oscillation (ENSO), changes in populations of predator and prey species of the species of interest, changes in pollution levels, and a broad array of other ecological changes and cycles. Under the most stable conditions, natural variability can mask the effects of economic exploitation, and the scale of human impacts is rapidly changing the global ecosystem. Science relies on replication and controls, and neither of these is possible when dealing with a unique species in a highly complex and rapidly changing ecosystem. We cannot scientifically estimate MSY accurately enough for use in resource management.[a]

Where the minimum viable population lies, and whether or not one exists for a given species or population, is similarly marked by extreme uncertainty. When a population becomes small enough, it is more susceptible to stochastic events and the negative impacts of inbreeding. For the North American passenger pigeon, probably once the most numerous bird on Earth, numbering in the billions but driven extinct in a matter of decades, it appears the minimum viable population was quite high, perhaps because of its colonial nesting habit. At the other extreme, the Mauritius Kestrel has rebounded from a known population of 6 in 1974 to a present population of over 600, though it would have almost certainly gone extinct without substantial conservation efforts, and inbreeding may make it highly susceptible to disease or other stochastic shocks. There is currently a debate over whether or not the blue whale and some populations of North Atlantic cod are below their minimum viable population. Obviously, experiments to scientifically determine minimum viable populations could not be replicated, as the first trial could wipe out the species in question.

On the other hand, uncertainty as to where the MSY is does not mean that we are not overshooting it, nor does it relieve policy makers of the responsibility to decide acceptable levels of offtake. As a general rule, the higher the uncertainty, especially in the presence of irreversibility (extinction), the more conservative (i.e., lower) the allowed offtake should be.

[a]D. Ludwig, R. Hilborn, and C. Walters, *Uncertainty, Resource Exploitation, and Conservation: Lessons from History,* Science *260:17, 36 (1993).*

numbers of eggs lead to insufficient recruits, reducing yield in spite of superabundant resources per individual within the population. This means that at some point there is a maximum sustainable yield, MSY on Figure 6.1. Again, we caution that the MSY may vary dramatically from year to year, and there is probably no way we can accurately estimate it in any given year. Although it is useful as a pedagogical device, it has very little value as a calculating device for setting annual quotas.[7] If harvests continue to be greater than annual yield, the population will continue to fall. Eventually, fish will become too scarce to find each other for breeding, and other ecological mechanisms (many only poorly understood) can break down. This means that at some population above zero, we may reach a minimum viable population, at which point the rate of growth is zero. Below this point the population enters into spontaneous decline, that is, death rates exceed birth rates.

Note that the sustainable yield curve enters into negative numbers below the minimum viable population. This means that in order to sustain the population, a negative harvest is needed; that is, recruits have to be placed into the system every year just to maintain the existing stock. Harvest is still possible anywhere below the 45-degree line, but this will simply lead to even more rapid extinction (or extirpation) of the population. Unfortunately, we do not know what the minimum viable population is and can only make very rough (and highly contentious) estimates of where the maximum sustainable yield occurs.

The same basic concepts explained here apply to plant species or even plant communities. For example, what happens when we clear some trees from a virgin forest for timber? Light, water, and nutrients become available to the other trees already there, accelerating their growth, and space becomes available for new seedlings to sprout. This process can initially speed up the growth rate and increase the sustainable yield of a forest. Soon, however, sources of new seeds become more distant from the cleared land, and recruitment slows. Nutrients are lost from the soil as trees are removed. Trees of the same species are too far apart to cross-pollinate, resulting in sterile seeds or the problems of inbreeding. The sustainable yield begins to fall. And as we described above, the removal of ecosystem structure can dramatically affect ecosystem function, with the potential of further reducing the capacity of the forest to reproduce itself. Thus, like animal populations, a stock of forest will show a maximum sustainable yield and a minimum viable size.

We will return to this analysis in Chapter 12 when we examine the microeconomics of biotic resource allocation.

---

[7]D. Ludwig, R. Hilborn, and C. Walters, Uncertainty, Resource Exploitation, and Conservation: Lessons from History, *Science* 260:17, 36 (1993).

Box 6-3	CASSOWARIES, MINIMUM VIABLE POPULATIONS, AND CRITICAL DEPENSATION OF ECOSYSTEMS

In the rainforests of northeastern Australia, up to 100 species of large-seeded fruit trees depend almost entirely on a single bird species for distribution. This bird, the cassowary, is a large ratite, an ostrich-like bird that lives in the forest. It is the only animal known in the region capable of swallowing and transporting very large seeds, up to 2 kg of which can be found in a single scat. Evidence suggests that some seeds must pass through the digestive tract of a cassowary before they can germinate.

Cassowaries need large home territories to survive, especially in the highland forests. As forests are cleared in a patchwork pattern, few areas remain that can sustain a viable cassowary population. Without cassowaries, many trees in the region will be unable to disperse, and some may not even be able to germinate. Eventually, these species are likely to go extinct. Other plants and animals depend on these species, and they too will go extinct, igniting a chain reaction of extinction in species that may in turn depend on them. The net result could be a dramatic change in forest composition, leading to a qualitatively different ecosystem. The entire process could take a very long time. It might not be noticed until centuries after it is too late.[a]

Such examples of mechanisms for critical depensation are just a few of the possibilities that have been proposed. Again, we must emphasize that we really have little idea where maximum sustainable yields or critical depensation points lie. Ignorance, uncertainty, and variability are our constant companions in the real world.

---

[a]J. Bentrupperbaumer, Conservation of a Rainforest Giant, Wingspan 8(Dec.): 1–2. (1992). Also extensive personal communications.

■ ECOSYSTEM SERVICES

In our discussion of ecosystem structure and function, we explained why forests need the functions generated by forests to survive, but we also hinted at the presence of extensive benefits that ecosystem functions provide for humans. *We call an ecosystem function that has value to human beings an ecosystem service.* For example, forested watersheds help maintain stable climates necessary for agriculture, prevent both droughts and floods, purify water, and provide recreation opportunities—all invaluable services for watershed inhabitants. But ecosystems provide many more services, of course. Unfortunately, we are unsure exactly how ecosystem structure creates ecosystem services, and we are often completely unaware of the services they generate. For example, prior to the 1970s, most people were unaware that the ozone layer played a critical role in making our planet habitable.[8] If we also take into account the tightly interlocking nature of ecosystems, it's safe to say that humans benefit in some way from almost any ecosystem function.

We just described forests as a stock of trees that generates a flow of trees. Now we want to look at the forest as a creator of services; as such, it is very different from a stock of trees. A stock of trees can be harvested at any rate; that is, humans have control over the rate of flow of timber produced by a stock of trees. Trees can also be harvested and used immediately or stockpiled for later use. Ecosystem services are fundamentally different. We cannot use climate stability at any rate we choose—for example, drawing on past or future climate stability to compensate for the global warming we may be causing today. Nor can we stockpile climate stability for use in the future. Nor does climate stability become a part of what it produces. If timber is used to produce a chair, the timber is embodied in that chair. If climate stability is used to produce a crop of grain, that grain in no way embodies climate stability. Furthermore, climate stability is not altered by the production of a crop of grain (unless perhaps the grain is grown on recently deforested land, but still it is the deforestation and not the grain that affects climate stability).

Intact ecosystems are funds that provide ecosystem services, while their structural components are stocks that provide a flow of raw materials. However, recall that stock-flow resources are used up, and fund-service resources are worn out. But when ecosystems provide valuable services, this does not "wear them out." The fact is, however, that ecosystems would "wear out" if they did not constantly capture solar energy to renew themselves. The ability of ecosystem fund-services to reproduce themselves distinguishes them in a fundamental way from manmade fund-services. Depreciating machines in a factory do not automatically reproduce new machines to replace themselves.

Examples of ecosystem services provided by a forest may help clarify the concept. Costanza et al. describe 17 different goods and services generated by ecosystems.[9] Forests provide all of these to at least some degree. Of these, food and raw materials are essentially stock-flow variables, though their ability to regenerate is a fund-service. The remaining fund-service variables included are described in Table 6.1.

---

[8]As further evidence of the extreme uncertainty concerning ecosystem function and human impacts upon it, in 1973 physicist James Lovelock, famous for the Gaia hypothesis, to his later regret stated that fluorocarbons posed no conceivable hazard to the environment. M. E. Kowalok, Common Threads: Research Lessons from Acid Rain, Ozone Depletion, and Global Warming, *Environment* 35(6):12–20, 35–38 (1993).

[9]R. Costanza et al., The Value of the World's Ecosystem Services and Natural Capital, *Nature* 387:256, Table 2 (1997).

■ Table 6.1

EXAMPLES OF SERVICES PROVIDED BY ECOSYSTEMS

Ecosystem Service	Examples from Forests
Gas regulation	Trees store CO_2 and growing trees create O_2; forests can clean SO_2 from the atmosphere.
Climate regulation	Greenhouse gas regulation; evapotranspiration and subsequent transport of stored heat energy to other regions by wind; evapotranspiration, cloud formation, and local rainfall; effects of shade and insulation on local humidity and temperature extremes.
Disturbance regulation	Storm protection, flood control (see water regulation), drought recovery, and other aspects of habitat response to environmental variability controlled mainly by vegetation structure.
Water regulation	Tree roots aerate soil, allowing it to absorb water during rains and release it during dry times, reducing risk and severity of both droughts and floods.
Water supply	Evapotranspiration can increase local rainfall; forests can reduce erosion and hold stream banks in place, preventing siltation of in-stream springs and increasing water flow.
Waste absorption capacity	Forests can absorb large amounts of organic waste and filter pollutants from runoff; some plants absorb heavy metals.
Erosion control and sediment retention	Trees hold soil in place, forest canopies diminish impact of torrential rainstorms on soils, diminish wind erosion.
Soil formation	Tree roots grind rocks; decaying vegetation adds organic matter.
Nutrient cycling	Tropical forests are characterized by rapid assimilation of decayed material, allowing little time for nutrients to run off into streams and be flushed from the system.
Pollination	Forests harbor insects necessary for fertilizing wild and domestic species.
Biological control	Insect species harbored by forests prey on insect pests.
Refugia or habitat	Forests provide habitat for migratory and resident species, creating conditions essential for reproduction of many of the species they contain.
Genetic resources	Forests are sources for unique biological materials and products, such as medicines, genes for resistance to plant pathogens and crop pests, ornamental species.
Recreation	Ecotourism, hiking, biking.
Cultural	Aesthetic, artistic, educational, spiritual, and scientific values of forest ecosystems.

THINK ABOUT IT!

Of the ecosystem services in Table 6.1, which are rival and which are excludable? Which would be impossible to make excludable?

Again we emphasize that the precise relationship between the quantity and quality of an ecosystem fund and the services it provides is highly uncertain and is almost certainly characterized by nonlinearities, thresholds, and emergent properties. We can say with reasonable confidence that the larger an ecosystem fund and the better its health, the more services it is likely to generate. As we deplete or degrade a complex ecosystem fund, we really cannot predict what will happen with any reasonable probability. Since we have defined *service* as an anthropocentric concept, we do know that it can be dramatically affected by human presence and use and not just by abuse. For example, a highly degraded forest in an urban setting may offer more water regulation and more recreational and cultural services (as measured by benefits to humans) than a pristine forest remote from human populations. Forests near orchards or other insect-pollinated crops may offer far more valuable pollination services.

Perhaps even more critical for the economic problem of efficient allocation of ecosystem services is their spatial variation. To use an example already described, large tropical forests can regulate climate at the local level, the regional level, and the global level. Flood control and water purification provided by forests may benefit only select populations bordering local rivers and floodplains, and the provision of habitat for migratory birds may benefit primarily populations along the migratory pathways.

Ecosystem services have some other characteristics that make them extremely important economically. Probably most important, it is unlikely that we can develop substitutes for most of these services, including the provision of suitable habitat for humans. We scarcely understand how these services are generated, and we are not aware of all of them. At the cost of some $200 million, a billionaire named Edward Bass initiated the Biosphere Two project in Arizona to see if he could develop substitutes for these services sufficient to sustain only eight people. The project failed. Imagine creating substitutes for billions of people! In addition, most ecosystem services are nonrival—if I benefit from a forest's role in reducing floods, providing habitat for pollinators, or regulating atmospheric gases, it does not diminish the quantity or quality of those services available to anyone else. Many ecosystem services (though certainly not all) are nonexcludable by their very nature as well.

The Relationship Between Natural Capital Stocks and Funds

In review, the structural elements of an ecosystem are stocks of biotic and abiotic resources (minerals, water, trees, other plants, and animals), which when combined together generate ecosystem functions, or services. The use of a biological stock at a nonsustainable level in general also depletes a corresponding fund and the services it provides. Hence, when we harvest trees from a forest, we are not merely reducing the stock of trees but are also changing the capacity of the forest to create ecosystem services, many of which are vital to our survival. The same is true for fish we harvest from the ocean, except we know even less about the ecosystem services produced by healthy oceanic ecosystems.

The relationship between natural capital stock-flow and fund-service resources illustrates one of the most important concepts in ecological economics: It is impossible to create something from nothing; all economic production requires a flow of natural resources generated by a stock of natural capital. This flow comes from structural components of ecosystems, and the biotic stocks are also funds that produce ecosystem services. Therefore, an excessive rate of flow extracted from a stock affects not only the stock and its ability to provide a flow in the future but also the fund to which the stock contributes and the services that fund provides. Even abiotic stocks (i.e., elements and fossil fuels) can be extracted and consumed only at some cost to the ecosystem. In other words, production requires inputs of ecosystem structure. Ecosystem structure generates ecosystem function, which in turn provides services. All economic production thus has an impact on ecosystem services, and because this impact is unavoidable, it is completely internal to the economic process.

■ WASTE ABSORPTION CAPACITY

But this is only half the story. The laws of thermodynamics ensure that raw materials once used by the economic system do not disappear but instead return to the ecosystem as high-entropy waste. They also ensure that the process of producing useful (ordered) products also produces a more than compensating amount of disorder, or waste. Much of this waste can be assimilated by the ecosystem. Indeed, waste assimilation and recycling are ecosystem services on which all life ultimately depends. However, as a fund-service, waste absorption occurs only at a fixed rate, while conversion of stock-flow resources into waste occurs at a rate we can choose. Waste absorption capacity is a sink for which we have control over the flow from the faucet but not over the size of the drain. The removal of ecosystem structure also affects the ability of the ecosystem to process waste. If we discharge waste beyond the ecosystem's capacity to absorb it, we can reduce the rate at which an ecosystem can absorb waste, which makes the waste accumulate more quickly. In time, the waste buildup will affect other ecosystem functions, though we cannot always predict which services will be affected and when.

A specific example can help illustrate these points. When we first begin to dump wastes, such as raw sewage and agricultural runoff, into a pristine lake, they will be heavily diluted and cause little harm. Higher waste loads may threaten humans who use the lake with intermittent health problems from bacteria and noxious chemicals contaminating the sewage, and water becomes unsuitable for drinking without prior treatment. Increasing nutrients allow bacterial and algal populations to thrive, increasing the ability of the system to process waste but reducing a number of other ecosystem services. Fish will begin to accumulate noxious compounds present in the waste stream and become inedible. Pollution-sensitive species will be extirpated. Yet more waste may make the water unsuitable for drinking even after extensive processing, and eventually it will become too contaminated for industrial use. Excess nutrients eventually lead to eutrophication, where algal and bacterial growth absorbs so much oxygen during the night[10] and during the decay process that fish, amphibians, and most invertebrate species die out. Birds and terrestrial animals that depend on the lake for water and food will suffer. With even greater waste flows, even algae may fail to thrive, and we have surpassed the waste absorption capacity of the system. Waste begins to accumulate, further decreasing the ability of algae to survive and leading to a more rapid accumulation of waste even if the waste flow is not increased any more. The system collapses.

Prior to the point where waste flows exceed the waste absorption capacity, a reduction in flows will allow the system to recuperate. After that point, it may not. Similar dynamics apply to other ecosystems. If the ecosystem in question provides critical life-support functions, either locally or globally, the costs of exceeding the waste absorption capacity of an ecosystem are basically infinite, at least from the perspective of the humans it sustains.

In general, ecosystems have a greater ability to process waste products from biological resources and a much more limited capacity to absorb manmade chemicals created from mineral resources. This is because ecosystems evolved over billions of years in the presence of biological wastes. In contrast, products such as halogenated cyclic organic compounds and plutonium (two of the most pernicious and persistent pollutants known) are novel substances with which the ecosystem has had no evolutionary experience and therefore has not adapted.

In contrast to many ecosystem services, waste absorption capacity is rival. If I dump pollution into a river, it reduces the capacity of the river

---

[10]While growing plants are net producers of oxygen and absorbers of CO_2, they also require oxygen for survival. During the day, photosynthesis generates more oxygen than the plants consume, but at night they consume oxygen without producing any. Average oxygen levels may be higher, but the lowest levels determine the ability of fish and other species to survive.

to assimilate the waste you dump in. It is also fairly simple to establish institutions that make waste absorption excludable, and many such institutions exist.

The bottom line is that the laws of thermodynamics tell us that natural resources are economic throughputs. We must pay close attention to where they come from and where they go.

Table 6.2 summarizes some of the important characteristics of the three biotic resources. We will discuss these characteristics and examine their policy relevance in greater detail in Chapter 12 and Part VI.

The points to take away from this chapter deserve reiteration. First, humans, like all animals, depend for survival on the ability of plants to capture solar energy in two ways: directly as a source of energy and indirectly through the life-support functions generated by the global ecosystem, which itself is driven by the net primary productivity of plants. There are no substitutes for these life-support functions. Second, every act of economic production requires natural resource inputs. Not only are these inputs being used faster than they can replenish themselves, but when these structural elements of ecosystems are removed, they diminish ecosystem function. Third, every act of economic production generates waste. Waste has a direct impact on human well-being and further diminishes ecosystem function. While the removal of mineral resources may have little direct impact on ecosystem function, the waste stream from their extraction and use is highly damaging to ecosystems and human well-being in the long run. As the economy expands, it depletes nonrenewable resources, displaces healthy ecosystems and the benefits they provide, and degrades remaining ecosystems with waste outflows.

Biotic resources are unique because they are simultaneously stocks and funds, and their ability to renew themselves is a fund-service. This means

▪ Table 6.2

ECONOMIC CHARACTERISTICS OF BIOTIC RESOURCES

Biotic Resource	Stock-Flow or Fund-Service	Can Be Made Excludable	Rival	Rival Between Generations	Substitutability
Renewable Resources	Stock-flow	Yes	Yes	Depends on rate of use	High at margin, ultimately nonsubstitutable
Ecosystem Services	Fund-service	For most, no	For most, no	No	Low at margin, nonsubstitutable
Waste Absorption Capacity	Fund-service	Yes	Yes	Depends on rate of use	Moderate at margin, nonsubstitutable

that ultimately economic scale is determined by the amount of fund-services provided in a given year, where one of those fund-services is the ability of renewable natural resources to renew. Biotic resources have a particularly large impact on scale because they ultimately have no substitutes, and we cannot survive without them.

BIG IDEAS to remember

- Ecosystem structure
- Ecosystem function
- Ecosystem services
- Stock-flow and fund-service resources
- Risk, uncertainty, ignorance

- Carrying capacity
- Minimum viable population
- Critical depensation
- Maximum sustainable yield
- Waste absorption capacity

CHAPTER

7

From Empty World
to Full World

Given the undeniable importance of entropy to the economic process and the resulting fact that "sustainable economic growth" is an oxymoron,[1] how do we explain the unwavering devotion to continuous economic growth by economists, policy makers, and the general public in the face of ecological and natural resource limits? Apparently, people believe that the economic system faces no limits to growth or that the limits are far off. The laws of thermodynamics ensure that there are limits to growth. Now we must briefly address the question of how close those limits are.

Certainly for most of human history, including the time when modern economic theory was being developed, human populations and levels of resource use were quite low. Material and energetic limits to growth appeared so far off that it seemed sensible to ignore them and concentrate on developing a system that efficiently allocated the much scarcer labor, capital, and consumer goods. But since the development of market economies and neoclassical explanations thereof, both human populations and per-capita levels of resource use have been increasing exponentially. The success of the market system reduced the relative scarcity of market goods and increased that of nonmarket goods and services provided by the

---

[1]To reiterate, we do not believe that sustainable growth of the "psychic flux" of satisfaction (what we would call development, not growth) is an oxymoron as long as that flux is not produced by ever-increasing natural resource consumption. However, as far as we know, economic growth as measured by GNP has never occurred without increased throughput. Even when each unit of GNP requires fewer resources, the net outcome has always been greater throughput. While this need not be the case, the free market economy seems poorly suited for promoting the activities that provide improved human well-being without increasing throughput.

111

Box 7-1	HOW CLOSE ARE WE TO A FULL WORLD?

Exponential growth occurs when a system keeps growing at a certain rate. For example, from 1900 to 2000, the per-capita material output of the global economy grew at about 2.3% per year. One can calculate the doubling time of a given growth rate by dividing 72 by that growth rate. This means that per-capita output doubled more than three times during the twentieth century. Over the same period, the human population has increased from 1.6 billion to 6.1 billion, almost a fourfold increase. The total material output increased more than 36 times in the twentieth century.[a] How many more times can our material output double?

Our situation may have parallels to a well-known riddle. If the area of a petri dish covered in bacteria doubles every hour, you inoculate the dish at noon on day one and it is completely full at noon two days later (and thereafter the population crashes because it has exhausted its food source and inundated the petri dish with waste), when is the dish half full? The answer, of course, is at 11 A.M. on the final day. At 9 A.M., $7/8$ of the resources available for continued growth are still present. The question right now for humans is: How close is it to noon?

Humans, of course, are very different from bacteria, and the Earth is different from a petri dish. Humans can control their rate of reproduction and, to an extent, the quantity of resources they use. The Earth hosts numerous ecosystems capable of providing renewable resources and processing wastes. However, human adaptation to resource scarcity requires taking time to develop new technologies, new institutions, and new ways of thinking—perhaps a great deal of time. Essentially, the closer it is to noon, the less time we have to develop and implement the necessary changes to show that we actually are substantially different from bacteria in a petri dish. (See Figure 7.1.)

[a]*Calculations by author from data found in Chapter 5 of J. B. Delong,* Macroeconomics, *Burr Ridge, IL: McGraw-Hill Higher Education, 2002.*

Figure 7.1 • How close are we to a full world?

sustaining system. What follows is a very quick assessment of how full the world is and how close we are to resource exhaustion.

THINK ABOUT IT!

The world is always "full" of some things and "empty" of others. In the "full world," what is it that the world is relatively full of? Relatively empty of? How is the fullness with respect to some things related to the emptiness with respect to others?

■ FOSSIL FUELS

As fossil fuels run the world economy and are among the most well studied of the resources required to sustain us, we will assess their limits first. At first glance, exhaustion hardly seems imminent. Economists tell us that price is a measure of scarcity, yet the price of crude oil averaged $24 per barrel between 1899 and 1999 (in 2008 dollars), when the price was only $23.60.[2] However, as we mentioned earlier, we can extract fossil fuels at virtually any rate we want, and it is the scarcity of flow that determines prices, not the scarcity of stocks (a point we return to in Chapter 11). In the regions with the vastest reserves, installed extraction capacity, more than the size of underground stocks, determines flow rates. Best estimates suggest that if we continue to extract oil at the same rate, we will exhaust probable stocks in about 40 years, yet the Energy Information Administration estimates that global demand for oil will increase by nearly 40% over the next 30 years.[3] As we said above, the net energy returns to fossil fuel exploration are declining dramatically. The same is true for new discoveries, which peaked in 1962 at 40 billion barrels per year[4] and fell to 6 billion barrels per year in the 1990s. Consumption currently (2008) stands at 31 billion barrels per year, exceeding new discovery rates by a factor of 2 to 6.[5] Although the rate of increase in global oil consumption began to decline after 1973, the world still used over twice as much oil

---

[2]British Petroleum, Statistical Review of World Energy 2009. Online: http://www.bp.com.

[3]Energy Information Administration, *Annual Energy Outlook 2006*, Washington, DC, February 2006.

[4]J. J. MacKenzie, Oil as a Finite Resource: When Is Global Production Likely to Peak? World Resources Institute, 2000. Online: http://www.wri.org/wri/climate/jm_oil_000.html.

[5]When a new oil field is discovered, it is very difficult to say exactly how much oil exists. Also, some sources report increased estimates of recoverable oil from a previously discovered source as a new discovery, while others do not, hence the discrepancy in estimates. MacKenzie (ibid.) cites a 2:1 ratio for 1996, and L. F. Ivanhoe cites a 6:1 ratio for major discoveries during the 1990s. Hubbert Center Newsletter #2002/2, M. King Hubbert Center for Petroleum Supply Studies, Petroleum Engineering Department, Colorado School of Mines, Golden, CO, 2002.

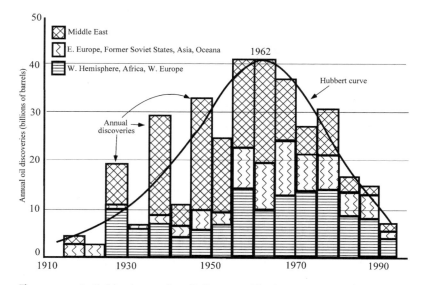

Figure 7.2 • A. Hubbert curve for oil discovery. The bars represent the average amount of crude oil discovered worldwide during each 5-year period from 1912 to 1992. The line known as the Hubbert curve is the weighted average of global oil discovered from 1915 to 1992. (Source: Adapted from L. F. Ivanhoe, King Hubbert, updated. *Hubbert Center Newsletter* #97/1. Online: http://hubbert.mines .edu/news/v97n1/mkh-new2.html.)

since 1973 than for all of human history prior to 1973.[6] What is the net result of all this?

M. King Hubbert, while working as a petroleum geologist for Shell Oil Company, developed a theory of nonrenewable resource extraction, graphically depicted in the **Hubbert curve**. Figure 7.2 shows a Hubbert curve for oil discoveries using actual data, and Figure 7.3 shows a Hubbert curve for oil production that includes estimates of future production. Hubbert hypothesized that peak production must follow peak discovery with a time lag. In 1954, Hubbert used this theory to predict that oil production in the U.S. would peak between 1967 and 1971—a prediction that was treated with considerable skepticism. In reality, it peaked in 1970. Applying Hubbert's methods, leading industry experts in the 1990s predicted that oil production would peak sometime between 2003 and 2020, followed by a decline.[7] Sophisticated analyses of oil prices accounting for both scarcity and information effects suggested that oil prices would rise

---

[6]The area under the curve in Figure 7.3 shows total oil production, which is almost identical to consumption. You can see that the area under the curve from 1973 to the present is nearly two and a half times greater than that from 1869 to 1973.

[7]C. J. Campbell and J. H. Laherrère, The End of Cheap Oil, *Scientific American*, March 1998.

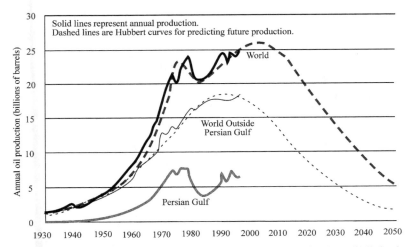

Figure 7.3 • A Hubbert curve for oil production. Global production of oil, both conventional and unconventional (solid black line), recovered after falling in 1973 and 1979. But a more permanent decline is less than 10 years away, according to the authors' model, based in part on multiple Hubbert curves (dashed lines). A crest in the oil produced outside the Persian Gulf region now appears imminent. (Source: Adapted from C. J. Campbell and J. H. Laherrère, The End of Cheap Oil, *Scientific American*, March 1998.)

suddenly and sharply.[8] Despite the highest global economic growth rates in decades and unprecedented demand for oil, oil production essentially stagnated from late 2004 to mid-2007 while prices doubled. Though production began increasing slightly from mid-2007 to July 2008, prices again doubled, leading to a simultaneous peak in production and prices. The onset of a global recession dragged prices and production back down, but in 2009 oil prices had begun to rise again even as the recession worsened. While oil exhaustion may not be imminent, we believe that oil production has already plateaued, and while fluctuations will occur, the trend in coming years will be toward steadily declining output and rising prices.

Given the vast supplies of solar power available as a substitute, does it matter if we exhaust oil supplies? Developing solar energy as a substitute will take considerable time. Also, because solar energy strikes the Earth as a fine mist, large areas of land are needed to capture that energy in significant quantities. With current technologies and without disrupting agriculture, forestry, or the environment, the amount of solar energy that could be captured in the U.S. would meet only 20–50% of our current energy demands. And we could not get around this problem simply by using

---

[8]D. B. Reynolds, The Mineral Economy: How Prices and Costs Can Falsely Signal Decreasing Scarcity, *Ecological Economics* 31(1):155–166 (1999).

lower-wattage light bulbs. Food production and transportation in the United States currently consumes at least three times as much hydrocarbon energy as it provides in carbohydrate energy.[9] One grain-fed steer "consumes" some 284 gallons of petroleum in the process of becoming our dinner.[10] Returning to animal traction to power our farms would require additional land devoted to fodder to feed the draft animals. We cannot sustain economic growth without cheap fossil fuels, but far more important, in the absence of radical changes in agricultural technology, we will not even be able to sustain food production.

■ MINERAL RESOURCES

Mineral resources are also growing scarcer. As noted earlier, the richest, most available ores are used first, followed by ores of decreasing quality. We previously used hematite ore from the Mesabi range in Minnesota, which is about 60% pure iron. That ore is now exhausted, and we must use taconite ore, at about 25% pure iron.[11] The situation with other ores is similar. At least metal can be recycled, and other materials are adequate substitutes. If we consider topsoil as a mineral resource, the situation looks more serious. Rates of topsoil depletion in the U.S. are currently 100 times the rate of formation.[12] Globally, experts estimate that 40% of agricultural land is seriously degraded, and this number is as high as 75% in some areas.[13] Currently, the most widely used substitute for declining soil fertility is petroleum-based fertilizers.

■ WATER

Among the more threatening of the imminent shortages is that of freshwater. While water is the quintessential renewable resource, thanks to the hydrologic cycle, global water consumption has tripled over the last 50

---

[9]D. Pimentel and M. Pimentel, Land, Energy and Water: The Constraints Governing Ideal U.S. Population Size, 1995. Online: http://www.npg.org/forum_series/land_energy&water.htm. Negative Population Growth, Inc. Forum Series. Georgescu-Roegen rightly pointed out that the notion of producing carbohydrates from hydrocarbons was absurd, and when this was first presented as an option, he also accurately predicted that we would move in the other direction first. N. Georgescu-Roegen, *The Entropy Law and the Economic Process*, Cambridge, MA: Harvard University Press, 1971. However, in the short term direct conversion of oil to food might be more efficient than Western agriculture.

[10]M. Pollan, Power Steer, *New York Times Magazine*, March 31, 2002.

[11]J. Hanson, Energetic Limits to Growth. Online: http://www.dieoff.com/page175.htm. Also appeared in *Energy Magazine*, Spring 1999.

[12]D. Pimentel and M. Pimentel. Land, Energy and Water: The Constraints Governing Ideal U.S. Population Size, 1995. Online: http://www.npg.org/forum_series/land_energy&water.html.

[13]World Resources Institute, *People and Ecosystems: The Fraying Web of Life*, Washington, DC: WRI, 2000.

years, and it continues to climb. Humans are pumping rivers dry and mining water from aquifers faster than it can be replenished. While global climate change may lead to a wetter climate overall as increased evaporation leads to increased rainfall, increased evaporation will also dry out the land much more quickly. Many climatologists believe the net result will be intense downpours interspersed with severe drying. In addition, global climate change is likely to affect where water falls, leading overall to greater risks of both flooding and drought.[14]

The dominant use of water (70%) is agriculture, so a water shortage will probably translate into hunger before thirst. The estimated water deficit (extraction of water greater than the recharge rate) in northern China is 37 billion gallons, which produces enough food to feed 110 million people.[15] The Ogallala aquifer in the United States has turned the arid Western plains into a breadbasket. Water levels have been in steady decline since the 1950s,[16] and diminishing rainfall in the American West[17] is likely to increase demand for aquifer water while reducing discharge rates.

The use of river water for irrigation has already led to one of the planet's worst environmental catastrophes in the Aral Sea. On every continent, important aquifers are falling at rates between 2 and 8 m per year.[18] Currently, nearly one billion people lack access to potable drinking water,[19] less than one-third of the world's population enjoys abundant water supplies,[20] and some studies suggest that nearly 50% of the world's population will be living in water shortage areas by 2025.[21] The World Bank warns that continued reduction in aquifers could prove catastrophic.[22] *Fortune* magazine suggests that water shortages will make

[14] C. J. Vörösmarty, P. Green, J. Salisbury, and R. B. Lammers, Global Water Resources: Vulnerability from Climate Change and Population Growth, *Science* 289:284–288 (July 14, 2000).

[15] L. Brown, Water Deficits Growing in Many Countries: Water Shortages May Cause Food Shortages. Earth Policy Institute, Eco-Economy Updates, August 6, 2002.

[16] V. L. McGuire, *Water-Level Changes in the High Plains Aquifer, Predevelopment to 2005 and 2003 to 2005*. Reston, VA: U.S. Geological Survey, 2007.

[17] E. Cook, C. Woodhouse, C. M. Eakin, D. Meko, and D. Stahle. Long-Term Aridity Changes in the Western United States. *Science* 306(5698):1015–1018 (2004).

[18] L. Brown, Water Deficits Growing in Many Countries. Earth Policy Institute, Eco-Economy Updates, August 6, 2002. Online: http://www.earth-policy.org/Updates/Update15.html.

[19] WHO/UNICEF Joint Monitoring Programme for Water Supply and Sanitation. Millennium Development Goals Assessment Report 2008: Country, Regional and Global Estimates on Water and Sanitation. New York: UNICEF; Geneva: WHO, 2008.

[20] Vörösmarty et al., op. cit.

[21] L. Burke, Y. Kura, K. Kassem, C. Revenga, M. Spalding, and D. Mcallister, *Pilot Analysis of Global Ecosystems: Coastal Ecosystems*, Washington, DC: WRI, 2000.

[22] Brown, op. cit.

water the oil of the twenty-first century, "the precious commodity that determines the wealth of nations."[23]

Projections concerning future water supplies are highly uncertain. First, we lack adequate data.[24] Second, consumption patterns and technology can dramatically change the demand for water. Third, as mentioned, climate change can have serious impacts on the hydrologic cycle, increasing evaporation rates and changing rainfall patterns.[25]

■ RENEWABLE RESOURCES

The fact that we live in a full world is even more obvious when it comes to "renewable" resource stocks. For virtually every renewable stock of significance, the rate of extraction is limited by resource scarcity, not by a lack of adequate infrastructure. It is a shortage of fish, not fishing boats, that has stagnated fish harvests over the last few years. The Food and Agriculture Organization (FAO) of the United Nations estimates that 11 of the world's 15 major fishing areas and 69% of the world's major fish species are in decline and in need of urgent management. For instance, cod catches dropped by 69% from 1968 to 1992. West Atlantic bluefin tuna stocks dropped by more than 80% between 1970 and 1993.[26] Similarly, it is a shortage of trees, not chainsaws, that limits wood production. As commercially valuable species are depleted, we turn to harvesting others that were formerly considered trash. As a result, for both fish and timber, the number of commercially valuable species has increased dramatically over recent decades.

Many economists cite this ability to substitute one species for another as evidence that there are no limits to potential harvests. However, when one fish species is exhausted because too many boats are going after too few fish, the whole fishing fleet is available to deplete any new stocks we identify. Having virtually exhausted rapidly reproducing species such as cod, we now pursue species such as orange roughy, which may take as long as 30 years to reach sexual maturity. We run the risk of harvesting

---

[23]N. Currier, The Future of Water Under Discussion at "21st Century Talks." *United Nations Chronicle* XL(1) (2003). Online: http://www.un.org/Pubs/chronicle/2003/webArticles/013003_future_ of_water.html.

[24]K. Brown, Water Scarcity: Forecasting the Future with Spotty Data, *Science* 297(5583): 926–927 (August 9, 2002).

[25]Vörösmarty et al., op. cit.

[26]FAO of the U.N. Focus: Fisheries and Food Security, 2000. Online: http://www.fao.org/focus /e/fisheries/challeng.htm.

such species to extinction before we even acquire sufficient data to esti-mate their sustainable yields.[27]

While there is serious cause for concern for the resource exhaustion of raw material inputs into the economy, these may pale in significance when compared to the dangers presented by the depletion and destruction of ecosystem services. Ecosystem services are destroyed directly by the har-vest of their structural components, primarily the renewable resources of which they are composed, and less directly by waste emissions. Forest cover is currently being depleted in the poorer countries at the rate of about 140,000 km^2 per year,[28] and if the World Trade Organization's ef-forts to liberalize trade in forest products go forward as planned, the rate of deforestation is expected to increase.[29] The Ramsar convention on wet-lands is an intergovernmental treaty providing a framework for the con-servation of wetlands and their resources, yet 84% of the wetlands supposedly protected by the treaty are threatened.[30] While we understand marine ecosystems less than terrestrial ones, it seems unimaginable that healthy fish populations do not play a vital role in these ecosystems and the scarcely understood mechanisms by which they provide ecosystem services. For example, biodiversity appears to enhance ecosystem pro-ductivity and stability along with other ecosystem services, and continued loss of oceanic biodiversity may lead to the total collapse of marine fish-eries by 2048.[31] Virtually all other ecosystems confront similar threats through depletion of their component stocks.

■ WASTE ABSORPTION CAPACITY

People have worried about resource exhaustion at least since the time of Malthus, but concern over the excessive accumulation of waste is more re-cent. Every economic activity produces waste. As humans overwhelm the waste absorption capacity of ecosystems at local and global levels, we suffer in two ways. First, accumulating toxins have direct negative effects on hu-mans. Second, pollutants damage ecosystems and degrade the ecosystem

---

[27]For example, one study found a 60–70% decline in total biomass of one stock of orange roughy in less than 10 years of fishing. P. M. Smith, R. I. C. C. Francis, and M. McVeigh "Loss of Genetic Diversity Due to Fishing Pressure," *Fisheries Research* 10(1991):309–316.

[28]World Resources Institute, *People and Ecosystems: The Fraying Web of Life*, Washington, DC: WRI, 2000.

[29]P. Golman, J. Scott, et al. *Our Forests at Risk: The World Trade Organization's Threat to Forest Protection* (Oakland, CA: Earthjustice, 1999).

[30]M. Moser, C. Prentice, and S. Frazier, "A Global Overview of Wetland Loss and Degrada-tion," Proceedings of the 6th Meeting of the Conference of Contracting Parties of the Ramsar Con-vention (1996), vol. 10. Online: http://www.ramsar.org/about_wetland_loss.htm.

[31]B. Worm et al., "Impacts of Biodiversity Loss on Ocean Ecosystem Services," *Science* 314(5800)(2006):787–790.

services on which we depend. Accumulating evidence suggests we are overwhelming the waste absorption capacity of the planet for several classes of wastes.

The most prominent category of waste in the news today is CO_2 emissions. In spite of an impressive ability of ecosystems to absorb CO_2, there is irrefutable evidence that it is currently accumulating in the atmosphere and near consensus in the scientific community that this has already contributed to global climate change. International recognition of the seriousness of the problem has led to international discussions, but at the time of this writing, the world's worst emitter of greenhouse gases has refused to participate in international accords. Even if the United States did participate, the reductions proposed under the Kyoto protocol would fail to limit CO_2 emissions to the waste absorption capacity of the environment, and would therefore at best merely slow the rate of global warming.[32] In the absence of major changes in human behavior, global warming will have dramatic impacts on global ecosystems. This is particularly true because so many remaining ecosystems are islands in a sea of humanity, and the species they contain will be unable to leave their islands in response to changing conditions.

Waste emissions from mineral resources also pose serious threats. Heavy metals are highly toxic to humans. As these metals are elements, there is no waste absorption capacity per se; once in the environment or in our aquifers, they remain indefinitely. These elements are normally highly diluted in nature or out of reach of living systems; humans have extracted and purified them and released them into the environment in dangerously high concentrations. Many of them tend to bioaccumulate; when ingested, they are not released, so predators retain all that has been consumed by their prey. Many fish species have dangerously high levels of mercury and other metals, which cause human birth defects and worse when consumed, not to mention their impacts on other species.

Nuclear wastes are also elements and far more toxic than the other heavy metals. Nuclear wastes do break down, but not on a human time scale. Plutonium, one of the most toxic substances known, has a half-life of 24,300 years. At minimum, we must sequester such waste for ten times that long—nearly fifty times as long as civilization has existed.

Halogenated hydrocarbons are another class of particularly dangerous manmade mineral wastes. Chlorofluorocarbons (CFCs) are the best known, and they are now banned. However, many countries continue to use hydrochlorofluorocarbons (HCFCs). While HCFCs have lower ozone

---

[32]Intergovernmental Panel on Climate Change, *Climate Change 2007: Synthesis Report. Summary for Policymakers.* Intergovernmental Panel on Climate Change. Cambridge, UK: Cambridge University Press, 2007.

depleting potential than CFCs, China and India have been increasing their use by as much as 35% annually. As a result, the greatest recorded decrease in the ozone layer occurred in 2006.[33] Ozone depletion threatens not only human health but also global plant and animal life. The Antarctic ozone hole poses a particularly serious threat to phytoplankton production in the southern seas. In addition to its key role at the bottom of the oceanic food chain, phytoplankton may play an important role in sequestering carbon dioxide, and its depletion may contribute to global warming.[34]

Other halogenated hydrocarbons are classified as persistent organic pollutants (POPs). International negotiators are currently calling for a ban on the most notoriously harmful POPs. These chemicals are now found in every ecosystem on Earth. Among their negative traits, some of them seem to mimic hormones and are capable of affecting the reproductive capacity of many species. As their name implies, POPs will continue to persist in the environment for many years to come, in spite of the ban. In the meantime, industry is busy introducing new chemicals, many with a very similar structure to the most toxic ones, at the rate of over 1000 per year. We often do not become aware of the negative impacts of these chemicals for years or even decades. And while it may be possible to perform careful studies about the damage caused by a single chemical, outside of the laboratory, ecosystems and humans will be exposed to these chemicals in conjunction with thousands of others.[35]

Pollution in some areas is becoming so severe that it threatens human health, ecosystem function, and even large-scale climate patterns. For example, a recent study has shown that a 3-km-thick layer of pollution over South Asia is reducing the amount of solar energy striking the Earth's surface by as much as 15% in the region yet preventing heat from the energy that does pass through from leaving. In addition to threatening hundreds of thousands of premature deaths, the pollution cloud is likely to increase monsoon flooding in some areas while reducing precipitation by as much as 40% in others.[36]

---

[33]K. Bradsher, The Price of Keeping Cool in Asia: Use of Air-Conditioning Refrigerant Is Widening the Hole in the Ozone Layer, *New York Times*. United Nations Environmental Program (UNEP), 2006 Antarctic Ozone Hole Largest on Record. New York: UNEP, 2006.

[34]R. C. Smith, B. B. Prezelin, K. S. Baker, R. R. Bidigare, N. P. Boucher, T. Coley, D. Karentz, S. MacIntyre, H. A. Matlick, D. Menzies, M. Ondrusek, Z. Wan, and K. J. Waters. Ozone Depletion: Ultraviolet Radiation and Phytoplankton Biology in Antarctic Waters, *Science* 255:952–959 (1992).

[35]A. P. McGinn, Why Poison Ourselves? A Precautionary Approach to Synthetic Chemicals, World Watch Paper 153, Washington, DC: World Watch, 2000.

[36]P. Bagia, Brown Haze Looms Over South Asia, *Science* 13 (2002).

In summary, it appears that the global sink is becoming full more rapidly than the global sources of natural resources are being emptied. This is understandable in view of the fact that sinks are frequently freely available for anyone to use (nonexcludable) but also rival. In contrast, sources are more often excludable resources, either privately or publicly owned and managed.

A rapid assessment of the resources on which the human economy depends suggests that we are now in a full world, where continued physical expansion of the economy threatens to impose unacceptable costs. Whereas historically people have been most worried about resource depletion, the source problem, it appears that the most binding constraint on economic growth may be the waste absorption capacity of the environment, the sink.

BIG IDEAS to remember

- Exponential growth
- Doubling time
- Hubbert curve

- Source and sink limits
- Measures of "fullness" of the world

Conclusions to Part II

In Part II we have examined the scarce resources upon which all economic production depends, applying insights from physics and ecology. The economic system, like all other known systems, is subject to the laws of thermodynamics. As it is impossible to create something from nothing, and equally impossible to create nothing from something, economic production must deplete natural resources and generate waste. On a finite planet, economic growth (and human population growth) must eventually come to an end. The real question is whether this must happen soon or in the far distant future. The previous summary suggests that we are indeed approaching a "full world" and that ecological economists' concern with scale is abundantly justified. While the ultimate scarce resource is low-entropy matter-energy, the different forms of low entropy have fundamentally different characteristics. In Part III, we will see how these characteristics affect the allocation process.

Part III will involve a shift in focus from the whole to the part, from the natural earth ecosystem to the human-dominated subsystem, from concepts of physical and biological science to concepts of social science, especially economics. But even as we focus on supply and demand, prices, national income, interest rates, and trade, we will not forget that all of these economic activities take place in an increasingly full world and that the whole system is governed more by thermodynamics and photosynthesis than by prices and GNP.

PART III

Microeconomics

CHAPTER

8

The Basic Market Equation

The purpose of microeconomics is to understand how decentralized decisions by thousands of independent firms and households are communicated, coordinated, and made consistent through prices determined by supply and demand in markets. In the absence of central planning, markets make it possible to have spontaneous order. This seems a bit miraculous, but an analogy will clarify. Even though no one designed a language, nevertheless a language is logically structured and ordered. Although the capacity for language is no doubt part of our genetic inheritance, an actual language is an ordered structure that evolved spontaneously through use. We can, after the fact, analyze the logical structure and grammar of a language, but that is description, not design. Even Esperanto, which is a designed language, was largely copied from Spanish, a natural spontaneously evolved language. Few people speak Esperanto. Like language, the market is also a communication system, though much more limited in what it can communicate. But the market, too, has a "grammar," a logic that can be analyzed even though it arose in a spontaneous, unplanned way. And the market does more than communicate—it allocates.

What are the grammar rules of markets? What is being communicated? In what sense is the market allocation efficient, and why? Is the outcome fair or sustainable? These are microeconomic questions. Our approach here is to develop the basic answers, the big ideas and conclusions, without getting lost in details and without resorting to calculus or indeed to any mathematics beyond ratios and simple equations. To do this we need five definitions and three principles. On the basis of these definitions and principles, we will derive a "basic market equation" and then interpret its meaning concretely. So a little patience will be needed for several pages before we see just how all the concepts and principles cohere into the main conclusion of price theory. Once we have the main conclusion and big picture, we'll show how basic supply and demand derive from that picture and will present the most basic "grammar rules" that govern supply and demand in individual markets. Then we will look behind supply and demand, first at the production function that underlies supply, then at the utility function that underlies demand. Finally, we will see the implications of ecological economic analysis with respect to the production and utility functions.

■ COMPONENTS OF THE EQUATION

Let's begin with the five definitions of concepts we'll need:

1. MU_{xn} = marginal utility of good x to consumer n. The marginal utility is the extra satisfaction one gets from consuming one more unit of the good, other things being equal. If x is pizza and you are the consumer, MU_{xn} is the amount of utility you get from consuming another piece of pizza.

2. P_x = the market price of good x (goods are x, y). What's the market price of a slice of pizza? $2.50?

3. P_a = the market price of factor a (factors are a, b). What factors of production go into making a pizza? The kitchen and stove are the capital (fund-service), the cook is the labor (fund-service), and the flour, tomatoes, and cheese are raw material stock-flows yielded by natural capital, with the assistance of cultivation. The cook and oven transform the raw ingredients into pizza.

4. MPP_{ax} = the marginal physical product of factor a when used to make good x. The marginal physical product is the extra output produced as a result of using one more unit of a factor as an input, all other inputs remaining constant. For example, if adding one more cook increases the pizza output of a pizzeria by 20 pizzas per night, then the marginal physical product of 8 hours of labor in terms of pizza is 20 pies.

5. Competitive market. A **competitive market** is a market in which there are many small buyers and sellers of an identical product. "Many" means "enough that no single buyer or seller is sufficiently large to affect the market price." Put another way: Everyone is a price-taker; no one is a price-maker. Everyone adjusts his or her plans to prices; no one has the power to adjust prices to plans. Since everyone treats price as a parameter (a given condition) rather than a variable (something one can change), this condition is sometimes called the parametric function of prices.[1]

---

[1] See O. Lange, On the Economic Theory of Socialism, *Review of Economic Studies*, for an excellent exposition of market prices and parametric function.

Here are the three principles:

Law of diminishing marginal utility. As one consumes successive units of a good, the additional satisfaction decreases, that is, total satisfaction increases, but at a decreasing rate. The marginal utility of one's first slice of pizza on an empty stomach is great. The marginal utility of the fifth slice is much less. How do we know that this principle is generally true? One way to make an argument is to assume the contrary and show that it leads to absurdity. Suppose that there were a law of increasing or even constant marginal utility—the utility of the fifth slice of pizza were equal to or greater than the first. Then what would we observe? A consumer would first purchase the good that gave him the highest marginal utility per dollar. But then the second unit of that good would, under constant or increasing marginal utility, give him the same or even more satisfaction per dollar than the first, and so on. With a law of constant or increasing marginal utility, consumers would spend all their income on only one good. Since the contrary of diminishing marginal utility leads to absurdity, we have indirectly established the reasonableness of diminishing marginal utility.

Law of diminishing marginal physical product. As a producer adds successive units of a variable factor to a production process, other factors constant, the extra output per unit of the variable factor diminishes with each addition, that is, total output increases at a decreasing rate. This is sometimes called the law of diminishing returns. Again we can convince ourselves of its reasonableness by assuming the contrary and showing that it leads to absurdity. Assume a law of increasing marginal physical product. We have a 10-acre wheat farm. We add one more laborer, and his marginal product is greater than that of the previous laborer. So we add another, and so on. The result is that all agricultural labor will be employed on a single farm. Indeed, we could grow the whole world's wheat crop in a single flowerpot. This is absurd, and thus we have indirectly established the reasonableness of the law of diminishing marginal physical product.[2]

The **equimarginal principle of maximization** was referred to earlier

---

[2]The law of diminishing marginal product should not be confused with economies or diseconomies of scale. An economy of scale occurs when a 1% increase in *all* the factors of production together leads to more than a 1% increase in output. This does not contradict the law of diminishing marginal physical product. With an economy of scale, we couldn't necessarily produce the world's wheat supply in one flowerpot, but we would want to grow it all on one very large farm. In reality, economies are likely to occur over a limited range of production, usually followed by diseconomies of scale. Over very limited ranges of production, an additional unit of a factor of production may have a higher marginal physical output than the first. For example, four carpenters building a house may finish a house more than four times faster than a single carpenter, as a single person simply cannot lift a large wall or maneuver a truss, but 16 carpenters on the same house are unlikely to get it built four times faster than four.

as the "when to stop" rule.[3] When does a consumer stop reallocating her income among different goods? When she has found an allocation that maximizes her total satisfaction or total utility. That point occurs when the marginal utility per dollar spent on each good is equal. Again, suppose the contrary—the marginal utility of good *a* per dollar spent was greater than that for good *b*. Then our consumer could increase her total utility by reallocating a dollar from *b* to *a*. Only when utilities were equal at the margin would it no longer be possible to increase total utility by reallocation of expenditure. Furthermore, the law of diminishing marginal utility guarantees that each reallocated dollar brings us closer to the optimum—buying more *a* reduces the marginal utility of *a*, buying less *b* raises the marginal utility of *b*, moving us toward equality at the margin.[4]

In a simple economy of shoes and pizzas, how would you spend your money? You buy pizza if a dollar spent on pizza provides more pleasure than a dollar spent on shoes, and you buy shoes if a dollar spent on shoes provides more pleasure than a dollar spent on pizza. To maximize pleasure, the last dollar spent on pizza must supply the same pleasure as the last dollar spent on shoes.

Similar logic applies to the producer who is maximizing her output by choosing a combination of factors such that the marginal product per dollar spent on each factor is equal. If the MPP of each factor were not equal, the same total output could be produced with fewer inputs, hence at lower total cost, thus yielding higher profit.

Think in terms of a pizza parlor. The owner can hire another cook for $1600/month who can produce an additional 20 pizzas per day, generating $1700 in additional net monthly revenue. Alternatively, if the owner spends $1600/month on payments for a better kitchen, she would be able to produce only 18 more pizzas per day. The owner will keep hiring cooks as long as the additional pizzas they produce generate net profits, and

---

[3]This basic rule in economics does have an important limitation. The rule says that the way to get to the top of the mountain is to take any step that leads upward. When you can no longer do that, when any step you take will move you downward, you know you are at the top of the hill, the maximum point. Or do you? If there is a temporary dip on the hillside, you might mistake a local maximum for the global maximum. The laws of diminishing marginal utility and diminishing returns are thought to guarantee hillsides with no dips, but the caveat is important to keep in mind.

[4]The "when to stop" rule assumes people are concerned only with maximizing their own utility and do not allocate resources for the sole purpose of making others happy. It also assumes people always make the utility-maximizing choice. Mainstream economic theory assumes that rational self-interest guides all allocation decisions, and someone who acts in this way is known as *Homo economicus*. Empirical studies and common observations show that in reality, people are not always "rational," as economists define the term, and sometimes act selflessly (helping others with no gain to themselves), vengefully (harming others even when it harms themselves as well), or in other "irrational" ways. Though the individual might maximize his or her own utility by harming or helping someone, applying the equimarginal principle of maximization to such actions will not lead to optimal outcomes for society.

those profits are greater than the profits from spending an equivalent sum on the kitchen. However, as she hires more cooks, the cooks get in each other's way, there are insufficient ovens, and their marginal productivity goes down. In contrast, with more cooks available, the marginal productivity of a better kitchen might increase. If the productivity of a bigger kitchen per dollar invested becomes greater than the productivity of another cook (and produces enough extra pizza so their sales will cover the costs of improvement), the owner will invest in the kitchen. The point is that to maximize profits in the pizzeria, the last dollar spent on hiring cooks should generate the same profit as the last dollar spent on expanding the kitchen.[5]

Now let's imagine a shoe store moves in next door to the pizzeria. Business grows, and the owner estimates that an assistant would generate an additional $1800 in net monthly revenue. The town is small, and the only available workers are employed by the pizza parlor. Because the shoe store owner profits more from an extra worker than the pizzeria, he can afford to pay more. One of the pizza cooks comes to work for him (assuming the skills are transferable). As the shoe store owner hires away more pizza cooks, his store gets crowded and production per laborer goes down, while the pizzeria gets less crowded and production per laborer goes up. The pizzeria owner can therefore raise wages to retain her employees. As long as the shoe store owner can make more profit from another laborer than the pizzeria owner, he can hire laborers away from her, but the decreasing productivity of more workers for him and the increasing productivity of fewer workers for her means that this cannot go on forever. Eventually, the marginal product valued in dollars of profit from a pizza cook and a shoe store assistant become equal, and neither store owner can outbid the other in hiring extra labor. Hence, the marginal physical product of labor for shoes and pizzas, as valued in dollars, will be equal. The same holds true for all factors of production across all industries.

Now we are in a position to state the basic market equation, then show why it must hold, and then interpret just what it means. The basic market equation can be written as

$$MUxn/MUyn = Px/Py = MPPay/MPPax$$

First, notice the central position of relative prices. On the left are conditions of relative desirability, reflecting the upper or ends part of the ends-means spectrum (see Figure 3.1). On the right are conditions of

---

[5]Of course, there are real differences between hiring cooks and improving a kitchen. Cooks can be hired or fired, can work longer or shorter hours, and in general may allow greater flexibility than kitchen construction. Improving a kitchen is a lumpier investment, harder to do in discrete units. The arguments presented here would make more sense if the restaurant owner could invest in cooks and kitchens in very small units.

relative possibility, reflecting the lower or means part of the ends-means spectrum. The intermediary role of prices is to bring about a balance between ends and means, an efficient allocation of means in the service of ends.

But how do we know that the basic market equation holds? We know the left-hand equality holds because it is just a restatement of the consumer's allocation rule of equal marginal utility per dollar, usually written as

$$\text{MU}xn/Px = \text{MU}yn/Py$$

The marginal utility per dollar spent on pizza should equal the marginal utility per dollar spent on shoes for consumer n. If you think about it and review the section on the equimarginal principle of maximization, you'll see that if the first ratio is larger than the second, n will increase her total utility by buying more pizza and fewer shoes.

Similarly, the right-hand equality must hold because it is the producer's equimarginal principle of maximization—all factors are employed in quantities such that the price of each factor equals the value of its marginal product. For all firms using a to produce x, we have

$$Pa = Px \, (\text{MPP}ax)$$

The price of labor equals the price of pizza times the number of pizzas an additional unit of labor can produce with all other factors of production held equal.

THINK ABOUT IT!
In real life, can more labor produce more pizza without additional pizza ingredients? We will return to this question.

If the marginal unit of labor costs more than the value of the pizzas it can produce, the pizzeria owner would earn higher profits by employing less labor.

Likewise, for all firms using factor a (e.g., labor) to produce good y (e.g., shoes) we have

$$Pa = Py \, (\text{MPP}ay)$$

Since Pa is the same for all firms, and things equal to the same thing are equal to each other, it follows that

$$Px \, (\text{MPP}ax) = Py \, (\text{MPP}ay)$$

The value of the additional pizzas a worker could produce in a pizzeria will equal the value of the additional shoes a worker can produce in a shoe store.

Reorganizing terms, we get

$$Px/Py = MPPay/MPPax$$

This is the right-hand equality in the basic market equation.

■ WHAT DOES THE MARKET EQUATION MEAN?

Now that we have derived the basic market equation, what does it mean? So what?

First, note that x and y are any pair of goods, a is any factor of production used by any firm, and n is any individual. The equation holds for all pairs of goods, all factors, all firms, and all individuals. We could string out marginal utility ratios to the left, one for each individual in the economy, not just n. For each individual, the ratio of his marginal utilities between x and y would equal the price ratio. Does this mean that all individuals consume the same amounts of x and y? Certainly not! People have different tastes, and in order to get equality at the margin, different consumers have to consume different total amounts of x and y. Unless each consumer is consuming amounts such that the ratio of marginal utilities is equal to the price ratio, that consumer is not maximizing his utility.

Likewise, we could string out marginal physical product ratios to the right, one for each factor of production ($a, b, c,$ etc.) used by any firm in making x and y. Does the equality of each marginal ratio imply that all firms use the same total amounts of factor a or b in producing x and y? No, because different firms have different production processes. But unless the ratio of MPP equals the price ratio, the firm in question is not maximizing profits.

We could write a similar basic market equation for every other pair of goods—one for x and z, one for y and z, and so on. So the equation holds for all relative prices.

The central role of Px/Py is worth emphasizing. It brings about an equality of the marginal utility ratios with the marginal productivity ratios. Things equal to the same thing are equal to each other—prices serve as a kind of sliding fulcrum on a seesaw that balances the weight of relative possibility with the weight of relative desirability, of means with ends (Figure 8.1). The rate at which consumers are *willing* to substitute one good for another (psychological rate of substitution) is equal to the rate at which they are *able* to substitute goods by exchange (market rate of substitution) and also equal to the rate at which producers are able to produce one good rather than another (essentially "transform" one good into another) by reallocating resources between them (technical rate of substitution or transformation).

Relative prices serve as a sliding fulcrum to bring about balance or equality between the utility and productivity ratios. But once that equality is achieved, what does it mean? To understand better, leave out the

---

The basic market equation is:

$$MUxn/MUyn = Px/Py = MPPay/MPPax$$

MU is the marginal utility of good x or good y to person n, and MPP is the marginal physical product of factor a used to produce good x or good y.

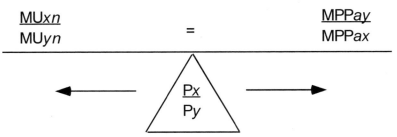

Figure 8.1 • The parametric or fulcrum function of relative prices.

intermediate price ratio and consider only the resulting equality of the marginal utility ratio and the marginal productivity ratio:

$$\text{MU}xn/\text{MU}yn = \text{MPP}ay/\text{MPP}ax$$

Rewrite this equation as

$$\text{MU}xn \times \text{MPP}ax = \text{MU}yn \times \text{MPP}ay$$

This states that the marginal utility derived from factor *a* when allocated to good *x* is just equal to the marginal utility derived from factor *a* when allocated to the production of good *y*, as judged by consumer *n*,[6] or in more concrete terms, the amount of pizza a worker produces in an hour provides the same utility to consumer *n* as the amount of shoes the worker could produce in an hour (assuming the same wage). Because *n* is any consumer, *x* and *y* are any goods, and *a* any factor, it follows that no consumer would want to reallocate any factor between any pair of goods. In other words, the **basic market equation** defines an optimal allocation of resources, one in which no one would want to reallocate any factor to any alternative use because doing so would only decrease that person's total satisfaction. No firm would want to reallocate any factor to any other use because doing so would lower profit.

Perhaps this seems a very big rabbit to pull out of a very small hat! We will come back to that, but for now, let's appreciate the result. Prices in competitive markets lead to an efficient allocation of resources in the sense that no one can be made better off in his own judgment by reallocating resources to produce a different mix of goods. Of course, individual *n* could be made better off if income or wealth were redistributed from individual *m* to himself. We all could be better off if we had a larger resource endowment. But this analysis assumes a given distribution of income and wealth among people and a given total resource endowment.

---

[6]A review of the units: MU*xn* is utility per unit of good *x*, or U/*x*. MPP*ax* is units of good *x* per unit of factor *a*, or *x*/*a*. The units of the product are U/*x* × *x*/*a* = U/*a*. The units are utility per unit of factor *a* when it is used to make *x* (utility as experienced by individual *n*). In other words, the utility yielded by factor *a* in its *x*-use is equal to that yielded in its *y*-use.

The optimal allocation of resources is what economists call a **Pareto optimum**: Everyone is as well off as they can be without making someone else worse off.

These qualifications shrink the rabbit back down to the dimensions of the hat, but it is still a nice trick to bring about a balanced adjustment of relative possibilities with relative desirabilities, to communicate and mutually adjust means to ends in an efficient way without central coordination. The technical possibilities of transforming one good into another by reallocating resources are balanced with the psychic desirabilities of such transformations as judged by individuals.

The key thing about this result is that it is attained by an unplanned, decentralized process. The problem solved by the price system is, in the words of F. A. Hayek, "the utilization of knowledge not given to anyone in its totality."[7] The psychological rates of substitution, the terms on which consumers are willing to substitute commodities, are known to the different individual consumers only. The technical terms on which producers are able to substitute or transform commodities are known only to various production engineers and managers. Yet all this piecemeal, scattered knowledge is sounded out, communicated, and used by the price system in the allocation of resources. No single mind or agency has to have all of this information, yet it all gets used.

> A *Pareto optimum* occurs when no other allocation could make at least one person better off without making anyone else worse off. This is also known as a Pareto efficient allocation (see Chapter 1), Pareto efficiency, or simply efficiency.

■ MONOPOLY AND THE BASIC MARKET EQUATION

The parametric or fulcrum function of prices depends on pure competition. It fails if there is a monopoly. Suppose the producer of good x is a monopolist, while y is still produced in a competitive market. The equimarginal rule of maximization tells both monopolist and competitive firms to produce up to where **marginal cost**—the additional expenditures required to produce one more unit—equals **marginal revenue**—the additional income from selling one more unit.[8] For the competitive firm, marginal revenue is equal to price (price is constant, and the extra revenue from selling one more unit of x is Px). But the monopolist is the only supplier and is definitely not too small to influence price by the amount he can produce. When the monopolist supplies more, it causes the price to fall. But the monopolist's marginal revenue is not equal to the price times the extra unit. Instead it is equal to the new lower price times the extra unit *minus* the fall in price times all previously sold units. For the mo-

---

[7] F. Hayek, The Use of Knowledge in Society, *The American Economic Review* 35(4):520 (1945).

[8] The pizzeria owner will keep hiring cooks as long as their marginal cost (their wage, plus the cost of the additional ingredients they use) is less than the marginal revenue they generate (the price of the pizzas they produce).

nopolist, marginal revenue is less than price, that is, $MRx < Px$. If we substitute MRx for Px in the basic market equation on the right-hand side, we have

$$MUxn/MUyn = Px/Py > MRx/Py = MPPay/MPPax$$

Therefore,

$$MUxn/MUyn > MPPay/MPPax$$

And furthermore,

$$MUxn \times MPPax > MUyn \times MPPay$$

This means that the marginal utility yielded by factor a in its x-use is greater than that yielded in its y-use. Consumer n would like to see some of factor a reallocated from y to x. But this is not profitable for the monopolist. The monopolist finds it profitable to restrict supply below what consumers would most desire. He does this to avoid losing too much revenue on previously sold units of x as a result of lowering the price a bit in order to sell another unit of x. The fulcrum is split, the balance between ends and means is broken, the invisible hand fails.

Neoclassical economics deserves a hearty round of applause for the interesting and important demonstration that competitive markets result in an optimal allocation of resources. If we do not rise to our feet in a standing ovation, it is only because conventional economists sometimes forget the assumptions and limitations of the analysis that led to the conclusion. A short list that we will address later includes the limiting assumptions that the analysis is independent of the distribution of income and wealth, that all goods are market goods (i.e., rival and excludable), that factors of production are substitutes for each other, that external costs and benefits are negligible, that information is perfect,[9] and that all markets are competitive.

■ Non-Price Adjustments

We need to consider two more results from the basic market equation: first, what it tells us about making adjustment by means other than price, as well as price adjustment; and second, how it relates to supply and demand, the most basic rules of market grammar.

In the seesaw diagram (Figure 8.1), the desirability conditions (MU ratios) can be altered only by substitution, by reallocation of consumers' expenditure (not by a fundamental change in preferences); likewise, the

---

[9]Perfect information requires that buyers and sellers can acquire information about a product at negligible cost and that one side in a negotiation does not have more information than the other (i.e., information is cheap and symmetrical).

possibility conditions (ratios of MPP) are alterable only by substituting factors (not by a fundamental change in technology). Prices, the sliding fulcrum, coordinate the substitutions and reallocations necessary to attain balance between the first and last terms of the equation. But what really defines the optimum is the equality of the first and last terms. Prices play only an adjusting and accommodating role.

Suppose that relative prices were fixed. Could we ever attain balance? Suppose the fulcrum position on a seesaw were fixed. How would we attain balance? By directly adjusting the weights on both sides. We could directly change the conditions of relative desirability by altering peoples' preferences through advertising. We could also directly alter the conditions of relative possibility by technological innovation. Vast sums of money are spent on advertising and on technological research. These efforts may be thought of as **non-price adjustments**. But regardless of whether adjustments to equality are by price or non-price mechanisms, the resulting equality of the end terms defines a Pareto optimum.

There are many Pareto optima, one for each distribution of income, set of technologies, and set of wants or preferences. However, if wants are created and preferences altered through advertising, and if advertising is a cost of production of the product, then production begins to look like a treadmill. If we produce the need along with the product to satisfy it, then we are not really making any forward motion toward the satisfaction of existing needs. The producer replaces the consumer as the sovereign. Then, the moral earnestness of production, as well as the concept of Pareto optimal allocation of resources in the service of such production, suffers a loss.

Indeed, even under price adjustment we have the parametric function of prices; each individual takes prices as given and adjusts his plans to prices rather than adjusting prices to his convenience. Yet the market price does change as a result of the market supply and demand conditions that result from each individual treating price as given. Yet if no individual can change the price, then how do prices ever change in the real world? Someone, somewhere has to be a price maker rather than a pure price taker if prices are ever to change. This puzzle has been met in two ways by economic theorists. One is to assume an auctioneer who takes bids and changes the price. This is fine for auction markets, but most markets are not auctions. The other solution is to say that markets really cannot be 100% competitive in the sense of total compliance with the parametric function of prices, or they would never be able to adjust prices. Someone has to have a bit of market power if prices are ever to change. So some real resources have to be dedicated to price adjustment, whether it be the salary of an auctioneer or the temporary monopoly profits of a price leader. But the point to remember is that there is non-price market adjustment as well as adjustment by prices, and the basic market equation helps to analyze all three types of adjustment: price adjustment, psychological adjustment, and technological adjustment. All will get us to Pareto optima, albeit different ones. There are many, many different Pareto optima. While it is good to know that the market will get us to some Pareto optimum, it is vital to remember that that is not enough. Some Pareto optima are heavenly, others are hellish. There is more to welfare than efficient allocation; there are distribution and scale, for example.

■ Supply and Demand

How do supply and demand relate to the basic market equation? This is important because supply and demand are the most useful tools of market analysis. Let's begin with demand. Take the left-hand side of the basic market equation, rewritten as

$$Mux_n/Px = MUy_n/Py$$

Let commodity y be money. Then MUy_n becomes Mum_n, the marginal utility of money, and Py becomes Pm, the price of money, which is unity—that is, the price of a dollar is another dollar. Then we have

$$Mux_n/Px = Mum_n/Pm$$

or

$$MUx_n/Px = Mum_n$$

or

$$Px = MUx_n/MUm_n$$

This is the condition for the consumer to be on his demand curve. To be on the demand curve is to be maximizing utility by substituting good x for money to the point where utility is a maximum—where the marginal utility of a dollar is just equal to the marginal utility of a dollar's worth of good x.

From the relation $Px = MUx_n/MUm_n$, and the law of diminishing marginal utility, we can see that the quantity of x demanded by the consumer will be inversely related to price.

If the consumer is always maximizing utility according to the relation $Px = MUx_n/Mum_n$, and if diminishing marginal utility rules, then we can see that the relation between Px and the amount of x demanded (Qx) will be inverse, as depicted in Figure 8.2. Suppose the left-hand side of the equation, Px, falls. Then, to reestablish equality the right-hand side will have to fall. The consumer makes the right-hand side fall by buying more x. More x means the numerator, Mux_n, will decline thanks to the law of diminishing marginal utility. That reduces the ratio. Also, as the consumer

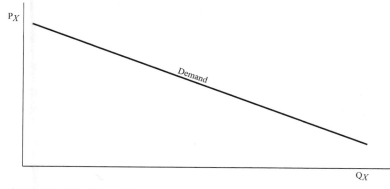

Figure 8.2 • The demand curve.

buys more x she has less money, so the marginal utility of money rises, increasing the denominator and again reducing the right-hand ratio.

> **THINK ABOUT IT!**
>
> *Market demand is nothing more than preferences weighted by purchasing power. In other words, markets allocate according to the principle of one dollar, one vote, which in a political system is called plutocracy. The preferences of the poor count for less than those of the rich. Do you think natural resources and ecosystem services should be allocated by markets? Is this appropriate in a democracy?*

Box 8-1	SPECULATION, MARKET DEMAND, AND THE UPWARD-SLOPING DEMAND CURVE

Obviously, the more of something you have, the less the use value of each additional unit of that thing. However, in the real world, many people, ranging from merchants to speculators, buy and sell things for their exchange value, in which case demand does not necessarily decrease with price. Consider wealthy investors who notice that stock prices are falling while real estate prices are on the rise. These investors cash in their stocks to buy real estate, decreasing the demand and price for the stocks while increasing the demand and price for land. With rising prices, other investors switch to real estate, driving up prices even further. Everyone decides to get in on the act, and banks readily lend money with real estate as collateral, escalating demand even further—if borrowers can't make payments, they can just sell their houses for profit. Rather than the negative feedback loop of rising prices leading to falling demand, speculation creates a positive feedback loop of rising prices leading to rising demand. But housing prices can't rise rapidly forever. The smartest speculators eventually sell off their investments and walk away with the profits. As more and more speculators sell, real estate prices plunge, leading to a frenzied sell-off—the bursting of a speculative "bubble." Thus, in the presence of speculators, the demand curve can slope upward.

Though conventional economics focuses almost entirely on negative feedback loops that lead to market equilibrium, the real economy is also characterized by positive feedback loops fed by speculators. Positive feedback loops are particularly likely to happen when the supply of something is fixed, such as land, subterranean oil supplies, or food grains (in the short run). The years 2001–2008 saw bubbles and crashes in oil, grains, and real estate.

How important is speculative exchange value relative to the value of goods and services bought and sold for their use? By some estimates, the former dwarfs the latter by as much as 20 to 1.[a] The more wealth concentrates in the hands of the few, the more liquid assets are available for speculation, and the more we can expect dangerously destabilizing positive feedback loops to overwhelm the stabilizing negative feedback loops of ordinary commerce.

[a]D. Korten, The Post-Corporate World: Life After Capitalism, *San Francisco: Berrett-Koehler Publishers, 1999.*

To go from the demand curve of the individual consumer to the demand curve of the whole market, we just add up all the individual demand curves; that is, for each price we add up all the q's demanded at that price by each consumer. Thus the market demand curve will be downward sloping, just like that for each individual. Only the units on the horizontal axis have changed, now Q instead of q.

Turning to supply, we know that at all points on the producer's supply curve he must be offering an amount at each price that would maximize profits. That would be when $Px = MCx$, where MCx is the marginal cost of producing x.[10] By definition, MCx is the cost of producing an additional unit of x. We can produce an additional unit of x by using an additional amount of factor a or b or some other factor. The marginal cost of producing x by using more factor a is $Pa/MPPax$—that is, the dollars spent to get one more unit of a (which is Pa) divided by the extra x that was produced by the extra unit of a gives the extra cost of a unit of x, or the marginal cost of x. We could do the same in terms of factor b, and so on. Whichever turns out to be the cheapest way to produce another unit of x (using more a or more b) is the marginal cost of x.

---

[10]Remember, we're assuming here that the marginal cost of producing x is increasing. As long as the marginal cost of production is less than the price, it pays to produce more. Therefore, the producer stops producing when the marginal cost of production equals the price.

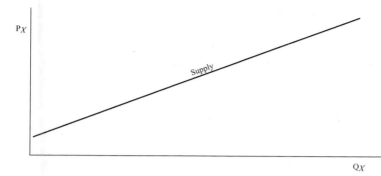

Figure 8.3 • The supply curve.

THINK ABOUT IT!

As you might guess, the marginal cost of x in terms of each factor would tend to equality. Can you explain why?

So for the profit-maximizing producer to be on her supply curve, the condition that must hold is

$$Px = MCx = Pa/MPPax \ldots = Pb/MPPbx \ldots$$

Let's consider only $Px = Pa/MPPax$ as the condition. From it we see that the relation between Px and Qx supplied must be positive, as shown in Figure 8.3. Suppose Px rises; then $Pa/MPPax$ will also have to rise to maintain equality. To bring this about, the producer makes more of x. By the law of diminishing marginal physical product, MPPax declines. Since it is the denominator of the right-hand ratio, that ratio rises to restore equality with the increased Px.

Once again we move from the individual's supply curve to that of the market simply by adding up the amounts supplied by each individual at each Px. The upward slope of the curve remains, and we simply have larger units on the horizontal axis, Q instead of q.

If we put the supply and demand curves together, their intersection will give the combination of Px and Qx such that buyers and sellers are both happy (Figure 8.4). Sellers are maximizing profits, and buyers are maximizing utility. It is as if the market were solving two simultaneous equations in two unknowns by a process of trial-and-error approximation.

At the equilibrium point we have the following:

$$MUxn/MUmn = Px = Pa/MPPax$$

The left-hand equality ensures that the buyer is on his demand curve. The right-hand equality guarantees that the seller is on his supply curve. The point of intersection satisfies both buyer and seller, so the market equilibrium is at P^* and Q^*.

Box 8-2	Can the Supply Curve Have a Downward Slope?

There are two important circumstances in which the supply curve can have a downward slope. The first is the case of increasing returns to scale. This occurs when a firm increases all inputs into production by X%, and output goes up by more than X%. However, microeconomic theory assumes that in the short run, some factors of production are fixed, which means that one cannot change the productivity of those factors, at least not by much. For example, it may take several years to build a new factory. While the firm is using the old factory, the marginal output from more workers or more raw materials is likely to be diminishing, leading to an upward-sloping supply curve. Over time, as new technologies (or new factories) change production systems, we frequently see downward-sloping supply curves.

The second case concerns the restoration of natural capital. As we will discuss at greater length in Chapter 12, many natural resources exhibit ecological thresholds. For example, some species may have a minimum viable population, and if the population falls below this level, it can no longer reproduce itself. An ecosystem can exhibit similar behavior. For example, the Amazon is said to recycle its own rainfall. If enough forest is cleared, it will no longer generate enough rainfall to sustain itself. Many critically important species and ecosystems have probably fallen below their minimal viable population or stock, but it may still be possible (and necessary, in many cases) to maintain or increase the stock through human intervention. The farther below the ecological threshold the stock falls, the more difficult and expensive it may be to preserve or restore it. Once we have re-crossed the threshold, the system is again capable of reproducing itself with no human intervention. In other words, the greater the supply, the cheaper it is to increase supply even further. This dynamic is limited, however. As we restore more and more natural capital, we must give up more important opportunities, including the forgone opportunity of harvesting the resource and using the land for other purposes. This opportunity cost is likely to rise as restoration displaces more and more valuable alternative uses.

We could do exactly the same analysis with the market for good y, and we would end up with similar supply and demand curves and a similar equation:

$$MU_{yn}/MU_{mn} = P_y = P_a/MPP_{ay}$$

If we divide each term in the x-market equation by each term in the y-market equation, we again arrive at the basic market equation previously derived. The terms for the marginal utility of money and the price of factor a cancel out, and we have again the basic market equation:

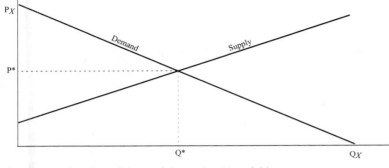

Figure 8.4 • Supply and demand determine P* and Q*.

$$\text{Mu}xn/\text{Mu}yn = Px/Py = \text{MPP}ay/\text{MPP}ax$$

Finally, it is worth commenting on the fact that the basic market equation states that the ratio of marginal utilities of good x and y are *inversely* proportional to the ratio of marginal physical products of factor a when used to make x or y. Why this inversion? What does it mean in terms of economics? Why not a direct proportionality instead of an inverse one? Actually there is a direct proportionality between the ratio of marginal utilities of x and y and the marginal costs of x and y. Prices, in effect, measure both the ratio of marginal utilities and the ratio of marginal costs. But we have to realize that the true marginal cost of x, the *opportunity cost* of a unit of x, is in fact the amount of y that has to be sacrificed to get the extra x. Thus MPPay is the amount of y given up when we use a to produce x instead of y. The amount of y sacrificed (MPPay) is, in real terms, the marginal opportunity cost of x. The best alternative sacrificed (in this case the only alternative) is by definition the opportunity cost. Therefore, the direct proportionality between prices and marginal opportunity costs is at the same time an inverse proportionality between prices and marginal physical products. The true marginal cost of x is *precisely* the amount of y you have to sacrifice to get it.

To summarize: We have derived a basic market equation, shown why it defines and how it brings about an optimal allocation of resources in the sense of Pareto (everyone as well off as they can be without making someone else worse off, i.e., without redistributing income or wealth), and shown how supply and demand derive from the equation. Since x and y represent any pair of goods, a and b any pair of factors, and n any consumer, our conclusions hold for all pairs of goods, all pairs of factors, and all consumers. In other words, we get a good insight into the meaning of general equilibrium yet without having to confront all the complexities of a general equilibrium model.

Box 8-3 DO DEMAND CURVES MEASURE UTILITY?

Economists usually assume that willingness to pay for a product accurately reflects its utility. Behavioral economists have tested this assumption. In one experiment, participants received a list of items to be auctioned and were asked to write down the last two digits of their Social Security numbers as a dollar amount at the top of the page. They were then asked to state whether they would be willing to pay that dollar amount for each item, followed by their actual bid for the item. The bids were collected, and the highest bidders then purchased the items at the price they bid. It turned out that those with higher Social Security numbers were, on average, willing to pay significantly more for the items than those with lower numbers. For example, people whose Social Security numbers ended in 80–99 were willing to pay 2 to 3.5 times more for the auctioned commodities than those whose numbers ended in 00–19. It's pretty hard to imagine that your Social Security number has any meaningful correlation to the utility you derive from different commodities or from money. A better explanation is that our willingness to pay is determined by reference points—in this case, the last two digits of a Social Security number—and that those reference points can be manipulated. Such experiments make it hard to assume that demand curves are an exact measure of utility and might lead us to question the inherent desirability of market equilibriums.

We have also repeatedly applied the equimarginal principle of maximization in determining how far consumers should substitute one good for another in their shopping basket and how far producers should substitute one factor for another in their production processes. We have taken it for granted that goods can be substitutes in the minds of consumers and that factors can be substitutes in the production processes of firms. This is sometimes called the **principle of substitution**. Goods (and factors) are not always related as substitutes. Sometimes they are complements, meaning that more of one makes the other more desirable (useful) rather than less. These relations of substitution and complementarity will play an important role in later chapters.

BIG IDEAS to remember

- Basic market equation
- Competitive market
- Law of diminishing marginal utility
- Law of diminishing marginal product
- Equimarginal principle of maximization
- Pareto optimal allocation
- Sliding fulcrum function of prices
- Monopoly and misallocation
- Marginal cost and marginal revenue
- Non-price adjustments
- Supply and demand

Supply and Demand

Since supply and demand are the most important rules of market grammar, it is worth saying more about them—how they are used as tools of analysis, and something about the theories of production and consumption that underlie them. Supply and demand are so important that some wags have said you could turn a parrot into an economist just by training it to repeatedly squawk "supply and demand"—no need to worry about grammar. However, as you will see in this chapter, there is a good deal more to supply and demand than even a genius parrot could master. But fortunately, it is not too difficult for human beings.

■ A Shift in the Curve Versus Movement Along the Curve

The amount demanded of a good can change for many reasons, and economists classify these causes into two categories: a change in the price of the good, and everything else. The effect of a change in price on the quantity demanded is shown by a movement along the demand curve. The effect of all other causes of a change in amount demanded is shown by a shift in the entire curve. What are these other causes? The most important are the consumer's income, his tastes, and the prices of related goods.

If the consumer's income rises, he will likely purchase more of every good including x, at every price. His demand curve for x shifts to the right. For a fall in income, it would shift to the left.

Goods are related as substitutes (ham and bacon) or as complements (bacon and eggs). If the price of bacon goes up, people will substitute ham and buy more ham at every price. The demand curve for ham will shift up. If the price of eggs goes up, the demand curves for both ham and

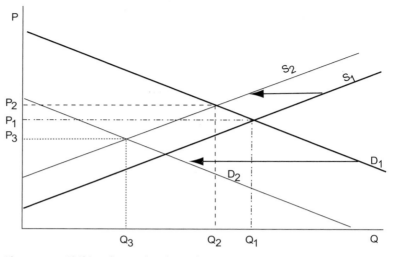

Figure 9.1 • Shifting demand and supply curves.

bacon are likely to shift down as people buy fewer eggs and eat bacon and eggs less often.

A change in tastes or information can shift the demand curve. If people worry more about cholesterol, they will buy fewer eggs at all prices (a shift downward in the demand curve).

To summarize: The demand curve is a relationship between P and Q. Within that given relationship, a change in P causes a change in Q, or vice versa, along the demand curve. But the whole relationship between P and Q can also change. That is a shift in the entire curve. The curve may change its shape and position, but it will always be downward sloping, at least in the absence of speculative demand (see Box 8.1).

Similar things can be said about a movement along the supply curve and a shift in supply. The discovery of a more efficient technology or a new deposit of resources, for example, would shift the supply curve outward so that more would be offered at each price.

The impacts of shifting supply and demand curves on equilibrium price and output are shown in Figure 9.1. A massive recall of *E. coli*–tainted beef and a closing of the guilty processing plant might shift the supply curve for beef from S_1 to S_2. In response to the shift, prices would rise along the D_1 curve, from P_1 to P_2, and supply would drop from Q_1 to Q_2. The recall leads to a series of investigative reports on conditions in meat-packing plants, showing that crowded conditions, rapid processing, and poor inspections make bacterial infection a regular and recurring problem. In response, the demand for beef might shift from D_1 to D_2. As a response to this shift, the amount suppliers would be willing to provide falls along the S_2 curve from Q_2 to Q_3, and price falls from P_2 to P_3.

A mild winter would cause the demand for natural gas to shift down. An increase in the price of electricity would cause it to shift up. Supply and demand analysis is not just finding the intersection but knowing where the curves are.

■ EQUILIBRIUM P AND Q, SHORTAGE AND SURPLUS

The intersection of supply and demand is important because it defines the equilibrium in which both buyers and sellers are satisfied, and, as we saw in the basic market equation, resources are optimally allocated.

In Figure 9.2, P^*Q^* is the equilibrium price at which both buyers and sellers are satisfied (maximizing utility and profits, respectively). It is called an equilibrium because once at it, there is no tendency to change, because everyone is satisfied.

Any movement away from equilibrium sets in motion forces pushing us back toward equilibrium (the equilibrium is stable). Suppose the price were P_1. Then at P_1 quantity supplied would be P_1B and quantity demanded would be P_1A. The excess of quantity supplied over quantity demanded (AB) is called a *surplus*, and if there is a surplus, the market is not in equilibrium. There are many eager sellers and few eager buyers. The many unsatisfied sellers will begin to compete with each other to sell to the few buyers. How? By offering a lower price. The price will be bid downward until the surplus disappears.

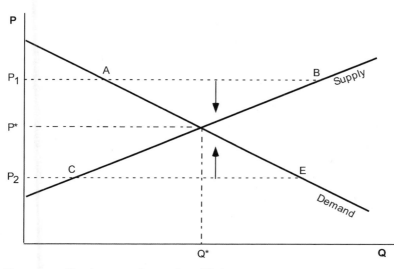

Figure 9.2 • Shortage, surplus, and equilibrium.

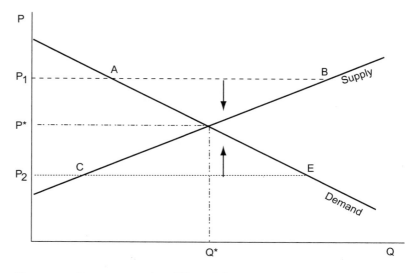

Figure 9.3 · Consumer surplus, differential rent.

At a price of P_2 we have a quantity supplied of P_2C and a quantity demanded of P_2E. The distance CE represents a *shortage*. It, too, is a disequilibrium situation. We have many willing buyers facing few willing sellers. Unsatisfied buyers will begin to compete with each other for the limited sellers. How? By offering a higher price. The price will rise, and as prices rise, producers will produce more, until the shortage disappears.

Note that shortage and surplus are defined with reference to a given price. When people complain about a shortage of petroleum or of labor, what they mean is a shortage of *cheap* petroleum or *cheap* labor. This has led some to claim that the free market can solve all problems. There can be no shortage of resources in a free market, nor any surplus of labor either. All we need to do is to let the price be free to find its equilibrium, and *violà*, no shortage, no surplus—and some would go on to say no resource scarcity, no involuntary unemployment! Hardly comforting if the shortage of a key resource disappears only at a near infinite price or if the surplus of labor disappears only at a wage below the level of subsistence. The market is a wonderful institution but not a magic charm.

In Figure 9.3, the equilibrium price is OC, and the equilibrium quantity supplied and demanded is CE. Notice that buyers pay the amount OC for *every* unit of x they buy, not just the last or marginal unit. Buyers would have been willing to pay a price of OA for the first unit of x rather than do without it. And for the second unit they would have been willing to pay almost as much, and similarly for the third, and so on. But for the last unit at CE, they were only willing to pay OC. The maximum the con-

Box 9-1	THE RATIONING AND ALLOCATION FUNCTIONS OF PRICE IN ACTION

The rationing function of prices apportions market products to whoever is willing to pay the most for them, ensuring that the product goes to the person who values it the most. This maximizes the monetary value of products (a proxy measure of their utility) across all consumers.

The allocative function of prices apportions factors of production to whatever industry is able to generate the greatest profits from those factors, thus maximizing monetary value generated by all producers.

By creating the greatest possible monetary value, the price mechanism balances what is possible with what is desirable—but only if we define desirability solely in terms of monetary value. Is monetary value what we actually want to maximize at all times? Does maximizing monetary value always ensure that resources are allocated to the most desirable use?

A few decades back, Aventis developed the drug eflornithine, which cures African sleeping sickness, a debilitating disease that threatens millions of Africans. Although the only other treatment for advanced sleeping sickness is extremely painful to administer and often ineffective or even lethal, Aventis could not profit from selling the drug to poor Africans. They had no market demand; their strong preferences for the drug were unfortunately weighted by negligible purchasing power. Aventis discontinued production for African sleeping sickness but licensed the drug to Bristol-Myers Squibb and Gillette for an alternative use: removing unwanted facial hair in women.

When a good like eflornithine can be put toward the vanities of the wealthy or the basic needs of the poor, the rationing function of price apportions it to the wealthy. When scientific research can be apportioned toward developing cosmetics for the wealthy or life-saving medicines for the poor, the allocative function of price apportions it toward cosmetics.

The story has a happy ending. After the NGO Médecins Sans Frontières threatened to publicize the issue, Aventis agreed to again produce eflornithine for the treatment of African sleeping sickness. Still, while markets can do an amazing job at allocation, we must not assume that they always apportion resources toward their most desirable use.[a]

The allocative function of price apportions few resources toward cures for lethal diseases that afflict the poor, and many towards the production of cosmetics for the rich. When something exists that can be used toward either end, like eflornithine, the rationing function of price apportions it towards the vanities of the wealthy rather than the basic needs of the poor.

---

[a]P. Gombe, Epidemic, What Epidemic? New Internationalist Spring 2003; P. Trouiller et al., Drug Development for Neglected Diseases: A Deficient Market and a Public-Health Policy Failure. The Lancet 359:2188–2194(2002); WHO Fact sheet no. 259: African Trypanosomiasis (Sleeping Sickness). Geneva: World Health Organization, 2006.

sumers would have been willing to pay rather than do without is AEHO. What they actually pay is OCEH. The difference, the triangular area ACE, is consumer surplus. It results from the law of diminishing marginal utility. The consumer enjoys the higher marginal utility on all infra-marginal units purchased but has to pay a price equal to the lowest marginal utility, that of the last unit purchased.

For necessities, consumer surplus is enormous. For example, I buy water at a price that equals my marginal utility of water—say, washing my car. But I get the benefits of the higher marginal utility uses on the infra-marginal units of water used to keep me from dying of thirst, from just being thirsty, from going without a bath, and so on. Many environmental goods and services are necessities that have a large or even infinite consumer surplus.

The producer is also getting a producer surplus, sometimes called **rent**, or differential rent. The producer sells all CE units at his marginal cost for making the last unit, or OC. But for the first unit he would have been willing to accept only OB rather than not sell it. And he would have been willing to accept just a bit more for the next unit, and so on. So thanks to the law of diminishing marginal physical product (law of increasing marginal costs) the producer can sell all his low-cost infra-marginal production for the same price as his highest marginal cost final unit. For example, the price of coal will be equal to the marginal cost of the most expensive coal that can be sold. That marginal coal will come from the worst, leanest, most inaccessible coal mine. But the coal from the rich, easy-to-dig coal mines will sell at the same high price (equal to marginal cost at the worst mine), so the good mines earn a surplus or rent. This is over and above the normal profits of the operation. Normal profit is reflected in the costs underlying the supply curve. **Normal profit** is defined as the opportunity cost of the time and money the entrepreneur has put into the enterprise— that is, what he could have earned from his time and money in his next best alternative. Rents are especially important in extractive industries.

The usual definition of *rent* in economics is payment over and above the minimum necessary supply price. The term is usually associated with land because any payment for the use of land is over and above its minimum supply price in the sense of its cost of production. The cost of production of land is zero. That doesn't mean that land is not scarce and that no charge should be levied on its use. But it does mean that such rent is unearned income, as the tax accountants so frankly call it. It is better to tax unearned income than earned income from the point of view of both fairness and economic efficiency, as Henry George argued over a century ago. Ecological economists have followed Henry George, generalizing his insight a bit to advocate Ecological Tax Reform: shifting the tax base from value added and onto that to which value is added, namely the through-

Rent (also known as differential rent or, in the case of natural resources, scarcity rent) is equivalent to producer surplus and is defined as payment over and above the minimum necessary supply price.

put flow. In bumper sticker form, "Tax bads, not goods!" The bads are depletion and pollution (throughput), and the goods are value added by labor and capital, that is, earned income. More about that later.

■ ELASTICITY OF DEMAND AND SUPPLY

How sensitive is a change in quantity demanded to a change in price? Does it take a big change in price to get even a small change in quantity demanded? Or does even a tiny change in price cause a big change in quantity demanded? Clearly it could be either, or anywhere in between. **Elasticity** is a measure of the responsiveness of a change in quantity demanded to a change in price.

The numerical measure of elasticity is defined as follows: Price elasticity of demand = percentage change in quantity demanded, divided by the percentage change in price. Or,

$$ED = (\Delta q/q) \div (\Delta p/p)$$

An exactly analogous definition holds for the price elasticity of supply.

Figure 9.4 shows the extreme values of elasticity of demand and helps give you a feel for the concept. In the case on the left, elasticity is infinite because even the smallest percentage change in price would cause an infinite percentage change in quantity demanded. This is the way the demand curve looks to a pure competitor, a pure price-taker (i.e., one whose production is too small relative to the market supply to have any noticeable effect on price). In the case on the right, even an infinite percentage change in price would cause no change in quantity demanded, which might approximate the demand for essential goods (e.g., water, food, energy, life-saving medicines, vital ecological fund-services) when in short supply. Most demand curves are neither horizontal nor vertical but are negatively sloped somewhere in between. For these in-between curves, elasticity varies along the curve in most cases. Demand is said to be elastic when a 1% change in price gives rise to a more than a 1% change in quantity demanded. If a 1% change in price causes a 1% change in quantity, then the formula gives an elasticity of 1, and consequently this case is called unitary elasticity.

A classic case of the importance of elasticity is the demand for agricultural crops in general. People need to eat no matter how high the price, and they will not eat much more than their fill no matter how low the price. The demand for food in general is rather inelastic, as shown in Figure 9.5. This means that the price (and total revenue) change drastically with small changes in quantity, putting the farmer in a risky position. This is one reason why governments often subsidize agriculture.

Figure 9.5 shows the impact of a shift in supply on the equilibrium

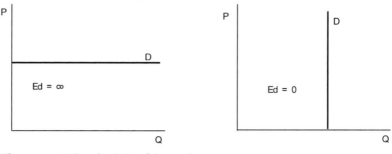

Figure 9.4 • Price elasticity of demand.

price and quantity of a good with inelastic demand, using food as an example. At a harvest of Q_1 the price is P_1 and the total revenue to the farmers is the area of the rectangle $P_1(Q_1)$. Next year comes a bumper crop due to good weather, and the supply shifts out to S_2, with demand staying the same. The farmers' total revenue has fallen to $P_2(Q_2)$, a much smaller amount, even though the harvest has increased from Q_1 to Q_2. The reason is that with inelastic demand, a small increase in quantity causes a large decrease in price. (Elasticity works in reverse, too!) When demand is elastic, price and total revenue move in the opposite directions; when demand is inelastic, they move in the same direction.[1]

THINK ABOUT IT!

What do you think would happen if the World Bank lent money to lots of developing countries to produce and export food crops?

What determines whether demand is elastic or inelastic? Mainly the necessity of the good for human well-being and the number of good substitutes available. If ham is easily available, the demand for bacon is likely to be elastic. Since there is no good substitute for food in general, the demand for agricultural goods in general is going to be inelastic.

There is an important caveat to this general rule. Elasticity is also inversely correlated with the share of one's budget spent on a particular good. For example, chewing gum is hardly a necessity, and there are many reasonable substitutes, but if its cost is negligible relative to your budget, a doubling in price may have little impact on how much you purchase. On the other hand, in poorer parts of the world, some people spend up to 50% of their income on food. When grain prices more than doubled

---

[1]Note the implications of this for GNP (to be discussed at length in Chapter 14). In 2008, grain and oil supply failed to keep pace with growing demand, leading prices to triple in response to supply constraints. The share of these commodities in GNP soared even as supply decreased or stayed the same. If essential resources become scarce enough, GNP could skyrocket.

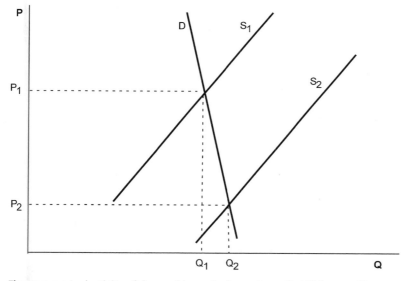

Figure 9.5 • Inelasticity of demand in agriculture. A small shift in quantity supplied leads to a large change in price.

between 2007 and 2008, many poor people were forced to buy much less
food, even though it is a necessity with no substitutes. Essential and nonsubstitutable resources susceptible to shifts in demand or variation in supply, such as food and oil (see Figure 11.4), can exhibit wild swings in
price. Moreover, high prices in one year may lead to large investments in
new output, which then result in price crashes. Price crashes can lead to
reduced investments and output, forcing prices back up. Such instability
presents a real challenge to the balancing function of price.

THINK ABOUT IT!

*Is the price mechanism the best means of determining production and
consumption levels of basic necessities? Are there better ways of
doing it?*

Finally, the longer the time period of the analysis (Q on the *x*-axis is really a flow, Q/t), the more time consumers have to adjust their habits, and
the more elastic demand will be.[2] The time period also greatly affects the
elasticity of supply. In the very short run, say daily, the fish supply is totally inelastic once the fishing boats have returned with the day's catch.
Weekly supply is more elastic because fishermen can respond to a higher

---

[2]Of course, it can be very difficult to adjust demand for food, water, and other things essential to life.

price by staying out longer or taking more crew. Over a year, elasticity is still greater because new fishing boats can be built. But then the supply could become totally inelastic as the limits of the fishery are reached. There is a tendency to neglect the last case in most microeconomics texts, but it is critical in ecological economics.

■ THE PRODUCTION FUNCTION

The **production function** is a relation that shows how factor inputs are converted into product outputs. It is basically a technical recipe for producing a good or service, or rather for *transforming* labor, capital, and resources into a good or service.

$$Q = F(a, b, c \ldots)$$

In plain English, the quantity produced is a function of the inputs (factors of production) *a*, *b*, and *c*. We earlier spoke of the marginal physical product of factor *a*. In terms of the production function, it is the increase in Q resulting from adding one more unit of factor *a* to the production process, holding factors *b*, *c*, and everything else constant. The main thing we know about production functions is that they follow the law of diminishing marginal physical product, as depicted in Figure 9.6. In the graph, the marginal physical product of *a* is the change in Q resulting from a change in *a* ($\Delta Q/\Delta a$), which is also the slope of the Q curve, when Δa is one unit. In Figure 9.6, Δa is the same in both cases, but the corresponding ΔQ is much smaller when *a* is larger. The slope diminishes as we increase *a*, and it finally becomes zero. A similar curve could be drawn for factors *b* and *c*. From the basic market equation we already know the significance of the law of diminishing marginal product, as well as the basic argument for why it is true. Another way of stating the relationship is to say that for each equal increment of Q we need to use increasing amounts of *a*. Stated this way, the relation is the law of increasing marginal cost.

The assumption behind the curve is that *a* is a substitute for *b*, *c*, etc., all the other factors held constant, but that it is an imperfect substitute. Therefore, the more we try to substitute *a* in the recipe, the less successful the substitution becomes in terms of producing more Q. The poorer a substitute factor *a* is for the other factors held constant, the sooner MPP*a* falls to zero. And if *a* is really a complement to the other factors held constant, then MPP*a* is zero from the start.[3] Production functions exhibit both **substitutability** and **complementarity** between factors. But standard economists tend to see mostly substitution, whereas ecological economists emphasize complementarity. Why is this so?

Probably it is because ecological economists put different things in the

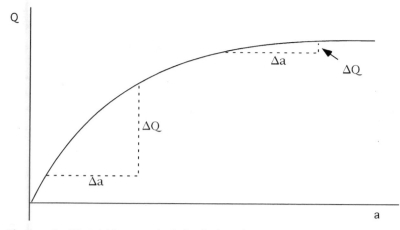

Figure 9.6 • Diminishing marginal physical product.

production function and assign different qualitative roles to the different factors in the production process. For instance, neoclassical economists treat all inputs the same: labor, capital, and resources. Ecological economists insist on a qualitative difference. Labor and capital are transforming agents, funds that transform the flow of resources into a flow of product, but are not themselves embodied physically in the product. Labor and capital are agents of transformation (efficient causes, or fund-services), while resources are that which is being transformed (material causes, or stock flows). The neoclassical production function abstracts from the difference between material and efficient causes of production and considers both to be equivalent. Ecological economists insist on the distinction. It is clear that while one **material cause** (resource) can often be substituted for another and one **efficient cause** (labor, capital) can often be substituted for another, the relation of efficient cause to material cause, of agent of transformation to material undergoing transformation, is mainly one of complementarity. Ecological economists emphasize this latter relation; neoclassical economics misses it entirely.

Ecological economists also insist that we account for the laws of thermodynamics in two other ways. First, we must recognize the importance of energy, which is essential to all production. Solar energy can be treated as an agent of transformation that is not physically embodied in what is produced and is not used up in the act of production.[4] However, fossil fuels account for the vast majority of energy use in today's world, and

---

[3]Note that the relations of complementarity and substitutability apply to factors of production as well as to goods used by consumers.

[4]Technically, any given photon of solar energy is used up in the act of production but does not reduce the number of photons available in the future.

while these are not embodied physically in the final product, the more important fact is that they are used up in the production process. Fossil fuels therefore are best treated as a material cause rather than an efficient cause. Food is the energy source used to power labor, and it is both used up and physically embodied in labor. Second, ecological economists recognize the law of entropy: All economic production inevitably generates waste, which is an unavoidable part of the production process. Furthermore, all economic products eventually become waste. Waste is not a factor of production, however, but rather an output of the production process.

The ecological economics production function embodies the fund-flow distinction (funds uppercase; flows lowercase) and accounts for energy use and waste emissions:[5]

$$q + w = F(N, K, L; r, e)$$

K and L are funds of labor and capital, r represents flows of natural resources, e flows of energy, q flows of products, and w flows of waste.[6] Funds and flows are basically complements. Substitution takes place within each category but usually not between them. N stands for natural capital, which exists both as a stock that yields a flow of resources (a forest yielding cut timber) and as a fund (a forest yielding the service of watershed protection or wildlife habitat). The stock function of yielding a flow of resources is already captured in r, so N should be taken here as representing the fund function of providing an indirect service that contributes to the transformation of r into q (and w as well), much as K and L provide direct services. For example, forest cover providing the service of water catchment to recharge aquifers used in irrigated agriculture is as much a capital fund-service to agriculture as is the fund of pipes and sprinklers used in irrigation. N also includes solar energy and waste absorption capacity, which helps break down w into forms that have negligible impact on human welfare or that can be reincorporated into r.

In most neoclassical economics textbooks, the production function is written as

$$Q = F(K, L)$$

In other words, waste and energy resources are neglected entirely! A flow of output is seen as a function of two funds, or two stocks that are not decumulated, or drawn down.

---

[5]Note that earlier we used a lowercase q for individual demand and supply and an uppercase Q for market demand and supply. Because conventional economics does not distinguish between funds and flows, we had to develop new notation, but we have simultaneously tried to maintain some continuity with conventional notation and so use uppercase letters to denote funds and lowercase to denote flows. See the next footnote for additional comments.

[6] q represents services and nondurable consumer goods (flows). Production functions for durable goods such as cars and houses (funds) should use Q.

Is it possible for stocks (or funds) that do not decumulate, by themselves, to yield a flow? An economist's first reaction is to say of course a stock by itself can yield a flow; consider the stock of money in the bank yielding by itself a flow of interest (the principal is not decumulated, yet the interest flow continues, perhaps even in perpetuity). True, but that is a convention of finance, not a physical process of production. How about a stock of cattle yielding a flow of new cattle in a sustained-yield fashion. Isn't that a physical stock yielding a physical flow? Not really. It is a stock (livestock) converting a *flow* of inputs (grass, grain) into a *flow* of outputs (new cattle and waste products). The resource inflow of grass, grains, and so on is *transformed* into a product outflow of new cattle (and replacement to the livestock herd for natural mortality or "depreciation"), plus waste. The correct description of "production" is transformation of a resource inflow into product outflows, with stocks (funds) of capital and labor functioning as the transforming agents. This is true not only for the living transformers in agriculture (plants and animals) but even more obviously for industrial processes of production where the transformation is visible within the factory at every stage.

Why are such basic facts excluded from neoclassical economics? Why does it choose to ignore them, to exclude natural resources from theoretical

Box 9-2   ARE E AND R SUBSTITUTES?

It's fairly obvious how labor and capital are substitutes for one another and how one natural resource can substitute another. Though it may be less obvious how stored solar energy and raw materials are substitutes, their substitutability has played an important role in economic activity. Fossil fuels are readily transformed into an enormous number of different raw materials ranging from fertilizers to plastics. Perhaps most important, fossil fuels have played an enormous role in avoiding Malthus' dire predictions of famine. Fossil fuels began to displace biomass as the dominant form of energy in the eighteenth century. This allowed more land to be converted to agriculture instead of forestry. As we moved from animal to mechanical traction and transportation, lands formerly dedicated to feeding oxen and horses could be dedicated to growing food. As we began to deplete the stocks of soil nutrients essential to agriculture, we learned to use natural gas to fix atmospheric nitrogen into bioavailable ammonia (NH_3). In recent years, society has sought to reverse this substitution process by converting agricultural products into biofuels. However, both energy and food are essential and nonsubstitutable. If we convert one resource into the other in response to rising prices, we are likely to simply shift the price increases to the other.

analysis right from the start? Perhaps it is a case of "money fetishism," assuming that what is true for money in the bank, the symbol and measure of wealth, must be true for the wealth it symbolizes.[7]

Recall that the production function is a recipe. Real recipes in real cookbooks always begin with a list of ingredients. They do not just say "take the labor of a cook and the capital equipment in a standard kitchen and make cherries jubilee." Real cookbooks give us a list of ingredients, followed by instructions about how to combine and transform the ingredients into the product. The cook and her kitchen are not physically transformed into an edible dish. They are the transformers, not the transformed.

Perhaps this latter consideration has led some neoclassical economists to include r in their production function. But they have done it in such a way as not to solve the problem. Most production functions are multiplicative forms—that is, the relationship F among the factors is one of multiplication (e.g., the Cobb-Douglass production function[8]). After all, what could be more natural than "multiplying" together things that we call "factors" to produce something that we call a "product"? Unfortunately there is nothing in the real-world process of production that corresponds at all to multiplication. There is only transformation. This means that substitutability is built into these production functions from the beginning as a mathematical artifact, including substitutability between r and K, and r and L (between funds and flows). In these multiplicative production functions, we can make one factor as small as we wish, while keeping the product constant, if we increase the other factor sufficiently. The only restriction is that no factor can be reduced to zero, but it can approach zero. But according to this logic, if our cook is making a 5-pound cake, he can increase it to a 1000-pound cake with no extra ingredients, just by stirring harder and baking longer in a bigger oven. The First Law of Thermodynamics (conservation of matter and energy) has been totally ignored.

Ignoring the necessary role of natural resources in production is part of a pattern in neoclassical economics that has the effect of denying that nature has any role in economic life. Value in their view is only value added—added by labor and capital. But added to what? In the neoclassical view, that to which value is added is thought to be merely inert stuff

---

[7]One current text included resources among its factors of production in discussion but on the next page wrote the production function as above, with only K and L as variables, and from then on forgot about resources. J. M. Perloff, *Microeconomics*, 2nd ed., Reading, MA: Addison Wesley, 2000.

[8]The Cobb-Douglas production function (in its simplest form, $Q = K^\alpha L^\beta$) states that production equals capital (raised to an exponent) times labor (raised to an exponent), and the sum of the exponents $(\alpha + \beta)$ equals one. The important point for us is simply that capital and labor are multiplied, and if R (resources) is added as a third factor, it too is multiplied in like manner.

having no independent value of its own. Ecological economics recognizes the contribution of nature in supplying that to which value can be added. It is by no means inert stuff. It is scarce low-entropy matter and energy, as we discussed in Chapter 4.

Moreover, since resource flows are complementary with manmade funds, they can become the limiting factor on production when they become more scarce than the manmade fund factors. This is exactly what is happening in the real world. As we cannot realistically increase our supplies of fossil fuels, any sustainable economy must ultimately be built around the fund of solar energy rather than the flow of fossil fuels. Economic logic tells us to maximize the productivity of the limiting factor in the short run and to invest in its increase in the long run. Economic logic has not changed, but the pattern of scarcity has. As we move from the empty to the full world economy, natural resource flows and services generated by natural capital stocks and funds become the limiting factor. Fish catches are no longer limited by the manmade capital of fishing boats but by remaining natural capital of stocks of fish in the sea and the natural funds that support their existence. We need to economize on and invest in the limiting factor. Economic logic has not changed, but the identity of the limiting factor has.

Because our basic neoclassical theory of production, when it considers natural resources at all, cannot distinguish funds from flows or recognize complementarity between them, we have been slow to recognize this change.

■ THE UTILITY FUNCTION

The **utility function** relates utility or want satisfaction to the flow commodities (goods and services) consumed by the individual:

$$U = F(x, y, z, \ldots)$$

In plain English, our happiness depends on what we consume (and not on our freedom, creativity, social relationships, etc.). The main thing we know about the utility function is the law of diminishing marginal utility, that as we consume increasing amounts of a single good, other things being equal, our additional satisfaction from each additional unit of the good in question declines—that is, total utility increases but at a decreasing rate.

THINK ABOUT IT!
Do you remember the argument demonstrating why it is reasonable to believe that this law is true?

The graph in Figure 9.7 simply shows that marginal utility declines with more units of x consumed in a given period, even though total utility in-

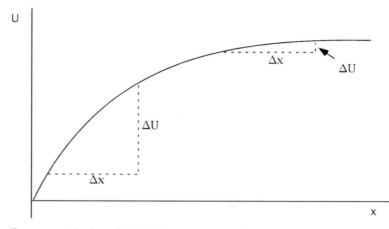

Figure 9.7 • The law of diminishing marginal utility.

creases (but ever more slowly). The first ice cream cone on a hot day is a delight, the third is not. To the extent that x is complementary with other goods that are being held constant, the marginal utility of x will decline faster, or may even be zero from the start. For example, a third ice cream cone without a complementary glass of water is not so enjoyable. And a right shoe without a left shoe has zero marginal utility to most people.

The neoclassical production function contains only flows of goods or services as variables. A more complete picture of consumer satisfaction would include the services of directly enjoyed natural capital funds: the provision of breathable air by a well-functioning atmosphere, drinkable water by a well-functioning hydrologic cycle, and so on. As we did previously, let's refer to these natural funds as N and include them in the utility function in a way analogous to our inclusion of the transforming funds L and K in the production function. Direct service from N is not just the pleasure of a beautiful landscape, although that is included. It also and primarily refers to the basic fitness of our environment to our organism that supports our life and consequently supports our consumption and want satisfaction from commodities. We have evolved over millions of years into a relation of fitness to our environment. But fitness is a reciprocal relation. If we are fit for our environment, then our environment is fit for us. The relation of fitness can be destroyed by either a change in us or a change in our environment.[9] This fitness of the environment implies

---

[9]Fitness-destroying changes in the environment can be induced by humans (e.g., global warming) or autonomously caused by nature (e.g., a volcanic eruption or earthquake). H. J. Henderson, *The Fitness of the Environment*, Gloucester, MA: Smith, 1913.

a relation of complementarity between N and us, including most of our artifacts, which are extensions of us.[10]

Let's rewrite the utility function as

$$U = F(N; r, w, x, y, z, \ldots)$$

Some natural resources such as wild strawberries or water from a spring enter directly into our utility function, as indicated by our inclusion of r. Waste, w, also enters directly into our utility function as a negative, ameliorated by waste absorption capacity (N). Commodities $x, y,$ and z are outputs of the production function. If x is a pair of hiking boots, then its utility depends on places worth hiking in (N). If y is a snorkeling mask, its utility depends on reefs and clean water (N)—not to mention prior dependence on breathable air, drinkable water,[11] sunlight filtered of enough of its UV rays so we won't get melanoma if we go snorkeling, and so on. N provides a complementary service without which the utilities of most consumer goods are not very great.

Consumers may be able to maintain the same level of satisfaction in the face of a reduction in x by simply consuming more y and z, which to some degree are substitutes. But N is a complement to $x, y,$ and z, and their increase will usually not compensate for a decline in N. In fact, their utility will fall with a decline in N. For example, you won't enjoy your new hiking boots very much if there are no pleasant hiking trails.

The production of more $x, y,$ and z may not make us any better off if accompanied by a decline in N. Indeed, it may make us worse off. The usual assumption is that N is superabundant and $x, y,$ and z are scarce. But that was back in the empty world. We are now in a full world. The use of N as a stock that yields r is likely being pushed beyond sustainable limits. Consequently we may lose not only N in the stock sense but also the services of N in the fund sense. And those services are complementary with most consumer goods, so reduced N will mean reduced utility from $x, y,$ and z.

We may argue that for rich people with a low MU for goods in general, the relative importance of N is high, but that for the poor, the MU of goods still outweighs the MU of N. That may be the case in certain instances, and it is certainly what institutions like the World Bank consider realistic. But it need not be. Hurricane Katrina showed us that the relative

---

[10]Cars and bicycles are extensions of our legs, clothes are extensions of our skin, telephones extend our ears, computers extend our brains, etc.

[11]It may seem like double-counting to include freshwater and breathable air as both flows (r) and funds (N). However, both the oxygen we breathe and the water we drink are stocks physically transformed through the process of consumption (r), while our ecosystems (N) function as funds that transform exhaled carbon dioxide and excreted water back into breathable air and drinkable water.

importance of healthy wetlands and barrier islands that reduce storm surges and a stable climate that reduces hurricane strength may have higher marginal utility for the poor than the rich, as the former were forced to ride out the storm, while the latter were able to flee. Even food bought by the poor has low marginal utility if the poor are being choked by poisonous air and fried by UV radiation.

Microeconomics is a big subject, and this section could turn into a book by itself. We have not tried to cover the whole topic but rather to explain the basic grammar of markets, why markets are efficient, and what conditions have to hold for markets to be efficient (and by implication what might possibly go wrong) and to give the rudiments of supply-and-demand analysis. We have also tried to point out the places where we think the ecological economics perspective improves microeconomics. We will now focus our attention more directly on N, the goods and services provided by nature, to determine how well they meet the criteria for efficient allocation by market forces.

BIG IDEAS to remember

- Shift in the supply or demand curve versus movement along the curve
- Shortage, surplus, equilibrium
- Consumer surplus
- Producer surplus, rent
- Elasticity
- Complementarity and substitutability

- Production function
- Utility function
- Diminishing marginal utility
- Diminishing marginal physical product
- Production as transformation
- Fitness of the environment
- Natural capital's role in both utility and production functions

CHAPTER

10

Market Failures

In the previous chapters, we saw how markets utilize individual self-interest to efficiently allocate resources (means) among alternative ends via the pricing mechanism. However, markets function efficiently only with a narrow class of goods. We have already shown how monopolies undermine the ability of the market to efficiently allocate, but monopolies are a type of market failure generated by a structural problem: the absence of competition. Markets also fail because of inherent characteristics of certain types of resources or because there are no institutions clearly defining property rights. The purpose of this chapter is to describe the characteristics that make a particular resource a market good and to examine what happens when resources do not have these characteristics. As we will show, none of the goods and services provided by natural capital has all of the characteristics required for efficient allocation by the market.

For simplicity, in this chapter we will boil down the various conditions for arriving at Pareto optimal allocations, described in Chapter 8, to the generic equation "marginal cost equals marginal benefit," or MC = MB. In addition, we will often use the term *Pareto efficient* or just *efficient* as the equivalent of Pareto optimal.

■ CHARACTERISTICS OF MARKET GOODS

Excludability

We first defined and discussed **excludability** in Chapter 4 and briefly review the concept here in recognition of its importance. An excludable good is one for which exclusive ownership is possible; that is, a person or community must be able to use the good or service in question and prevent others from using it, if so desired. Excludability is virtually synonymous with property rights. If a good or service is not owned exclusively by

165

someone, it will not be efficiently allocated or produced by market forces. The reason for this is obvious. Market production and allocation are dedicated solely to profits. If a good is not excludable, someone can use it whether or not any producer of the good allows it. If people can use a good regardless of whether or not they pay for it, they are considerably less likely to pay for it. If people are unwilling to pay for a good, there will be no profit in its production, and in a market economy, no one will invest in producing it, or at least not to the extent that the marginal benefit to society of producing another unit is equal to the marginal cost of production.

Of course, many nonexcludable goods, such as fish in the ocean, are produced by nature, not by humans. In this case, "investment" is simply leaving smaller fish to grow larger, or maintaining a high enough population stock to ensure future production. The "cost" of investment is opportunity cost, the profit that would have been earned by catching those fish today. If a fisherman throws back a small fish to let it grow larger, it is more than likely that a different fisherman will catch that larger fish, and in a market economy, people rarely invest when others will reap the returns.

Excludability is the result of institutions. In the absence of institutions that protect ownership, no good is truly excludable unless the possessor of that good has the physical ability to prevent others from using it. Some type of social contract, be it government or less formal social institutions, is required to make any good excludable for someone who lacks the resources to defend her property. Excludability therefore is not a property of the resource per se but rather of the regime that controls access to the resource. It is fairly easy to create institutions that provide exclusive property rights to tangible goods such as food, clothing, cars, and homes. Slightly more complex institutions are required to create exclusive property rights to intangibles such as information. Patent laws protecting intellectual property rights are ubiquitous in modern society, but it remains difficult to enforce such property rights. For example, have you ever recorded copyrighted music or installed unpurchased, copyrighted software on your computer?

However, many goods and services, such as the majority of the fund-services produced by ecosystems, have physical characteristics that make it almost impossible to design institutions that would make them excludable. As we suggested in Chapter 6, it is pretty much impossible to conceive of a workable institution that could give someone exclusive ownership of the benefits of the ozone layer, climate regulation, water regulation, pollination (by wild pollinators), and many other ecosystem services. It is often possible to establish exclusive property rights to an ecosystem fund (e.g., a forest) but impossible to establish such rights to the services the fund provides (e.g., regional climate regulation). If, like a forest, the fund is simultaneously a stock that can supply a flow (e.g., of timber), market allocation will account only for the stock-flow benefits of the resource. When there are no excludable property rights to a good or service, that good or service is nonexcludable.

■ RIVALNESS

A second characteristic that a good or service must have if it is to be efficiently produced and distributed by markets is **rivalness**. We defined a rival good or service in Chapter 4 as one for which use of a unit by one person prohibits use of the same unit at the same time by another. Rivalness may be qualitative, quantitative, or spatial in nature. Again, food, clothing, cars, and homes are rival goods.

A nonrival good or service therefore is one whose use by one person has an insignificant impact on the quality and quantity of the good or service available for another person to use. Among nonrival goods produced by humans, streetlights, information, public art and eradication of contagious diseases come to mind. Climate stability, the ozone layer, beautiful views, and sunny days are a few of the nonrival goods produced by nature.

Note that all stock-flow resources are quantitatively rival. If I eat food (a stock), there is less for you to eat. In contrast, fund-service resources may be rival or nonrival. When a fund-service is rival, it is spatially rival at each point in time and qualitatively rival over time. If I wear clothes, drive a car, use a machine that makes cars, or use a house (all fund-service), they are not available for you to use at the same time I do, and if you use them afterwards, they are just a bit more worn out. As we pointed out in Chapter 4, all nonrival resources are fund-service.

As discussed in Chapter 9, market efficiency requires that the marginal cost to society of producing or using an additional good or service be precisely equal to the marginal benefit. However, if a good is nonrival, an additional person using the good imposes no additional cost to society. If markets allocate the good, it will be sold for a price. If someone has to pay a price to use a good, he or she will use the good only until the marginal benefit is equal to the price. A price is by definition greater than zero, while the marginal cost of additional use of nonrival goods is zero.[1] Therefore, markets will not lead to efficient allocation of nonrival goods, or conversely, a good must be rival to be efficiently allocated by the market.

---

[1]This does not necessarily imply that providing a nonrival good free of charge is efficient either. We will return to this topic later.

> **THINK ABOUT IT!** *Are nonrival resources scarce? Does it make sense to ration their use to those who can afford them? What price for a nonrival resource maximizes its value to society? How much of a nonrival resource would the private sector produce or preserve at this price? What options are available for producing and preserving nonrival resources?*

Resource Scarcity and Abundance

Some fund-service resources such as roads, beaches, and golf courses appear to be nonrival at times and rival at others. For example, if I drive my car down an empty road, it does not diminish your ability to drive down that same road. However, if thousands of people choose to drive down the same road at the same time, it results in traffic jams, and the ability of the road to move us from point A to point B is seriously diminished. Economists typically refer to such resources as nonrival but congestible, or simply as congestible. On closer inspection, however, congestibility is not a case of nonrival resources becoming rival but rather one of abundant resources becoming scarce.[2] If my car occupies a place on the highway or my towel a place on the beach, that physical location is no longer available to you. As long as there is an abundance of other equally good locations, the fact that a given spot is unavailable is irrelevant since there is no competition for that particular spot. Only when the resource grows scarce must we compete for its use. A congestible resource is one that is bordering on scarcity; at times of light use it is abundant with no competition for use, while at times of heavy use it becomes scarce, and potential users must compete for it. Note that **congestibility** is an issue of scale: as scale increases, as the world becomes more full, formerly abundant resources become scarce. Truly nonrival goods cannot become scarce through use, though they can of course be depleted.

The Interaction of Excludability, Rivalness, and Congestibility

What happens when goods and services are nonrival, nonexcludable, or both? The simple answer is that market forces will not provide them or will not efficiently allocate them. However, we need to be far more precise than this if we are to derive policies and institutions that will lead to the efficient allocation and production of nonrival and nonexcludable resources. Effective policies must be tailored to the specific combination of excludability, rivalness, and congestibility that characterizes a particular good or service. The possible combinations are laid out in Table 10.1 and described in some detail next.

---

[2]*Abundant* means that there is enough of the resource for all desired uses; scarce means that one must choose between alternative competing ends.

■ Table 10.1

THE MARKET RELEVANCE OF EXCLUDABILITY, RIVALNESS, AND CONGESTIBILITY

	Excludable	Nonexcludable
Rival	Market goods; food, clothes, cars, houses, waste absorption capacity when pollution is regulated.	Open access regimes ("tragedy of the commons"), e.g., ocean fisheries, logging of unprotected forests, air pollution, waste absorption capacity when pollution is unregulated
Nonrival	Potential market good, but if so, people consume less than they should (i.e., marginal benefits remain greater than marginal costs); e.g., information, cable TV, technology.	Pure public good, e.g., lighthouses, streetlights, national defense, most ecosystem services
Congestible	Toll or club goods: Market goods when scarce, zero marginal value when abundant. Greatest efficiency occurs when price fluctuates according to usage, or if clubs are formed that prevent the resource from becoming scarce; e.g., ski resorts, toll roads country clubs.	Open access regimes: Only efficient to make them excludable (i.e., to limit access) during periods of high use; e.g., non-toll roads, public beaches, national parks

■ OPEN ACCESS REGIMES

The first class of goods and services we will examine are open access resources, those that are nonexcludable but rival. The use of such goods commonly leads to what Garret Hardin has called "the tragedy of the commons."[3] The classic example Hardin used was the grazing commons once common in England. Say a village has a plot of land that anyone in the community can use for grazing cattle. There are 100 households in the community, and the plot of land is sufficient to support 100 head of cattle indefinitely without being overgrazed. In the terminology of Chapter 6, if we think of cattle as grass harvesters, then 100 cows will harvest the maximum sustainable yield of grass (see Figure 6.1). If one person adds one more cow to the commons (as might happen when there is no institution preventing her from doing so), not only does the grass need to be shared among more cows, but the grass yield declines, and each cow will be just a bit thinner. One person will gain the benefits of having two cows but will share the costs of all the cattle being thinner with everyone else in the community. If everyone thinks in the same manner, households will keep adding cattle to the commons until its productive capacity has de-

---

[3]G. Hardin, The Tragedy of the Commons, *Science* 162:1243–1248 (1968).

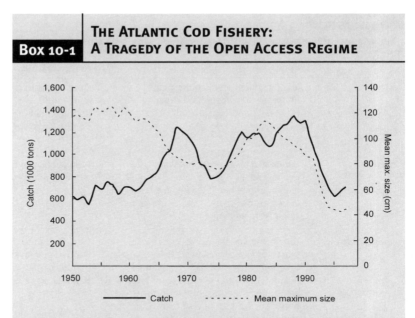

Figure 10.1 • Time series of total catch and mean maximum size of species in catch for Canada, northwest Atlantic. (Source: FAO Statistics. Online: http://fishbase.sinica.edu.tw/manual/fishbasefao_statistics00002679 .htm.)

The Atlantic cod fishery was once capable of providing enormous sustained yields at a low harvest cost, but years of overharvesting have been blamed for dramatically reduced yield.[a] *Sustainable* yields may now be close to zero,[b] and scarcity has made those that remain more expensive to harvest. For cod and many other commercial species, selective fishing pressure has also drastically reduced the age and size at maturity. Smaller mature females produce far fewer and smaller eggs, threatening the ability of the species to re-establish.[c] All the fishermen as a group would clearly have been better off if they had limited their harvest to sustainable levels. However, in any given year, any given fisherman was better off continuing to harvest more fish as long as a profit was to be made. Lacking institutions to keep harvests sustainable, if one fisherman reduced his catch, those cod would simply be caught by another fisherman.

In the case of renewable resources, reducing harvests in one year is passive investment in future production. The costs of investment are the opportunity costs of not harvesting now. As we pointed out in our discussion of excludability, if a good or service is nonexcludable, the market provides no incentive to invest in it. In the case of Atlantic cod, each fisherman pursuing his own rational self-interest has virtually wiped out the stock, and other fishermen are rapidly following the same path for the majority of fish species worldwide.[d] Economists tell us that if we wipe out one species, we can always replace it with another. Fishermen have done this around the world, harvesting fish lower and lower on the trophic chain (i.e., lower-level predators), yet in spite of this substitution, harvests in many places are still plunging.

---

[a]L. Burke, Y. Kura, K. Kassem, C. Revenga, M. Spalding, and D. Mcallister, Pilot Analysis of Global Ecosystems: Coastal Ecosystems, Washington, DC: WRI, 2000.

[b]It is actually quite possible that cod populations in the North Atlantic have fallen below the minimum viable population stock (point of critical depensation) and will gradually diminish, even in the absence of further harvests.

[c]F. Saborido-Rey and S. Junquera, Spawning Biomass Variation in Atlantic Cod (Gadus morhua) in Flemish Cap in Relation to Changes in Growth and Maturation, Journal of Northwest Atlantic Fishery Science 25:83–90 (1999).

[d]As mentioned earlier, the U.N. estimates that 11 of the world's 15 major fishing areas and 69% of major commercial fish species are in decline. The only reason that total harvests have not dropped accordingly is that favored species high on the food chain have been replaced by others lower down (Burke et al., op. cit.).

clined, and it is no longer capable of generating the biomass it once did. Each person acting in what appears to be rational self-interest degrades the commons, and everyone is worse off than if he or she had stuck with one cow per person. Under these circumstances, rational self-interest does not create an invisible hand that brings about the greatest good for the greatest number but rather creates an invisible foot that kicks the common good in the rear.

It is extremely important to note that the "tragedy of the commons" is a misnomer. Common property is property for which a community, not an individual, controls the property rights. Those who are not members of the community are not allowed access to the resource. In many cases, communities have developed institutions that prevent individuals within the community from overexploiting the resource, and there is no problem with the "tragedy of the commons." A better term, therefore, is the "tragedy of open access regimes" or simply "the open access problem."

There are many goods characterized by the open access problem. Hardin originally wrote his classic article to describe the problem of population growth. Especially in labor-driven agrarian economies, large families can be great assets. However, if everyone has a large family, the land must be divided up among all the children, and it eventually grows too scarce to sustain the population. People harvest soil nutrients faster than the system can restore them, and sustainable production declines. Other resources plagued by this tragedy include ocean fisheries (the Atlantic cod is a classic example) and the planet's waste absorption capacity for unregulated pollutants, such as carbon dioxide.

Many economists have correctly pointed out that the open access problem results from a lack of enforceable property rights (i.e., excludability). If the English commons in the first example had been divided up into 100 equally productive excludable private lots, the rational individual would graze only one cow in each lot, and the tragedy would be avoided. Unfortunately, for many of the resources of concern to us, the ability to bestow individual property rights is more the exception than the rule, and in other cases we will describe later, private property rights will not lead to efficient outcomes.[4]

For now, we will draw attention only to the difficulty of establishing property rights in the fairly simple case of open access resources. Analysis of oceanic fisheries provides a good starting point. Most of the oceans are international waters over which there is little or no institutional control. There are treaties limiting harvests, prohibiting certain harvest techniques, or prohibiting harvests of certain species altogether, but countries can choose whether or not to sign those treaties, and little in the way of enforcement is available even when they do sign. For example, most nations have agreed to cease or drastically reduce the harvest of many species of whale, but countries such as Norway, Japan, and Iceland do not always follow these regulations, and little can be done to force them to do so.[5]

Nations now enjoy 200-mile zones of exclusion in coastal waters, where they can prohibit boats from other nations from harvesting marine species and physically enforce this exclusion if necessary. Exclusion zones at least allow the potential for the regulation of fisheries within these waters, and we will discuss effective mechanisms in Chapter 20. Unfortunately, fish are generally pretty disrespectful of such boundaries, and once outside of those bounds, they are fair game to all. In addition, many species of fish migrate from the coastal exclusion zone of one nation to the exclusion zone of its neighbor. This is the case with many salmon populations off the coasts of Canada and the U.S. These two nations, which enjoy some of the best relations of any two nations in the world, are in the midst of a bitter dispute over who is entitled to what share of the catch. In the meantime, salmon populations continue their rapid decline.

---

[4]Nonetheless, sloppy analysis and a lack of rigor on the part of too many economists have led to a widespread belief that establishing private property rights is the answer to most, if not all, of our environmental problems.

[5]The regulations actually put a moratorium on harvesting certain species of whales for commercial purposes but still allow harvesting for scientific research. Japan now harvests endangered whale species for "scientific research," selling the carcasses commercially afterwards. CNN.com, Japan Whaling Fleet Returns Home Amid U.S. Dispute, *Nature* (2000). Online: http://www.cnn.com/2000/NATURE/09/21/whaling.japan.reut/, posted September 21, 2000.

If we are unable to establish defensible property rights to a resource such as fish, how are we going to address the far more difficult "tragedy of the commons" problem of overpopulation?[6]

■ EXCLUDABLE AND NONRIVAL GOODS

A second class of goods of great interest is those that are excludable but nonrival and noncongestible. The prime example of this type of good is information. In the not-too-distant past, most information was relatively nonexcludable as well as being nonrival. In Adam Smith's time, firms would jealously guard their trade secrets, but if such secrets got out, there was nothing to prevent others from using them. As Smith[7] pointed out, trade secrets were equivalent to monopolies, and "the monopolists by keeping the market constantly understocked, by never fully supplying the effectual demand, sell their commodities much above the natural price. . . . The price of monopoly is upon every occasion the highest which can be got" (p. 164). "Monopoly, besides, is a great enemy to good management" (p. 251). (We will say more about patents, which are monopolies on information, in our discussion of globalization in Chapter 19.) In more recent times, of course, trade secrets have been protected by patents,[8] an institution that makes them legally excludable and hence marketable. The justification for this is the assumption that without excludable property rights, people would not profit from inventing new things. Inventors would have no incentives, and the rate of advance of technology would slow, to the detriment of society. Once a patent expires, the knowledge embodied in it becomes a pure public good.

The problem is that one person's use of information not only has no negative impact on someone else's use, it can actually lead to improvements in quality—in the words of one computer programmer, "the grass grows taller when it's grazed on."[9] Intellectual progress is invariably a col-

---

[6]Kenneth Boulding actually proposed a solution to the overpopulation problem based on awarding all women the "property right" to 2.1 children (replacement fertility level) in the form of tradable permits. Needless to say, many people object to such a system. Can you suggest a better solution? See K. Boulding, *The Meaning of the Twentieth Century*, New York: Harper & Row, 1964.

[7]A. Smith, *The Wealth of Nations: Books I–III* (with an introduction by Andrew Skinner), Harmondsworth, Middlesex, England: Penguin Books, 1970.

[8]Trade secrets do still exist in the traditional form. Patents provide only exclusive ownership to information for a fixed time period. To avoid making information public knowledge at the end of this time span, some companies prefer not to patent certain processes or recipes, instead keeping them hidden from potential competitors. See J. E. Stiglitz, "Knowledge as a Global Public Good." In I. Kaul, I. Grunberg, and M. A. Stern, eds., *Global Public Goods: International Cooperation in the 21st Century*, New York: Oxford University Press, 1999.

[9]Cited in D. Bollier, *Silent Theft: The Private Plunder of Our Common Wealth*, London: Routledge, 2002.

THE LINUX OPERATING SYSTEM AND OPEN SOURCE: THE EFFICIENCY OF THE INFORMATION COMMONS

BOX 10-2

Tux is the penguin mascot of Linux. He was designed by Larry Ewing using GIMP, a free software graphics package. Like Linux, Tux is free for anyone to use and modify, as long as they acknowledge authorship.

Much information is covered by patents designed to make it excludable. Open-source software, in contrast, is protected by licenses designed to keep it nonexcludable. While many open-source licenses allow people to sell open-source software, they insist that it also be legal to redistribute the software for free. Even more important, the licenses must allow distribution of the source code, that is, the computer program written in a language intelligible to humans, compared to the compiled binary code intelligible only to computers. This practice enables other programmers to find and remove bugs in the software, to modify the software, and to incorporate the software into their own work on the condition that that work remain open-source as well. The philosophy behind this approach is that when many programmers are free to improve the source code for a piece of software, it evolves and improves at an astonishing rate. The grass grows taller the more it's grazed on.

While some economists would tell us that invention requires the incentive of the profit motive, empirical evidence suggests otherwise. Take the example of the Linux operating system. Linux is an open-source operating system for computers invented by Linus Torvald. Computer experts around the world have worked on this operating system free of charge, and as a result it has become stable, powerful, and adaptable.[a] IBM has contributed to the Linux code, and both IBM and HP use the

Linux operating system on their high-end mainframe servers. The use of Linux continues to increase rapidly. Apple has turned to open-source for its Mac OS-X operating system. Open-source Apache runs more than half of the world's Web servers, and hundreds of thousands of other open-source packages exist.[b] Certainly this proves that neither profits nor patents are always needed to spur innovation.

[a]*The Great Giveaway.* New Scientist *(2002). Online: http://dsl.org/copyleft/dsl.txt; D. Bollier,* Silent Theft: The Private Plunder of Our Common Wealth, *London: Routledge, 2002.*

[b]*Open Source Initiative Website (http://www.opensource.org/).*

lective process. In academia, people have freely shared and built upon each other's ideas for centuries. The Internet and much of its associated software were primarily the result of freely shared knowledge. In many ways, the free flow of information and ideas creates an "efficiency of the commons," not a tragedy.

Patents, on the other hand, may slow the rate at which we develop new knowledge and use it. Existing knowledge is the most important input in the production function of new knowledge. Keeping existing knowledge artificially expensive during the life of the patent also makes the production of new knowledge more expensive. In addition, corporations often patent scientific methods and even mathematical algorithms, thereby making it much more expensive to conduct research using those methods. Many researchers are engaged in research for the sake of advancing knowledge and not for making profits, and any additional costs are likely to reduce their ability to advance knowledge. For example, a new virus-resistant strain of rice cannot be distributed because there are as many as 34 separate patent holders with competing claims on the knowledge that went into its invention.[10]

The costs of intellectual property rights have become a serious issue. For example, the U.S. Constitution authorizes Congress to issue copyrights and patents "for limited times" to "promote the progress of science and the useful arts." Both copyrights and patents were initially awarded for 14 years, with the possibility of a single 14-year extension on copyrights if the author was still alive. Under pressure from corporate lobbies, Congress has gradually increased copyright longevity, and corporate copyrights are now good for 95 years, while individual copyrights are good for the life of the individual plus 70 years. Two hundred years ago, technol-

---

[10]Ibid.

ogy advanced slowly, and 14 years was typically only a small fraction of a technology's useful life. Technologies now change so fast that many are obsolete within a decade or less, yet we have extended patents to 20 years, frequently keeping technologies out of the public domain until they are useless. Many studies have shown that the proliferation and prolongation of patents and copyrights actually slows the progress of science and the useful arts.[11]

Patents can also generate serious inefficiencies on the consumption side. Consider the case of AIDS medicine. A currently available drug cocktail can dramatically reduce the level of HIV in the human bloodstream, potentially decreasing the risk of transmission. The benefits of controlling contagious and deadly diseases are nonexcludable. Currently drug companies hold patents on these medicines, making them prohibitively expensive for poorer countries,[12] decreasing their ability to control the disease, and increasing the risk of everyone contracting it. Of course, from the perspective of corporations that profit from these medications, total elimination of the disease would be a very unprofitable outcome. The argument in favor of patents is that without profits, corporations would not have the incentives to invent new drugs. The irony is that patent rights are protected in the name of the free market, yet patents simply create a type of monopoly—the antithesis of a free market.

So we see that while there may be a solid rationale for allowing patents, there also exist compelling arguments against them. If information is free, it will presumably be used until the marginal benefits of use are just equal to the marginal costs of additional use, which is zero. This is a prerequisite for efficient allocation. On the other hand, if a good is nonexcludable, the market provides no incentive to invest in it. Patent laws recognize this problem by imposing artificial excludability on information, at least for the time period of the patent, creating artificial scarcity. Nonetheless, Linux (see Box 10.2) and many other examples show that patents are not necessary to spur invention, so the belief that patents will result in a faster rate of technological advance is little more than an assertion. Widespread recognition of this problem has led to the "copyleft" movement, a general method for making a program or other work free and requiring all modified and extended versions of the program to be free as well.[13]

We believe that public provision and common ownership of informa-

---

[11]See M. Heller and R. Eisenberg, Can Patents Deter Innovation? The Anticommons in Biomedical Research. *Science* 280:698–701 (1998); L. Lessig, *Free Culture: How Big Media Uses Technology and the Law to Lock Down Culture and Control Creativity,* New York: Penguin Press, 2004.

[12]In spring 2001, a number of drug companies agreed to drop their suit against the South African government and its policy of producing and selling the drugs without paying full royalties. Numerous other drug patents exist that still illustrate the basic principles explained here.

[13]Free Software Foundation, What Is Copyleft? 2009. Online: http://www.gnu.org/copyleft/

tion, especially information needed to protect, provide, and restore public goods, is likely to be more sustainable, just, and efficient than private ownership; making nonrival resources excludable may create a tragedy of the non-commons.

Still, goods such as information and knowledge present difficult issues. We will return to these issues later in this chapter, and again in our discussion of trade and development in Chapter 18, with the issue of so-called trade-related intellectual property rights.

Congestible Resources: On the Edge of Scarcity

What about excludable, congestible goods? As discussed earlier, congestible goods are abundant at low levels of use and scarce at high levels of use. We used roads and traffic jams as an example, and recreational resources such as beaches, swimming pools, parks, and wilderness hiking trails are similar (though for the gregarious, crowding may actually add value). When goods or resources have these properties, positive prices may produce efficient outcomes for high levels of use, while at low levels of use, pricing will lead to inefficient outcomes. This suggests that under certain circumstances, it may be reasonable to treat congestible goods as market goods during peak usage and nonmarket goods at other times.

Multi-tier pricing structures are one possible solution. **Multi-tier pricing** involves charging different prices at different times or for different users. In this case, prices could be charged when congestion occurs (e.g., rush hour tolls on a bridge), but the good or service would remain free while uncongested. Such pricing structures can be expensive to implement, and whether the strategy is reasonable generally depends on the specific case. Whether the strategy is possible depends on excludability.

■ PURE PUBLIC GOODS

As most economists readily admit, the market is not capable of optimally producing or efficiently allocating pure public goods, which are both nonrival and nonexcludable. We add the adjective *pure* only because many people are careless in their use of the term *public goods*. As we explained in Chapter 8, in a market setting, each person can purchase a good or service until the marginal benefit from purchasing one more unit of that good or service is just equal to the marginal cost. As long as anyone is willing to pay more for a good than it costs to produce that good, the supplier will supply an additional unit. If a public good exists, however, anyone can use it regardless of who pays for it. An additional unit of a market good is worth producing only as long as at least one individual alone is willing to pay at least the cost of producing that unit. In contrast, a public good is worth producing as long as all individuals together are willing to pay the cost of producing another unit.[14] Look again at Figure 9.4 for supply and demand. When we moved from the individual demand curve to the market or social demand curve, we added up the quantities each individual would be willing to pay at a given price because we were talking about market goods. This is because the goods were rival, and what one person consumed another could not. However, public goods are non-rival, so one person consuming the good does not leave any less for others. In this case, we obtain the social demand curve by adding up the prices each individual is willing to pay for a given quantity to find out how much society as a whole is willing to pay for that quantity.

For market goods, each person consumes exactly as much as they purchase, so people's consumption preferences (weighted by their income, of course) are revealed by how they spend their money in the market. For public goods, in contrast, each person consumes as much as all of society purchases. This leads to problems.

For example, assume a nice forested park in the middle of a big city would cost $100 million for land purchase, landscaping, and infrastructure. Imagine that if we added together everyone's demand curve for the park, we would find that for a park of the proposed size, society is cumulatively willing to pay $150 million. Therefore, if everyone in the city contributed two-thirds as much money to building the park as he or she thought it would be worth, the park would be built. The problem is, how do we get everyone to contribute toward the park the amount that the park would be worth to him or her? Would market forces (i.e., the private sector) build it? Assume a corporation builds the park, fences it off, then charges admission. Knowing that the average person should be willing to pay $150 for a lifetime pass to the park, the corporation decides to sell such passes to recoup its investment. But problems arise. Not everyone values the park equally. Some people would be willing to pay much more than $150 for the pass, while others would be willing to pay very little. Those who are only willing to pay less than $150 will in effect no longer value the park at all if there is a $150 fee, and the corporation fails to recoup its investment. The corporation runs into similar problems if it charges an entrance fee for each use, say $1. In this circumstance, even those who value the park the most will use the park less than they would have if it were free. Since they will use it less, they will value it less. Again, the corporation will be unable to recoup its investment, and the park will not be built. If the park were free, more people would use it, increasing the total welfare of society while imposing no additional costs on society, but then, of course, the corporation would not build the park. Therefore,

---

[14]P. Samuelson, The Pure Theory of Public Expenditure, *Review of Economics and Statistics* 36: 387–389 (1954).

BOX 10-3   THE FREE-RIDER EFFECT

What would happen if some institution solicited voluntary donations to build a public park in my neighborhood? I am trying to decide how much to donate. If I meet standard neoclassical economic assumptions, I want to maximize my own utility. I live close to the proposed park site and would value it more than most people. I decide that I am indifferent between a park of the proposed size and $1000 and prefer the park over any cost less than $1000. However, I rationalize that if I contribute nothing to the park and others contribute what it's worth to them, that will reduce the size of the park by only one one-hundred-thousandth. I would vastly prefer a park 99.999% of the size of the proposed park at zero cost to myself than the proposed full-size park for $1000. Alternatively, if I contribute what the park is worth to me and others contribute less, the resulting park will be smaller because of insufficient funds and therefore no longer worth $1000 to me.

From this narrow perspective of self-interest, my best strategy, regardless of what others choose, is to contribute nothing and instead rely on the contributions of others. Unfortunately, if everyone else also makes a similarly rational calculation in his or her own self-interest, the city ends up with no park whatsoever, and everyone is worse off than they would have been if the park had been built. This is known as the **free-rider effect**, and it is a serious obstacle to the provision of public goods. In this case, rational self-interest has created an invisible foot that kicks the common good in the rear.

the market will not provide the park even as a private good, and if it did, it would not be used efficiently.

Let's examine another example of the clash between markets and public goods. A small sharecropper in southern Brazil is kicked off his land share so that the landowner can grow soybeans under a heavily mechanized system requiring little labor. The soybeans are exported to Europe as cattle feed for higher profits than the landowner could make using sharecroppers to produce rice and beans for the local market. The sharecropper heads to the Amazon and colonizes a piece of land. Researchers have "guess-timated" the value of the ecosystem services sustainably produced by this land at roughly $1660/hectare/year.[15] These ecosystem services are primarily public goods. If the colonist deforests the land, he may make a one-time profit of $100/hectare for the timber (the timber is,

---

[15]R. Costanza, R. d'Arge, R. de Groot, S. Farber, M. Grasso, B. Hannon, S. Naeem, K. Limburg, J. Paruelo, R. V. O'Neill, R. Raskin, P. Sutton, and M. van den Belt, The Value of the World's Ecosystem Services and Natural Capital, *Nature* 387:253–260 (1997). The land also produces a number of goods, such as timber and marketable nontimber forest products, which are valued in the cited paper, but those values are not included in this estimate.

of course, worth much more on the market, but the market is far away, and middlemen and transportation costs eat up the profits) and an estimated $33 annualized net profits per year from slash-and-burn farming.[16]

In terms of society, there is no doubt that the annual flow of $1660/year far outweighs the private returns to the farmer. However, the ecosystem services are public goods that the farmer must share with the entire world, and there is no realistic way of giving the farmer or anyone else meaningful private property rights to the ecosystem services his forests supply.[17] In contrast, the returns to timber and agriculture are market goods that the farmer keeps entirely for himself, and existing institutions give him the right to do as he pleases with his private property. Clearly, both the farmer and society could be better off if the beneficiaries of the public goods paid the farmer to preserve them. As long as the farmer receives more than $150/year, he is better off, and as long as global society pays less than $1660/ha/year, it is better off.

Unfortunately, a number of serious obstacles prevent this exchange from happening, and we'll mention three. First, most people are ignorant about the value of ecosystem services (more on this later). Second, the free-rider effect means that many beneficiaries of public goods will pay little or nothing for their provision. Third, we currently lack institutions suitable for transferring resources from the beneficiaries of ecosystem services to the farmer who suffers the opportunity cost of not deforesting. Thus, from the farmer's point of view, in a market economy deforestation is clearly the rational choice, and society suffers as a result.

Public Goods and Scarcity

Anyone who accepts the basic premise that global ecosystems create life-sustaining ecosystem services must believe that public goods are critically important. Yet market economic theory offers little advice concerning the production and allocation of public goods.

As we have repeatedly stressed, it is impossible to make something from nothing and nothing from something. The production of market goods requires raw materials and generates waste. Raw materials are stock-flow resources taken from ecosystem structure, which therefore deplete ecosystem fund-services. Waste returned to ecosystems further depletes these services. Thus, if our economic system provides incentives solely for producing and allocating market goods, it will systematically

---

[16]A. Almeida and C. Uhl, Developing a Quantitative Framework for Sustainable Resource-Use Planning the Brazilian Amazon, *World Development* 10 (1995).

[17]This does not mean we cannot develop mechanisms for compensating the farmer for providing ecosystem services; it simply means that if the farmer provides them, they are provided for one and all.

undermine the production of absolutely invaluable public goods—and life-sustaining functions of our planet. One of the underlying assumptions of ecological economics is that many of the scarcest and most essential resources are public goods (services provided by natural resource funds), yet the existing economic system addresses only market goods.

Let's return now to the question of knowledge and information, presented above. If information is a private good, it will not be efficiently allocated; if it's a public good, it will not be produced in sufficient quantity by market forces. If we set theory aside momentarily and simply look at the rate of technological progress, we might believe we have little to complain about. Technological progress is extremely rapid. While it is true that patents create legal monopolies, they do so for only a limited time, after which knowledge becomes a public good. It is not difficult to believe that it is the lure of temporary monopoly profits that brings new inventions onto the market faster than would otherwise be the case. Why worry about a system that works?

One reason is that the creation of this knowledge imposes an opportunity cost on society. There is a limited pool of resources (e.g., money, scientists, laboratories) for conducting research, and if it is being used in one task, it is simply not available for another. If new inventions are driven primarily by the pursuit of profits, then we have a serious bias against the invention of public goods or technologies that preserve or restore public goods. For example, the pharmaceutical industry employs legions of scientists and spends billions on research and development for noncommunicable diseases afflicting the wealthy.

On the other hand, the control of communicable diseases is a public good, and from a societal perspective, we should channel resources toward it. An excellent example is found with the case of tuberculosis treatments. Tuberculosis is a highly contagious disease that is difficult to treat. Effective treatments were developed in the 1950s, but they require close monitoring of patients for 6 months to a year. Many people who suffer from tuberculosis are not sufficiently responsible to treat themselves, and governments throughout the world have spent enormous amounts of public money to track down people and force them to take their medicine. In response to declining infection rates, federal funding in the United States targeted for tuberculosis treatment was slashed in the 1970s, and public health expenditures suffered further cuts in the 1980s. As a result, many tuberculosis sufferers did not receive treatment or began to take their medicine only erratically. This contributed to a resurgence of tuberculosis in the 1980s, including multiple-drug-resistant varieties. In New York City alone, it cost over $1 billion in government spending to bring this epidemic back under control.

Tuberculosis affects primarily the poor, which reduces the profitability of any cures and explains the lack of investment in new treatments by drug companies.[18] (It is no coincidence that only 13 of the 1240 new drugs licensed between 1975 and 1996 dealt with lethal communicable diseases that afflict primarily people from developing countries.[19]) However, even if drug companies did develop new treatments, they would need to patent the medicine and sell it for a profit to recoup their investments. Patents increase the prices of medicines to cure contagious diseases, while from the perspective of society their cost to patients should actually be *negative*. In other words, it would be efficient for the government to pay people to use such medicines because their use provides positive benefits to the rest of society.

Most research scientists working today are employed by the private sector, which retains rights to whatever they produce. The private sector is increasingly responsible for funding research in universities as well. It will logically concentrate on research with market potential. Corporate scientists would presumably work for a public organization for the same salaries. In this case, the resulting knowledge could be free for all to use, a prerequisite for efficient use (as defined by neoclassical economics) of nonrival goods. We are not suggesting here that all research be government funded.[20] But unless some nonmarket institutions fund research into public goods, technological advance will tend to ignore nonmarket goods.

As the great Swiss economist Sismondi argued long ago, not all new knowledge is a benefit to humanity. We need a social and ethical filter to select out the beneficial knowledge. Motivating the search for knowledge by the purpose of benefiting humanity, rather than by securing monopoly profit, provides a better filter—a filter more likely to give us a cure for AIDS or tuberculosis or malaria than a new liposuction or heart transplant technique.

If the market is extremely effective at producing market goods but very poor at producing or preserving public goods, then over time, public goods inevitably become more scarce relative to private goods, giving rise

---

[18]L. Geiter, "Ending Neglect: The Elimination of Tuberculosis in the United States," Committee on the Elimination of Tuberculosis in the United States, Division of Health Promotion and Disease Prevention, 2000. Online: http://www.nap.edu/books/0309070287/html/.

[19]L. Garret, *Betrayal of Trust: The Collapse of Public Health*, New York: Hyperion, 2000.

[20]It is worth noting that the government does fund enormous amounts of primary research with taxpayer dollars yet subsequently allows private corporations to establish patents to products derived from that research. This allows corporations to earn monopoly profits from taxpayers on research paid for by those very taxpayers.

to a problem of what we call **macro-allocation**, which is the allocation of resources between market and nonmarket goods and services.

Public Goods and Substitution

In previous chapters, we discussed the issue of substitution, pointing out that ecological economists believe we cannot substitute efficient cause for material cause, except at a very small margin, while neoclassical economists (NCEs) argue that manmade capital is essentially a perfect substitute for natural capital. After all, haven't people been arguing that resource exhaustion is imminent since the time of Malthus, and haven't they consistently been proven wrong? NCEs (and many other people) argue that as a resource grows scarce, the price increases, encouraging the invention and innovation of substitutes. It is true that some civilizations in the past appear to have disappeared from exhaustion of their natural capital, but NCEs assert that the market has averted such collapses since the advent of capitalism. This, of course, is tantamount to claiming that the profit motive is more powerful than the survival motive.

One can certainly find numerous examples where the profit motive has apparently produced substitutes for scarce resources, but that's no guarantee that there will be adequate substitutes for every vital resource. Moreover, even if the profit motive does provide a marvelous spur to our creative processes, what happens when the resources becoming increasingly scarce are public goods? Such goods have no price, and there will therefore be no price signal telling our entrepreneurs that we need substitutes, nor is there any profit to be made by creating such substitutes.[21] What happens then? Conventional market economics does not address this question.

The Distribution of Public Goods Through Space

Another complication arises with some public goods, particularly those produced by ecosystem function, which is highly relevant to policy choices. We pointed out earlier that ecosystems can provide different public goods and services for different populations. For example, water regulation and storm surge protection provided by intact mangrove forests are local public goods, the role of mangroves as a fish nursery is a regional public good, and global climate stability promoted by forest carbon storage is a global public good. Individuals are ultimately responsible for how ecosystem stock-funds are treated, they will prefer market flows over public good services, and the two are often mutually exclusive. Unlike individuals, society in some circumstances should prefer public goods

---

[21]Not to mention, as we pointed out in Chapter 6, that it is probably much easier to create substitutes for ecosystem structure (stock-flow resources, raw material) than for ecosystem services provided by the wickedly complex interaction of structural elements in an ecosystem.

over the production of private goods that deplete them. However, local communities may show little concern for providing national public goods. Sovereign nations may show little concern for providing global public goods. Thus, decision makers at different levels (individual, local, national, international, intergenerational) will have different incentives for preserving or destroying ecosystem functions, and these incentives must be understood in order to develop effective policies that meet differing needs at all levels. Unfortunately, political systems are based largely on the nation-state or smaller political units and hence are inadequate for addressing global issues.

The inadequacy of existing political and economic systems for managing public goods is particularly problematic in light of the fact that many ecosystem services are public goods that provide vital services. On the global level, such functions include protection from excessive solar radiation, global climate regulation, and the role of biodiversity in sustaining the web of life. On the local level, ecosystems provide microclimate regulation (often critical for successful agricultural production), buffering from storms, and maintenance of water quality and quantity, all of which may be essential for community sustenance.

■ EXTERNALITIES

Another important type of market failure is known as an externality. An **externality** occurs when an activity or transaction by some party causes an unintended loss or gain in welfare to another party, and no compensation for the change in welfare occurs. If the externality results in a loss of welfare, it is a negative externality, and if it results in a gain, it is positive. The **marginal external cost** is the cost to society of the externality that results from one more "unit" of activity by the agent.

The classic example of a negative externality is a coal-fired utility plant that moves in next door to a laundry service that air-dries its wash. The soot from the coal plant dirties the laundry, and the laundry service receives no compensation from the coal utility. Both air and water are great conveyors of externalities. If a farmer allows his cattle to defecate in a stream flowing through his property, all those downstream from him suffer the negative externality of polluted water. Alternatively, a farmer might reforest his riparian zone, reducing access by cattle. The canopy shades the stream, killing in-stream vegetation. Water can now run faster, allowing it to scour sediments out of buried springs in the stream, thereby increasing water flow.[22] Shaded water is cooler, reducing the ability of some harmful

---

[22]Note that the outcomes of reforestation are highly dependent on both the system in question and the techniques and species used for reforestation. In some cases, reforestation may reduce water quantity.

bacteria to thrive, thereby increasing water quality. Downstream landowners benefit from these positive externalities. Another example is the pollution we spew every time we drive a car, which decreases air quality and contributes to global warming.

Because the agent conducting the activity in question is not compensated for positive externalities and pays no compensation for negative ones, she does not take into account these costs or benefits in her decision to pursue the activity. In the case of negative externalities, the agent carries the activity too far. With positive externalities, the agent engages in too little of the activity. If the agent conducting the activity were to be appropriately compensated or charged, there would be no more externality; the activity would be carried out until marginal benefits equaled marginal costs, not only for the agent conducting the activity but also for society.

As in the case of public goods, economists have suggested that assigning property rights will eliminate the externality problem. If the laundry has the right to clean air, then the coal utility will be forced to pay the laundry service for dirtying its laundry.[23] Once compensation is paid, the externality is gone. Alternatively, it would be possible to assign the right to pollute to the coal utility. In this case, the laundry would have to pay the coal utility not to pollute.[24] In perhaps the most widely cited article ever written on externalities, Ronald Coase argued that under certain conditions it doesn't matter whether the utility is assigned the right to pollute or the laundry is assigned the right to clean air.[25] In either case, the negotiated outcome will lead to an identical amount of pollution, precisely at the level where marginal costs of pollution to the laundry are just equal to the marginal benefits to the utility. The implication is that the externality issue requires no government intervention aside from the assignment and enforcement of property rights; market forces are perfectly capable of sorting it out. This is known as the **Coase theorem**.

A graphic analysis may help make this a bit clearer. Figure 10.2 shows pollution on the x-axis and marginal costs and benefits on the y-axis. The coal-fired utility benefits from polluting, while this pollution imposes costs on the laundry service. There are several technologies available for

The *Coase theorem* states that the initial allocation of legal entitlements does not matter from an efficiency perspective as long as they can be exchanged in a perfectly competitive market.[26]

---

[23]In reality, this will not necessarily lead to an efficient solution in a dynamic setting. For example, if the payment makes the laundry service profitable, another laundry may locate nearby, which would also be profitable with a subsidy from the utility. For fairly obvious reasons, it is inefficient if the promise of a subsidy from the utility attracts businesses that are otherwise harmed by the utility's presence.

[24]In this case, we would have to look at installation of pollution reduction equipment as generating a positive externality for which the laundry service must compensate the public utility.

[25]R. Coase, The Problem of Social Cost, *Journal of Law and Economics* 3:1–44 (October 1960).

[26]R. Cooter, "Coase Theorem." In *The New Palgrave: A Dictionary of Economics*, New York: Macmillan, 1987, pp. 457–459.

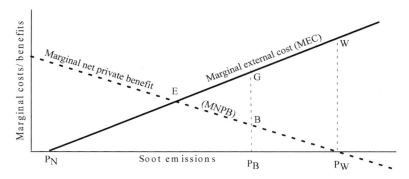

Figure 10.2 • "Optimal" pollution levels. "Optimal" pollution levels are theoretically determined by the intersection of the marginal external cost (MEC) curve and the marginal net private benefit (MNPB) curve.

reducing pollution, and firms are likely to use the cheapest technologies first. For example, the firm might install low-cost smokestacks to reduce some local pollution, more costly scrubbers to reduce more pollution, and very expensive conversion to natural gas to reduce it even further. If the marginal cost of reducing pollution is increasing, then the marginal net private benefits (MNPB) of pollution are decreasing, as depicted by the downward-sloping MNPB curve.

For the laundry service, the cost of a small amount of pollution is negligible. As the amount of pollution increases, however, drying the laundry outside makes it noticeably dirtier, resulting in fewer customers and reduced profits. If the pollution gets worse, the laundry service will have to move its laundry lines indoors, install electric dryers, and perhaps even use an air filtration system if the pollution gets bad enough. Each of these options is more expensive than the last, so the marginal cost of pollution to the laundry is increasing, as depicted by the upward-sloping marginal cost curve (MEC). In reality, it is unlikely that either the MNPB or the MEC curves would be smooth. Technologies are "chunky"; one cannot purchase smokestack or scrubbers in small, incremental units. External costs often exhibit thresholds beyond which costs increase dramatically. The assumption of smooth curves, however, does not affect the discussion.

Economic efficiency demands that MNPB = MEC (a variant of our basic MB = MC rule of efficiency). In the absence of laws preventing the coal-fired utility from polluting, it will produce until the MNPB of additional pollution is zero, at point P_W. However, at pollution level P_W, the laundry service suffers very high costs from the soot—the area EWP_w measures the net loss to society. The laundry service could increase its profits by paying the coal utility anything less than $P_W W$ to reduce pollution by one unit from point W, while any positive payment for a unit reduction will increase the utility's profits. If the laundry pays the coal utility to reduce pollution to P_B, there is still room for mutually beneficial exchange—any payment for pollution reduction less than P_BG will benefit the laundry, and any amount greater than P_BB will benefit the utility. The possibility for mutually beneficial exchanges (payments from the laundry service to the coal utility to reduce pollution) continues until we reach point E, where MNPB = MEC, and we have achieved a socially efficient outcome in the absence of government intervention. The same result applies if the laundry has the right to clean air, and we start at pollution level P_N. In this case, the coal utility will keep paying the laundry service for the right to pollute until reaching point E.

Three serious problems with this analysis are that it assumes that both the laundry and the utility are able to pay (i.e., that there are no wealth effects), that individuals are rational maximizers of self-interest, and that there are no real transaction costs. If the laundry earns insufficient profits to pay the utility to decrease pollution and must go out of business if the utility is assigned the right to pollute, that is an example of the *wealth effect*. Remember that efficiency of allocation is defined only for a given distribution. Since vesting property rights in the polluter implies a different distribution (wealth effect) than does vesting them with the pollutee, we simply cannot say that the two situations envisaged by Coase are equally efficient, because they are based on two different distributions of wealth. "Efficiency" in the Coase theorem is from the viewpoint of society, and different legal entitlements also have significant direct impacts on the well-being of the parties involved.

We will return to the assumption of human rationality in Chapter 13 but for now will simply quote Coase on the subject: "There is no reason to suppose that most human beings are engaged in maximizing anything unless it be unhappiness, and even this with incomplete success."[27] Probably the most important obstacle to real-life applications of the Coase theorem, however, is **transaction cost.** This is simply the cost of thrashing out an agreement, which can include legal fees, the cost of gathering information and, locating the interested parties, the time cost of bargaining, and so on. Even in the simplest case of one laundry service and one utility, transaction costs can be quite high. Estimates suggest that transaction costs account for nearly half of the gross national product in the wealthier nations. And this is primarily for market goods. Pollution from a coal-fired utility in Ohio not only pollutes the surrounding community but also contributes to smog and acid rain regionally and to global warming, which affects everyone on the planet. Transaction costs increase with the

---

[27]R. Coase, *The Firm, the Market and Law*, Chicago: University of Chicago Press, 1988, p. 4.

number of people affected by an externality and are particularly high for public goods. As the most important externalities we face today affect public goods such as climate stability and other ecosystem services, transaction costs cannot be ignored.

Coase himself was very clear on the importance of transaction costs. In his own words, "Without the concept of transaction costs, which is largely absent from current economic theory, it is my contention that it is impossible to understand the working of the economic system, to analyze many of its problems in a useful way, or to have a basis for determining policy."[28] The "perfectly competitive market" of the Coase theorem requires a world of no transaction costs and is therefore of theoretical interest only. In the real world, institutions, laws and policy are critically important in addressing externalities, topics we will address at length in Part VI.

Yet again we must stress that all economic production requires raw material inputs and generates waste outputs, thus depleting ecosystem services. All economic production inevitably generates "externalities." Indeed, *externalities* is a misnomer, since there is an unbreakable link (throughput) between resource depletion, production, and waste emissions, so these "externalities" are actually 100% *internal* to the economic process. If converting a forest to farmland imposes negative externalities at the local, national, and global levels, transaction costs for an efficient solution would be prohibitively expensive. When externalities affect future generations, we must accept that transaction costs between generations are infinite, and the market will not solve the "externality" problem unaided.

■ MISSING MARKETS

For a market to function optimally, everyone who would want to produce or consume the goods being marketed must be able to participate. For example, if the *Mona Lisa* were to be auctioned off and only people from Waco, Texas, were allowed to participate, it might not fetch as high a price as it would on the international market. Yet the fact is that future generations cannot possibly participate in today's markets, and therefore today's market prices will not reflect their preferences. The market can therefore "efficiently" allocate resources only if we assume that future generations have no rights whatsoever to the resources being allocated.

How could we provide future generations with property rights to resources? One way to bring this about would be to impose sustainability criteria. For example, we might decide that the rights of future generations to certain resources, such as the ecosystems responsible for generating life-

---

[28]R. Coase, *The Firm, the Market and the Law*, Chicago: University of Chicago Press, 1988, p. 6.

support functions, are inalienable,[29] much like human and political rights, where entitlements are not decided by efficiency criteria. As we deplete nonrenewable resources, we could invest a sufficient percentage of the profits in renewable substitutes to replace the depleted resource (we'll describe this option in greater detail later). For renewable resources (including waste absorption capacity), we could make sure that they were never depleted beyond their capacity to regenerate. If renewables were depleted below their maximum sustainable yield (MSY), we would need to bequeath some substitute that would compensate for the reduction in future harvests. Or we could lower offtake (passive investment in natural capital) enough to replenish the renewable resource to at least the MSY level.

How we handle intergenerational gambles with unknown reward structures is an ethical issue, but it certainly seems that most ethical systems would demand at the very least that we do not risk catastrophic outcomes for the future in exchange for nonessential benefits today. Given our ignorance of ecosystem function, this means we would have to stay well back from any irreversible ecological thresholds. Such sustainability criteria would essentially distribute resources between generations, and the market could then function to allocate them within a generation.

Alternatively, we could just continue to act on the ethical assumptions of neoclassical economics. If we are indeed rational maximizers of self-interest, and Pareto efficiency is an objective criterion for allocation, then the rights of future generations can be completely ignored. After all, as Kenneth Boulding once asked: What have future generations ever done for us? We certainly cannot increase our own consumption by redistributing resources to the future.

In reality, conventional economists do not disregard future generations entirely, but in their analyses they do systematically discount any costs and benefits that affect future generations. In Chapters 11 and 12, we'll look at how the convention of discounting can affect decisions concerning natural resource use.

Intertemporal Discounting

Do conventional economists really ignore future generations? In a standard economic analysis where they have to compare costs and benefits in the future with costs and benefits in the present, conventional economists will systematically discount any costs and benefits that affect future generations. There are some very plausible reasons for giving less weight to resources in the future than resources in the present, and we will explore the topic in some detail in Chapter 16. Here we offer only a brief intro-

---

[29]D. Bromley, *Environment and Economy: Property Rights and Public Policy*, Oxford, England: Blackwell, 1991.

duction to help you understand, in the following chapters, how the convention of **intertemporal discounting** can affect decisions concerning natural resource use. When evaluating present and future values, intertemporal discounting is the process of systematically weighting future costs and benefits as less valuable than present ones.

Why should resources in the future be worth less than resources today? If I have $100 today, I can invest it in some profit-making venture, and I will have more than $100 next year. In perhaps the simplest example, if I can safely invest money in the bank at 5% real interest (i.e., at an interest rate 5% greater than inflation), then I will always prefer $100 today to anything less than $105 a year from now for the simple reason that if I have the money today I have the option to spend it now or allow it to become $105 next year. Next year, of course, I would again have the option to spend the money, or leave it again to grow at 5% to become $110.25, then $115.76, then $121.55, and so on indefinitely. Conversely, $100 in the future is worth less than $100 today because of the *opportunity cost* involved (the lost opportunity to invest), and the farther in the future we look, the less the money is worth. Most conventional economists assume that money is an adequate substitute for anything, and therefore anything in the future is worth less than the same thing today. In general, the present value (PV) of a sum of money t years in the future, X_t, when the interest rate is r, will be given by

$$PV = X_t/(1 + r)^t$$

If we have a stream of money at different dates in the future, we can calculate the PV for each yearly amount and sum them. This is basically what is done in the more complicated formula below.

A standard cost-benefit analysis (CBA) will tell us the **net present value (NPV)**—the value to us today—of a given stream of costs and benefits through time. The farther off in time that a cost or benefit occurs, the more we discount its present value. The basic equation is

The discount rate is r, and the discount factor is $1/(1 + r)$. If we let r =

$$NPV = \sum_{t=0}^{T} (Benefits_t - costs_t)\left(\frac{1}{1+r}\right)^t$$

5%, as in the earlier example, then the discount factor is 1/1.05, which is less than one. The letter t represents time, and $benefits_t - costs_t$ is simply net benefits in period t. As t increases, the discount factor is raised to a larger and larger power, and because it is less than one, raising it to a higher power makes it ever smaller, reducing the net present value by ever more the farther in the future the benefit or cost is. The symbol • tells us to sum together the net benefit stream from time 0 to time T.

THINK ABOUT IT!

An economic analysis of global climate change conducted by Nicholas Stern and others[30] found that we could avoid catastrophic climate change and the potential loss of hundreds of millions of lives by investing 1% of global GNP in mitigation activities. Since GNP grew by about 3% per capita in 2006, this would require that we revert to our living standards of 4 months ago, with slower rates of increase in consumption. William Nordhaus,[31] in contrast, argued that the costs of such a strategy would dramatically outweigh the benefits. The major difference between the two studies was that Stern used a lower discount rate. Do you think that discounting the future costs and benefits of climate change is appropriate? Why or why not?

■ SUMMARY POINTS

What are the most important points you should take home from this chapter? Markets only balance supply and demand, possibility with desirability, under a very restrictive range of assumptions. Among others, goods and resources must be both excludable and rival (where *excludable* implies well-defined and enforced property rights), market actors must be able to make transactions with zero cost (which would automatically eliminate most transactions), and people must have perfect information concerning all the costs and benefits of every good. Even if all of these conditions are met, markets will not account for future generations. In reality, these conditions are never met, though many excludable and rival goods meet these criteria well enough that the market is a very useful allocation mechanism. When resources are nonrival or nonexcludable, the specific combination of these characteristics has much to tell us about how the resources should be allocated. You should clearly understand the implications of these various combinations. Remember also that social institutions are needed to make resources excludable, but some resources are nonexcludable by their very nature, and rivalness is a physical property.

In particular, we must recognize that the "optimal" production of pure public goods cannot be based on the criterion of Pareto efficiency. The public good problem appears to be beyond the scope of market allocation. You might think about policies and institutions that could be effective mechanisms for allocating public goods and the ecological fund-services that provide many of them. One possibility worth considering is a participatory democratic forum that captures a broader spectrum of human values than self-interest and does not weight participant values solely by the purchasing power at their disposal.

BIG IDEAS to remember

- Excludable
- Rival
- Congestible
- Public goods
- Open access regimes
- Nonrival, excludable resources
- Externalities

- Coase theorem
- Transaction costs
- Wealth effects
- Missing markets
- Intertemporal discounting and net present value
- Inalienable rights

CHAPTER

11

Market Failures and Abiotic Resources

Now that we have seen the circumstances under which the market is an effective allocation mechanism, we can examine how the market performs with respect to the goods and services provided by nature. There are eight classes of these, as discussed in Chapters 5 and 6:

1. Fossil fuels (nonrenewable stock)
2. Minerals (partially recyclable, nonrenewable stock)
3. Water (nonrenewable stock, or fund, depending on use, recyclable)
4. Solar energy (indestructible fund)
5. Ricardian land (indestructible fund)
6. Renewable resources (renewable stock)
7. Ecosystem services (renewable fund)
8. Waste absorption capacity (renewable fund)

If a resource is excludable, its market allocation is possible. If it is rival, we understand all the impacts of its use, and production and consumption generate no externalities, then market allocation is also efficient within the current generation. If the well-being of future generations is not affected by the use of the resource, then market allocation may also be intergenerationally fair. As we will see, however, no good or service provided by nature meets all of these criteria. In this chapter we examine abiotic resources and lay some of the groundwork for subsequent discussion of policies that can improve the allocation of these resources.

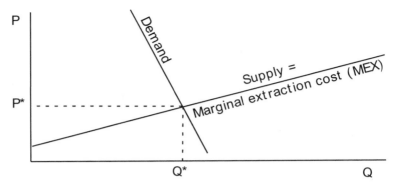

Figure 11.1 • Optimal extraction (Q*, P*) of fossil fuels in the absence of scarcity and market failures.

■ FOSSIL FUELS

Fossil fuels are both rival and excludable and thus can be allocated by market forces. If we ignored resource scarcity (or the future) and market failures, the optimal allocation of fossil fuels would simply be the intersection of the demand curve with the supply curve, where the supply curve would equal **marginal extraction costs (MEX)**,[1] as depicted in Figure 11.1.

External Costs

However, problems arise. First of all, the production and consumption of fossil fuels generate serious externalities at the local, regional, and global levels. Most of these externalities are in the form of public bads. Examples of these externalities are categorized in Table 11.1 according to their spatial and temporal characteristics.

Many of these externalities have different impacts at different spatial levels. Optimal extraction of fossil fuels would need to include these marginal external costs, as shown in Figure 11.2, where MEC shows the cost of extraction including externalities. Because these externalities are so widespread, affecting not only virtually everyone in the world alive today but future generations as well, transaction costs for resolving these externalities through the market would be infinite. Given the inability of the unfettered market to address these externalities, extramarket institutions (e.g., government regulation) will be needed.[2] However, these institutions

---

[1]MEX include all costs, i.e., the costs of equipment and labor and the opportunity cost of money invested.

[2]This is not to say that government regulations cannot create market incentives for reducing externalities, as we will see in Chapters 18–19.

■ Table 11.1

SPATIAL AND TEMPORAL CHARACTERISTICS OF SELECTED EXTERNALITIES ASSOCIATED WITH FOSSIL FUEL EXTRACTION AND CONSUMPTION

Externality	Local	Regional	Global	Intergenerational
Global warming			X	X
Acid rain	X	X		X
Oil spills	X	X		X
Damage from extraction (see Table 11.2)	X			X
War[a]	X	X		X
Water pollution	X	X		X
Soil pollution	X			X
Air pollution (gaseous)	X	X	x	x
Air pollution (particulate)	X			
Heavy metal emissions	X	X	X	X

[a]The number of wars that have been fought and are currently being fought over the control of fossil fuels argues for the treatment of some wars, or at least some military expenditures, as an externality of fossil fuel production. See, for example, M. Renner, WorldWatch Paper 162: The Anatomy of Resource Wars. Washington, DC: WorldWatch, 2002.

must be on the scale of the problem they address, as there is little incentive for governments (generally the most relevant institutions) to deal with externalities beyond their borders. At present, appropriate institutions for addressing international externalities either do not exist or are inadequate.

User Costs

Another problem is that fossil fuels are a nonrenewable resource upon which the well-being and even the survival of future generations is highly dependent. Even ignoring future generations, economists agree that the use of a nonrenewable resource now increases scarcity (decreases supply) in the future. As supply goes down, price should go up. Therefore, if the owner of a nonrenewable resource extracts that resource today, she loses the option of extracting it in the future when the price is higher.

The more of a resource we extract today, the greater the current supply and the lower the current price. Also, greater extraction now means greater scarcity in the future and a higher future price. All else being equal, the **marginal user cost (MUC)**—the opportunity cost of producing *one more unit* of the resource today instead of in the future—should therefore be increasing with total production.[3]

The marginal user cost is a real cost of production, and it must be added to MEC and MEX to give the full cost per unit that represents all

User cost is the opportunity cost of nonavailability of a natural resource at a future date that results from using up the resource today rather than keeping it in its natural state.

Marginal user cost is the value of one more unit of the resource in its natural state. In a perfectly competitive economy, marginal user cost would in theory equal the price of a resource minus its marginal extraction cost.

[3]We caution that some theoretical studies suggest that rising marginal extraction costs or the presence of an inexhaustible substitute (a backstop technology) may lead to declining marginal user costs over time. Different sets of assumptions in mathematical models lead to different results.

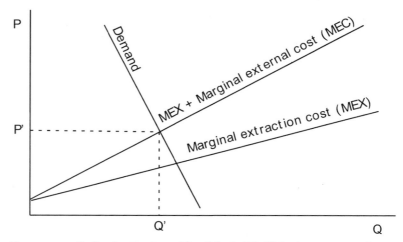

Figure 11.2 • Optimal extraction of fossil fuels (Q', P') in the presence of negative externalities (global marginal external costs), without scarcity.

marginal opportunity costs. The individual producer takes prices as given and therefore should produce up to the point where marginal benefit (price) equals marginal cost (MEC + MEX + MUC), as shown in Figure 11.3. Of course, when the producer does not have to pay marginal external costs, she is likely to ignore them.

User cost can also be thought of as the value of the resource in its natural state, the in-ground value before it has been extracted. Marginal user cost

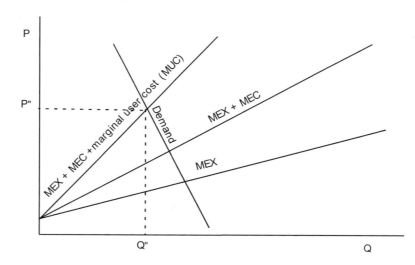

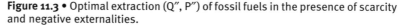

Figure 11.3 • Optimal extraction (Q", P") of fossil fuels in the presence of scarcity and negative externalities.

is the value of one more unit of the resource in its natural state. Theoretically, in a competitive market, producers would pay resource owners a per-unit fee for extraction rights precisely equal to the marginal user cost.

> **THINK ABOUT IT!**
>
> *Can you explain why, in a competitive market, producers would pay resource owners a per-unit fee equal to the MUC for the right to extract a resource?*

This fee is known as a **royalty**; the marginal user cost should be equal to the price of the resource minus the marginal extraction cost. Because no human effort is needed to produce a natural resource in its natural state, user cost is unearned economic profit, also known as economic rent. As explained in Chapter 9, **rent**, or scarcity rent, is defined as unearned profits, or the payment to a firm for a product above and beyond what is needed to bring that product to market.

Royalty is the payment to the owner of a resource for the right to exploit that resource. Theoretically, in a competitive market, the per-unit royalty should be equal to the marginal user cost.

We have explained that marginal user cost should be rising with increasing scarcity, but how fast will it increase? We must remember that user cost is the opportunity cost of extraction today, while the discount rate is the opportunity cost of leaving a resource in the ground instead of extracting it and investing the profits in the most profitable alternative productive activity. Conventional economists argue that, all else being equal, the optimal rate of resource extraction is one at which increasing in-ground scarcity drives the marginal user cost up at a rate equal to the usual rate of return on alternative (aboveground) investments.[4] This is known as the **Hotelling rule**, after economist Harold Hotelling, who first stated it. As a result, extraction rates should decrease through time, causing the price to increase.

If we imagine that MEC and MEX are zero, then the price under a profit-maximizing extraction regime will increase at the discount rate. This is an intuitive result: if the price increased more slowly than the discount rate, the resource owner would maximize profits by extracting the resource faster and investing the profits, which would then grow faster than the value of the resource in the ground. Alternatively, if the price of fossil fuels is growing faster than the discount rate, then leaving the resource in the ground to appreciate in value generates the greatest profits. In other words, if the opportunity cost of leaving the resource in the ground (the discount rate times the current value of the resource) is greater than that of extracting it (the user cost), we extract, and vice versa.

In theory, the market mechanism automatically (through the invisible hand) incorporates marginal user cost into the market price and is equal to the market price minus the extraction cost. In reality, as natural resource

---

[4]H. Hotelling, The Economics of Exhaustible Resources, *Journal of Political Economy* 2: 137–175 (1931).

markets in general are highly imperfect because of cartels, an absence of competition, poorly defined property rights, and imperfect information, they will not reveal true user costs. As an alternative measure, economists can estimate the time of total depletion of the resource, the time when we'll have to turn to the best available substitute, assuming there is one. If there is a reasonable renewable substitute (e.g., solar power) or extremely abundant substitute (e.g., hydrogen fusion) for the resource in question (known as a backstop technology), the price of the resource can never rise above the price of the substitute. This reduces the opportunity cost of using the resource in the present—that is, it lowers the user cost, thereby leading to more rapid extraction and a lower price. To determine the user cost, the extra unit cost of the best substitute, over and above that of the depleted resource, is estimated. That amount is then discounted from the future date of exhaustion back to the present to tell us the marginal user cost.

In summary, the user cost will be low if (1) the discount rate is high, (2) exhaustion is far in the future because either reserves are large or annual usage rates are low, and (3) good substitutes are expected to be available, or high if opposite conditions exist. We see once again the im-

---

| **Box 11-1** | **MARGINAL USER COST IN SUSTAINABLE INCOME ACCOUNTING** |

It is interesting that in discussing user cost, John Maynard Keynes, arguably the most influential economist of the twentieth century, was interested mainly in applying the concept to depreciation of the fund of manmade capital (in order to arrive at a proper measure of income). He made reference to the more "obvious" case of accounting for user cost in a copper mine (natural capital) as a way of clarifying his argument.[a] Nowadays, if texts discuss user cost at all, they refer to the more "obvious" necessity of accounting for depreciation of manmade capital as an argument for applying the same logic to natural capital. Perhaps this reversal is a measure of how much we have recently come to neglect natural capital.

In any case, recalling our terminology of previous chapters, it is clear that user cost is a necessary charge for the depletion of stocks (natural or manmade inventories) and the depreciation of funds (natural and manmade productive equipment). In proper accounting usage, both inventories and machines are capital. Inventories are depleted; machines are depreciated. Both require the accounting of user cost. If user cost is not deducted in calculating income, then income will be overstated and will not be sustainable. In most national accounts around the world, user cost is erroneously counted as income.[b] Keynes, as the major architect of modern macroeconomics, was very interested in getting a correct measure of national income.

Few nonrenewable natural resources are fund-service in nature. We will examine the implications of user cost for the stock-flows and fund-services provided by renewable resources in Chapter 12.

---

[a]*J. M. Keynes,* The General Theory of Employment, Interest, and Money, *Orlando, FL: Harcourt Brace, 1991, p. 73.*

[b]*See S. El Serafy, The Proper Calculation of Income from Depletable Natural Resources. In Y. J. Ahmad, S. El Serafy, and E. Lutz, eds.,* Environmental Accounting for Sustainable Development, A UNEP–World Bank Symposium, Washington, DC: World Bank, 1989; and S. El Serafy, Green Accounting and Economic Policy, *Ecological Economics 21:217–229 (1997).*

portance of the discount (interest) rate, expectations about substitutes, and uncertainty about stocks in the ground.

Flaws in the Analysis

From the perspective of ecological economics, however, this analysis of fossil fuels is inadequate. First, it looks only at the net present value of the resource for the existing generation, ignoring any ethical obligations to leave some of the resource for future generations; that is, it focuses

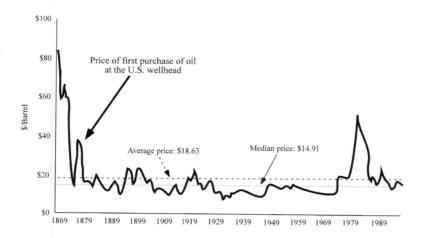

Figure 11.4 • Oil prices from 1861 to 2009 (estimated), in inflation-adusted 2008 dollars. The volatile nature of prices in the short run is obvious (note particularly the surge in prices in response to reduced OPEC production in the 1970s and the price shock of 2007–2008), but perhaps more noteworthy is the relative stability of prices for most of the period. (Source: British Petroleum. 2009. Statistical Review of World Energy, Full Report 2009. Online: http://www.bp.com; 2009 estimate of oil prices from Energy Information Administration, Short-Term Energy Outlook, 2009. Online: http://www.eia.doe.gov/steo/)

on intragenerational efficiency, ignoring scale and distribution. Second, neither producers nor consumers currently pay marginal external costs. Third, empirical evidence contradicts the conventional theory; as Figure 11.4 shows, oil prices were fairly stable for much of the last 140 years. We'll return to this last point following our discussion of mineral resources.

■ MINERAL RESOURCES

Mineral resources are also rival and excludable and amenable to market allocation. As in the case of fossil fuels, their production and consumption generate serious externalities. In fact, mining accounts for nearly half of all toxic emissions from industry in some countries, such as the United States.[5] As many of these negative externalities are less well known than those associated with fossil fuel use, we have summarized them in Table 11.2.

Although many of these externalities are fairly localized compared to problems from fossil fuel emissions, they can be very persistent and severe. For example, acid mine drainage still occurs on mine sites worked by Romans over 1500 years ago. Over 500,000 abandoned mine sites exist in the U.S. alone,[6] with estimated cleanup costs of $32–$72 billion.[7] Again, transaction costs for market resolution of these externalities will be extremely high to infinite, depending on our concern for future generations. Unregulated markets will not solve the problem.

It is worth noting here an interesting anomaly. Within a generation, for the market to efficiently allocate resources, they must be rival. However, future generations cannot participate in today's markets. Thus, if a good is rival between generations—that is, its use by one generation prohibits use by another—the market will still not allocate it efficiently because future generations cannot participate. Fossil fuels are rival between generations. Mineral resources, to the extent they can be recycled, are rival within a generation, but less so between generations. Thus, if mineral resources were efficiently recycled and had no negative externalities associated with their production and consumption, market allocation could be both intragenerationally efficient and intergenerationally

[5]P. Sampat, From Rio to Johannesburg: Mining Less in a Sustainable World. World Summit Policy Brief #9. World Watch Institute News Releases. Online: http://www.worldwatch.org/press/news/2002/08/06/.

[6]Center for Streamside Studies. Online: http://depts.washington.edu/cssuw/Publications/FactSheets/minec.pdf.

[7]Cleanup costs for some mines have reportedly been greater than the value of minerals extracted. Environmental Media Services, Mining Companies Profit from Public Lands While Taxpayers Pay for Cleanup, 2002. Online: http://www.ems.org/mining/profits_costs.html.

■ **Table 11.2**

PRODUCTION EXTERNALITIES OF MINERAL RESOURCE EXTRACTION, SPECIFICALLY FROM HARD ROCK MINES

Externality	What Is It?	What Does It Affect?
Acid mine drainage	Metal sulfides are common in mineral ores and associated rocks. When these rocks are mined and crushed, exposure to air and water oxidizes these sulfides, generating acids and toxic heavy metal cations.	Water required by oxidation also washes products into nearby surface water and aquifers; in addition to acidification effects, heavy metals build up in animal populations and humans.
Erosion and sedimentation	Heavy machinery, strip mines, and open pit mines destroy surface vegetation holding soils in place; water washes away small particles from erosion and waste materials, depositing it elsewhere.	Major impacts are on wetlands and other aquatic habitats; soil organisms, vegetation, and restoration efforts are also affected.
Cyanide and other chemical releases	Cyanide and other toxic chemicals are regularly used to help extract minerals.	Cyanide released into ecosystem has adverse impacts on water, soil, aquatic organisms, wildlife, waterfowl, and humans.
Dust emissions	Ore crushing, conveyance of crushed ore, loading bins, blasting, mine and motor vehicle traffic, use of hauling roads, waste rock piles, windblown tailings, and disturbed areas all generate dust.	Dust can be an air pollutant and may also transport toxic heavy metals.
Habitat modification	Mining can have dramatic impacts on landscape and uses enormous amounts of water.	Ecosystem structure and function are affected.
Surface and groundwater pollution	Mining uses massive quantities of water, pumping water from mines affects water tables, and mine wastes pollute water.	Altered surface and groundwater flows, with accompanying impacts on wetlands and other water-dependent habitats.

Source: EPA Office of Waste Water Management, Hardrock Mining: Environmental Impacts, http://cfpub.epa.gov/npdes/docs.cfm?view=archivedprog&program_id=14&sort=date_published. The Web site was archived shortly after the G.W. Bush administration took office.

fair. The more efficient the recycling process, the lower the marginal user cost, and the less the theoretically efficient price would need to rise over time. However, the record shows low levels of recycling, enormous increases in demand (from 93 million metric tons in 1900 to 2900 million metric tons in 1998 in the U.S. alone), and substantial decreases in real prices.[8]

---

[8]D. E. Sullivan, J. L. Sznopek, and L. A. Wagner, Twentieth Century U.S. Mineral Prices Decline in Constant Dollars. Open File Report 00-389, Washington, DC: U.S. Geological Survey, U.S. Department of the Interior, 2000. Between January 2002 and January 2008, the UNCTAD index metal and mineral prices rose by 285%. Prices fell significantly in late 2008, then began to climb again in 2009.

Do Prices Reflect Scarcity?

How do we explain the major anomaly between the empirical fact of falling prices of nonrenewable resources for the last century and the theoretical prediction of rising prices? Conventional economic theory generally assumes that prices increase as a function of scarcity, and it is an unalterable physical fact that extraction has reduced the quantity of in-ground stocks of nonrenewable resources. This does not necessarily mean that prices do not reflect scarcity, as long as we assume that scarcity is defined not only by the physical quantity of a resource remaining. Scarcity is also determined by new discoveries[9] and by the availability of substitutes. Prices equilibrate supply and demand, and if supply increases from new discoveries or demand falls because substitutes are invented, scarcity is reduced, and prices fall as well. For example, fiber optic cables dramatically decreased the demand for copper in telephone lines, which might explain the fall in prices.

However, as we pointed out previously, oil discoveries peaked in 1962, production surpassed new discoveries in 1982, and consumption currently exceeds new discoveries by a factor of two to six. What's more, while we do have more potential substitutes for oil, relative to 100 years ago we have created far more technologies that depend on oil (complements) than technologies that substitute for oil. Just as substitutes reduce resource scarcity, complements increase it. Nonetheless, steady increases in the demand for oil apparently failed to affect the price of oil for most of the twentieth century.

What is the explanation? First of all, we must recognize that if prices reflect in-ground scarcity, they do so very poorly, and for obvious reasons. There is considerable debate even among the experts about the precise amount of oil and minerals left in the ground (though less about ultimately recoverable reserves than about "proven" reserves), and estimates of "proven" reserves have changed dramatically over the years, often increasing substantially even in the absence of new discoveries.[10] If the experts do not know how much remains underground, how can prices tell us?

While prices cannot effectively equilibrate unknown in-ground supply with demand, they can equilibrate the available aboveground supply with demand. Available aboveground supply is determined solely by the rate of

---

[9]Remember, new discoveries do not increase the amount of resources in the ground; they just make it easier for us to get them. In the short-run market sense, new discoveries decrease scarcity, while in the long-run physical or geological sense, any increase in extraction increases the scarcity of resources remaining in the ground.

[10]In January 1988, Iran, Iraq, and Venezuela each reported a doubling of their reserves, presumably to earn higher quotas under OPEC. In spite of continuous extractions since then, their reported reserves have scarcely changed. C. J. Campbell, Proving the Unprovable, *Petroleum Economist*, May 1995.

extraction, which depends on known deposits, existing infrastructure and technology, as well as the resource owner's decision of how much to extract. Hotelling suggested that a rational producer will limit current production to take advantage of higher prices in future years. However, if real prices are not increasing, the owner has no incentive to leave the fossil fuels in the ground and would rationally extract the resource as long as the marginal extraction cost remains lower than the price. Essentially, the producer ignores MUC, as in the simplest static analysis depicted in Figure 11.1.

Even if the producer ignores MUC, we would still expect MEC to increase. As we suggested earlier, economic analysis typically assumes that nonrenewable resources will be mined from the purest, easiest-to-access sources first. As these are depleted, we then move on to sources that are more expensive to extract, again putting upward pressure on prices. However, there are two serious problems with this argument. First, as Norgaard has noted, when we begin to exploit a new resource, we typically know very little about where the best fields are. A great deal of chance is involved with the initial discoveries. Norgaard compared this to the *Mayflower*. If people always exploited the best resources first, the first pilgrims would have settled on the best land in America. However, prior to their arrival, the pilgrims knew virtually nothing about land resources in North America and ended up where they did largely by chance.[11]

Second, as we exploit a new resource, we diminish the total stock, but we gradually acquire more information about where to find it and how to extract it, and more of the resource becomes accessible. Thus, there are two effects at work. The scarcity effect decreases the total amount of resource available, but the information effect increases the amount that is accessible and reduces the costs of extracting it. Thus, as long as the information effect is dominant, the price of the resource should decrease. Eventually, however, the scarcity effect must come to dominate, and the price must then increase. Rather than predicting a gradual price increase in a resource, this model suggests the likelihood of decline followed by sudden, rapid increases. In the first edition of this textbook, we combined this analysis with the estimates of petroleum geologists and predicted a sudden and dramatic increase in oil prices in the next 2–20 years, which we indeed saw from 2005 to July 2008. Prices then plunged but began to climb again in spite of recession. This analysis suggests that any return to high growth rates of throughput is likely to be accompanied, and perhaps stifled, by sharply rising oil prices.[12]

---

[11]R. Norgaard, Economic Indicators of Resource Scarcity: A Critical Essay, *Journal of Environmental Indicators and Management* 19(1):19–25 (July 1990).

[12]D. B. Reynolds, The Mineral Economy: How Prices and Costs Can Falsely Signal Decreasing Scarcity, *Ecological Economics* 31(1):155–166 (1999); C. J. Campbell and J. H. Laherrère, The End of Cheap Oil, *Scientific American*, March 1998.

This result is particularly important if we are concerned with sustainability. As we pointed out earlier, economists assume that price increases will trigger innovation and generate substitutes for any given resource. If resource owners are optimists, they believe new discoveries will be made and substitutes invented. This means that their resource will not become scarce and its price will not go up (and may even go down). Under such circumstances, it makes sense to extract the resources as quickly as possible and invest the returns. If the resources are being extracted quickly, aboveground supply is large, and the price is low. This reduces the incentives for exploration and the development of substitutes. The problem is that developing substitutes requires technology, technological advance requires time, and the less warning we have of impending resource exhaustion, the less time there is to develop substitutes. Perversely, then, in a world of optimists, the pessimist is most likely to be correct, and vice versa.[13] While these arguments are far from the only ones discrediting the belief that we can ignore resource exhaustion, they are important.

■ FRESHWATER

As the economically relevant characteristics of freshwater depend on the specific use to which it is put, and because it is used in most economic and ecological processes, the economics of water could fill a textbook on its own. Some relevant characteristics of specific water uses can be gleaned from discussions in other sections of this text. Specifically, water in fossil aquifers is a nonrenewable resource similar to fossil fuels with fewer externalities.[14] Water as an ecological fund-service is similar to other ecosystem services discussed in Chapter 12. In this section, we'll limit our discussion to some unique attributes of water as a stock-flow resource. Specifically, we address the facts that water is 100% essential to human survival, that it has no substitutes,[15] and that water distribution systems generally show substantial economies of scale.

The scarcity of clean and available water has traditionally been experienced as a local matter, but international disputes over access to water are increasing, indicating a global dimension to water scarcity. These characteristics have important implications for water markets. Though traditionally supplied by the public sector at least since early civilization's first large-scale irrigation projects, water is a rival good that can be made ex-

---

[13]P. Victor, Indicators of Sustainable Development: Some Lessons from Capital Theory, *Ecological Economics* 4:191–213 (1991).

[14]If aquifers rise to the surface, feeding streams and rivers, aquifer depletion can have serious negative externalities.

[15]Though technology can, of course, increase the efficiency of water use in some applications.

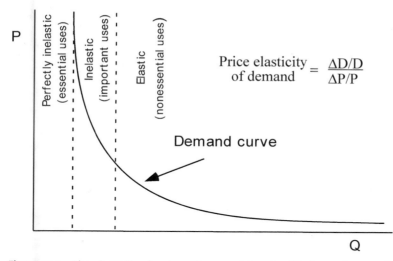

Figure 11.5 • The elasticity of water with respect to price (% change in quantity demanded with respect to % change in price) for different quantities of water.

cludable under most circumstances and therefore technically amenable to market allocation. Indeed, in recent years, more and more cities, states, and even countries are turning their water supplies over to the private sector in the name of greater efficiency, and many neoclassical economists applaud this trend.[16] However, the fact that water is nonsubstitutable and 100% essential means that there are serious ethical implications to the market allocation of water, and it therefore makes a good case study of why just distribution precedes efficient allocation in ecological economics.

In many places, water is very abundant and is used for fairly unimportant activities; some 90% of industrial and household water is simply wasted.[17] Higher prices for water would reduce this waste. However, because in its most important uses water has no substitutes and is essential to our survival, as water supplies become scarce or prices increase, the demand for water becomes extremely inelastic with respect to price. A 1% increase in price will lead to less than a 1% decrease in demand (Figure 11.5). When water is abundant, we use it for nonessential activities, and demand is price elastic. As water becomes scarcer, we use it only for more important activities, and it becomes inelastic with respect to price. As water becomes still scarcer, it will be used exclusively for essential activi-

---

[16]For example, the International Monetary Fund and the World Bank often make privatization of water supply a requirement for loans.

[17]I. A. Shiklomanov, World Water Resources: Modern Assessment and Outlook for the 21st Century, 1999. Federal Service of Russia for Hydrometeorology & Environment Monitoring, State Hydrological Institute, St. Petersburg.

ties, such as raising food and drinking, and demand becomes perfectly inelastic.

This situation leads to two very serious problems. First is the distribution issue. In the market economy, the most "efficient" use is the use that creates the highest value, and value is measured by willingness to pay. In a world with grossly unequal income distribution and a growing relative scarcity of water, many people have very limited means to pay. "Perfect" market allocation of water could easily lead to circumstances in which rich people use drinking water to flush their toilets, while poor people must drink from water contaminated by sewage. While economically sparkling toilet bowls might be more efficient, ethically most people would probably agree that survival of the poor should take precedence.

The second issue is efficiency. Markets are rarely perfect, and in the case of water, they are likely to be less perfect than most. Providing water requires substantial infrastructure that would be very costly to duplicate. For this reason it makes sense to have only one provider, so even where water is privatized, there is typically not a competitive market but rather a natural monopoly. A natural monopoly occurs when there are high fixed costs to production (e.g., building a reservoir and water main) and low and constant marginal costs (e.g., hooking up another house to the water main), so that average costs fall with increasing use. Many public utilities are natural monopolies. Dealing with inelastic demand, the monopoly provider knows that a 10% increase in price will lead to less than a 10% decrease in quantity demanded, leading to higher revenue and lower costs. Moreover, everyone needs water and cannot exit the market no matter how inefficient and expensive the monopoly supplier is. With no threat to their market share, firms bent on maximizing short-term profit may delay needed improvements in infrastructure. Only extensive regulation will deter the private supplier from increasing prices and decreasing quality. With no competition to drive down prices or regulation to control costs, private sector provision of water is likely to be less efficient than public sector provision—as well as less just.

■ RICARDIAN LAND

As we explained in Chapters 4 and 5, by *Ricardian land* we mean land simply as a physical space capable of capturing sunshine and rainfall, and not the various productive qualities inherent to the land itself. The latter qualities, such as soil fertility, we class as ecosystem services. Within a generation, Ricardian land is both rival and excludable and hence can be allocated by markets. Between generations, it is nonrival, which suggests that market allocation of land might meet the criterion for both efficiency within a generation and fairness between generations.

Before we reach this conclusion, however, we must ask: What is it that makes Ricardian land valuable? Certainly in market terms, the most valuable land in the world is found within the borders of big cities, where prices may pass $100,000 per square meter, and the least valuable land is generally found in the most deserted areas. What makes land valuable appears to be proximity to other humans. Some might reply that the low value of land in uninhabited areas is due to other factors, such as extreme cold or extreme heat, and those same factors prevented people from settling there in the first place. But if we look at some of our planet's less-inviting habitats, we find that where they are inhabited, land prices are highest at the sites of densest habitation and lowest where population is thinnest, even if the sites are otherwise virtually identical.

Why would the presence of other humans make land more valuable? Humans are social animals, dependent on each other for both psychic and physical needs. Living near others allows individuals to specialize, and the economic benefits of specialization are common knowledge. Empirically, in a growing economy with growing populations, land appreciates in value even in the absence of improvements by the landowner. As cities grow, the land on their peripheries becomes more valuable. If the government builds a subway system or road, the value of adjacent land can skyrocket. Proximity to new infrastructure, such as sewage systems, electric grids, highways, and subways, can similarly increase land value.

The truth is that land attains value as a positive externality of the decisions of others. Land values thus result from a market failure, and we cannot simply assume that markets are the best means for allocating even Ricardian land. These insights into land value were first popularized by nineteenth-century economist Henry George.[18]

The origin of land value is not just an academic argument; it is directly related to important policy debates. For example, some time in the early 1990s, a case was widely discussed in which an elderly woman on the outskirts of Chicago owned land worth some $30 million. An endangered butterfly was found on her land, imposing serious restrictions on development and causing the value of the land to plummet. Many people argued that this was entirely unfair, and the government should be forced to compensate the woman for "taking" the value of her land. However, what was it that caused her land to be worth so much in the first place? The woman had owned the land for decades, during which time Chicago had grown considerably. Government-built roads, sewage, and electric grids had gradually expanded, making her formerly remote piece of land highly valuable. Government action created the value in her land, and a different government

---

[18]H. George, *Significant Paragraphs from Henry George's* Progress and Poverty, *with an Introduction by John Dewey*, Garden City, NY: Doubleday, Doran, 1928.

action subsequently reduced its private value through an effort to meet important public needs. This raises an important question: Are individuals entitled to wealth created by society or by nature rather than through individual effort, or should this wealth belong to society as a whole?

In addition to the market failure associated with land values, there is another reason that market magic does not work with land. Land is present in a fixed amount, and supply is perfectly inelastic—it does not respond to changes in price. With fixed supply and increasing demand (as populations and wealth increase), land prices trend upward. Thus, whoever manages to acquire land will in general see the value of that land grow through no effort of her own. This makes land the subject of speculative investment. Land purchased for speculation is often left idle, but the demand for land for speculative purposes must be added to demand for land for productive purposes, driving up the price even further and reducing the ability of people to buy land for production. In other words, under certain circumstances, speculative markets in land can reduce the production from land. None of this means that land ownership and land markets are necessarily bad; it simply means that we should not automatically attribute all the theoretical virtues of markets to markets in land.

■ SOLAR ENERGY

Primarily for completeness, the final abiotic resource we will consider is the flux of a solar energy that continuously warms the Earth and turns its biogeochemical cycles. Obviously, neither human institutions nor human invention (short of giant mirrors or umbrellas in space) can directly change the allocation of sunlight on the Earth or the supply. Indirectly, however, market forces can have significant impacts on the scale, distribution, and allocation of solar energy.

In terms of scale, human impacts on ecosystems appear to be degrading their capacity to capture solar energy; for example, forests sickened by acid rain capture less solar energy than healthy ones. In terms of distribution, land is an essential substrate for the capture of solar energy in most forms other than heat, and markets thus determine indirectly who can use the solar energy striking the Earth. Allocation—to what uses sunlight is put—is also determined in part by who owns the land it strikes. In this sense, solar energy is both rival and excludable. Also something of an allocation issue, land uses affect the spatial distribution of solar energy; recall how the evapotranspiration from the Amazon transports solar energy in the form of heat to the temperate zones. In terms of policy considerations, however, these issues can be treated as attributes or externalities of the use of other resources.

BIG IDEAS to remember

- Classification of eight natural goods and services: fossil fuels (nonrenewable stock), minerals (partially recyclable, nonrenewable stock), water (nonrenewable stock, recyclable), solar energy (indestructible fund), Ricardian land (indestructible fund), renewable resources (renewable stock), ecosystem services (renewable fund), waste absorption capacity (renewable fund)

- Stocks (depletion), funds (depreciation)
- Renewable, nonrenewable
- Rival, nonrival, excludable, nonexcludable
- Extraction cost, marginal extraction cost
- External costs, marginal external costs
- User cost, marginal user cost
- Royalty
- Hotelling rule
- Price as problematic measure of resource scarcity
- Henry George

Market Failures and Biotic Resources

In contrast to abiotic resources, biotic resources are renewable, if degrad-able, and are valuable as much for the services they provide as for any goods that can be derived from them. In this chapter, we examine whether specific natural resources meet the criteria for market allocation, turning our attention to biotic stock-flow resources (ecosystem structure) and eco-logical fund-service resources (ecosystem function), paying special atten-tion to waste absorption capacity. However, we must never forget that the extraction of any part of an ecosystem and the emission of wastes affect vital ecosystem functions. Species interact with each other in unknown and unpredictable ways. Studying species, or other components of these complex systems, can provide useful insights, but the integrated system is far more than the sum of its parts.[1]

■ RENEWABLE RESOURCE STOCKS AND FLOWS

Renewable resource stocks and flows are rival and potentially excludable, depending on whether institutions exist that can regulate access to them. If depleted at a rate no faster than they regenerate, they are nonrival between generations. Unfortunately, unless we explicitly take future generations into account, economic incentives are quite likely to lead us to deplete many of these resources faster than they can regenerate and may eventually threaten them with extinction. What's more, as we have pointed out repeatedly, the use of renewable resource stocks and flows unavoidably depletes ecosystem

---

[1]Chapters 6 and 7 of the workbook focus on synthesis, the integration of the parts in order to better understand the whole.

funds and services as an "externality" of their production. This dramatically complicates economic analysis of these resources.

In our earlier discussion of renewable resource stocks and flows, we looked at their physical properties with little if any discussion of the economics involved. Recall the sustainable yield curve from Chapter 6, here reproduced as Figure 12.1.[2] At first glance, it appears that the goal of economists would simply be to make the resource as productive as possible. If this were the case, we should strive to maintain a population that produces the maximum sustainable yield, or MSY. However, this ignores two major issues. First, there are costs to harvesting, which we will call P_EE (price of effort times effort, where effort includes all the resources needed to harvest a stock), and these costs are likely to increase per unit harvested as the population in question grows smaller. Obviously, the smaller the population of fish that remains, the harder they are to catch. Even for forests, the most accessible timber will be harvested first, and as forest stocks decrease, it will cost more to bring the less accessible stocks to market.

Second, and even more important, if we were to consider all resources substitutable, as many economists do, and money is the perfect substitute for any resource, then the economic goal would be to maximize not the sustainable harvest of any specific resource but rather the monetary sum of annual profits yielded by the resource. But even that is incomplete, as we will discuss shortly, for the market goal is in fact to maximize present value, the monetary sum of discounted future profits. Maximizing present value is probably not a desirable goal, or even an achievable one, given the uncertainties inherent to biotic resources (see Chapter 6). However, it is important to understand how and why conventional economists pursue this goal because it provides some useful insights both into how biotic resources are and should be managed.

To simplify analysis, we can assume a linear relationship between effort, stock, and harvest, known as the **catch-per-unit-effort hypothesis**. For any given effort, more stock leads to a larger harvest in a linear fashion, and for any given stock, more effort yields a larger harvest. However, unless harvests are less than the annual increase in stock (i.e., below the sustainable yield curve), an increase in effort in any given year, all else being equal, implies a smaller stock and thus a smaller harvest from the same level of increased effort in the following years.[3]

---

[2]We changed the scale of the figure to make it clearer. Specifically, the y-axis has been stretched, so that the 45-degree line appears to be steeper than 45 degrees.

[3]When population stocks are large, a large harvest can lead to a larger sustainable yield in the following year, but this sustainable yield will still be smaller than the harvest needed to reduce the population to that stock that provides that sustainable yield.

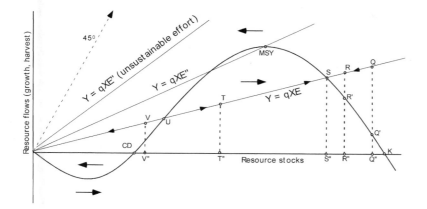

Figure 12.1 • The sustainable yield curve and catch-per-unit-effort curves. Recall from Chapter 6 that the curved line represents the rate of growth of stock for any level of stock, which is the same as sustainable yield. The straight lines Y = qXE represent the harvest for a given effort at any level of stock. A steeper slope indicates greater effort.

Using simple algebra, we could assume that Y = qXE, where Y is harvest, X is stock (e.g., of fish or trees), E is effort (e.g., the number of fishing boats or sawmills), and q is a constant we can think of as the harvestability coefficient.[4] As those of you familiar with math will recognize, this is the equation for a line starting at the origin with slope equal to effort, E. In Figure 12.1, we have drawn in lines Y = qXE, Y = qXE′, and Y = qXE″, where E″ > E′ > E. Say the stock is Q″ in year 0, and effort is E (e.g., 100 boats). Our harvest then is Q. Of this harvest, Q″Q′ corresponds to the annual growth for that year, and Q′Q must therefore reduce the stock, to R″. (Note that this figure is not drawn with the same scale on both axes; i.e., resource flows are in smaller units than resource stocks.) At R″ and effort E, the harvest will be at point R, reducing the stock by R′R. As long as effort remains constant, this process continues until we reach stock S″ and point S on the sustainable yield curve.

Now imagine an external shock, such as El Niño, pushes populations down to T″ in a given year. As long as effort stays the same, the annual harvest will be less than the growth increment or net recruitment, and the stock will gradually recover, until we again reach S″. S is therefore a *stable equilibrium point*. However, if another El Niño year occurs before the fish population has recovered, it could push the population down to V″. At V″, the same effort will lead to a harvest greater than the growth in-

---

[4]C. Clark, in *Mathematical Bioeconomics: The Optimal Management of Renewable Resources*, New York: Wiley, 1990, uses the phrase *catchability coefficient* in discussing fisheries. We use *harvestability coefficient* because we are taking about many different types of stocks.

crement, and the population will not recover. Recall from Chapter 6 that any harvest below the sustainable yield curve leads to a higher stock in the following year, and any harvest above the curve leads to a lower stock. The arrows on the Y = qXE curve illustrate this dynamic. Thus, point U, where the catch-per-unit-effort curve intersects the sustainable yield curve, is an unstable equilibrium of no practical interest in a dynamic world and hereafter ignored.

Maximizing Annual Profits

Suppose the goal is to maximize sustainable annual profits (π) from exploiting the fishery. This requires that we figure out where on the sustainable yield curve profits are maximized. Unfortunately our graphical analysis using the effort curve does not directly show profit. To analyze the question from the perspective of annual profit is worthwhile and will require a somewhat different graph (Figure 12.2). The axes and the yield curve remain the same, except that we multiply the vertical (flow) axis by an assumed constant, P_F, the price of fish.[5] This converts the yield curve into a total revenue (TR) curve without changing its shape, since we are multiplying by a constant. Since profit, π, is equal to TR–TC (total cost), we need to add a TC curve. If we define effort as all the equipment, labor, and other resources that go into fishing, then TC is equal to the amount of effort times the price of effort and therefore can be derived from a series of catch-per-unit-effort curves.[6] We may think of TC as a curve starting at maximum population and rising to the left as more fish are caught. TC will increase as more fish are caught, both as a result of stock depletion and as a result of harvesting a larger sustainable yield (at least up to MSY). But even beyond MSY the TC will probably still rise because stocks have become so sparse that the fish have become hard to find.[7] At some point, the level of effort in the fishery is more than fish populations can sustain, and harvest levels become unsustainable. This is depicted by the dotted lines extending the total cost curves in Figure 12.2.

Focusing on the flow dimension only, π = TR–TC, and maximum ≠ for the industry occurs at π^* where MR = MC, where the slope of the tangent

---

[5]Note that the assumption of a constant price assumes perfect competition: one small, managed fishery among many in the world. We'll discuss the implications of relaxing this assumption shortly.

[6]Figure 12.2 is not actually derived from Figure 12.1. The linear yield effort curves of 12.1 would not transform into a linear total cost curve as depicted in 12.2. However, the assumptions of linearity in either case are simplifications designed to facilitate analysis.

[7]Note that Figure 12.2 is not the same shape as Figure 12.1. Figure 12.2 depicts a more resilient population that can sustain higher levels of effort without causing extinction, to better illustrate the points we are making here.

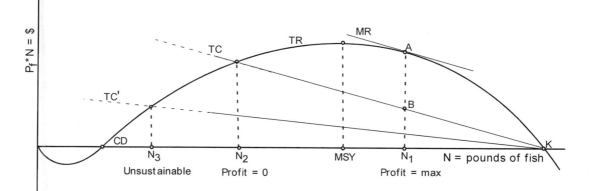

Figure 12.2 • Maximizing annual profit from renewable resources.

to the TR curve (MR) equals the slope of the tangent to the TC curve (MC). Remember that we have left the stock dimension out of our concept of total revenue; we are analyzing profit as a sustainable flow, not as the result of unsustainable stock reduction. Some stock reduction is necessary to arrive at the profit-maximizing stock—for example, if S were the profit maximizing stock in Figure 12.1, then Q'Q and R'R are the one-time stock reductions that lead to that stock—but just assume for now that the stock-reducing, one-time catch of fish are thrown away. We'll deal with them later.

In Figure 12.2, we can see that the maximization of annual profit (AB) occurs at N_1, which is larger than the stock corresponding to MSY. In other words, in this analysis an owner of a small, competitive fishery seeking to maximize sustainable annual profits will not even reach MSY, much less drive the population to extinction. If total costs were zero (or constant), marginal costs would be zero, and profit maximization would occur where marginal revenue was also zero, namely at MSY, where the tangent to the TR curve is horizontal (slope equals zero). So even zero harvest cost would not lead the capitalist to exploit beyond MSY.

> **THINK ABOUT IT!**
> *Can you find the tacit assumption responsible for the happy result that profit-maximizing exploitation does not require much stock depletion?*

We have assumed a single capitalist exploiting the fishery, a single owner or decision maker. Instead of private property with excludability, suppose the fishery were open access, as most are. Under an open access regime (in which you'll recall the resource is nonexcludable), new fishermen will enter as long as there are profits to be made.[8] New entrants will push the stock down to N_2, at which π = zero, or (TR = TC). At N_2 many more resources are going into fishing, but the sustainable catch is less than at N_1, and no one is making a profit.[9]

Curve TC′ depicts lower harvest costs, which might come about from a technological advance such as sonar devices for locating fish schools. At these lower costs, in an open access fishery it would be profitable for new fishermen to keep entering the fishery even after harvests become unsustainable. This is the case where the tragedy of open access resources may well lead to extinction and may be a realistic depiction of what was happening with North Atlantic cod and many whale populations before regulation began. Competition in this case leads to a race to catch the remaining fish before someone else can.

And what happens if we relax our assumption about constant prices and instead assume that prices increase as harvests decrease? However, the total revenue curve would still equal zero at minimum viable population and at K but would otherwise shift up where harvests were low and down where harvest are high. As stocks became too low to sustain higher harvests, profits would increase, attracting more fishermen into the industry and increasing the risk of unsustainable levels of effort. However, the single annual profit-maximizing owner would end up harvesting fewer fish from a larger stock.

Box 12-1	ANNUAL PROFIT MAXIMIZATION IN WILD VS. BRED POPULATIONS

Before leaving the annual profit-maximizing approach, we'll use it to distinguish between the exploitation of wild and bred populations. The fishery example we have used is, of course, the case of a wild population. A catfish pond would be an example of a bred population. In both cases, the biological population growth function is similar. But the cost functions are very different. For a wild population, costs are mainly costs of capture. For a bred population, costs of capture are minimal, but the costs of feeding and confinement are high.

---

[8]Profits, in economic parlance, are returns above and beyond the cost of production, where the cost of production obviously includes wages. For a small fishing crew sharing returns, profits would mean higher wages than they could find elsewhere.

[9]Note that if total costs are high enough, it is possible that the open access equilibrium might be at a greater yield than the profit-maximizing equilibrium, though with a lower stock and still with zero profit.

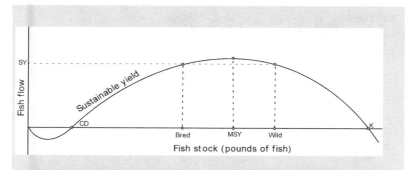

Figure 12.3 • Sustainable harvests from wild vs. bred populations.

Figure 12.3 shows two equal sustainable yields on a given population growth curve, one corresponding to a wild mode of exploitation and the other a breeding mode. In the wild mode, the TC curve (not shown) would start from the right and rise to the left (increasing with capture, as in our fishery example). In the breeding mode, the TC curve would start from the left and rise to the right (increasing with feeding and confinement costs of the larger population). The graph depicts a case in which the same SY could be had from exploiting the population in either a wild or a bred mode. Note that just as the annual-profit-maximizing owner of a wild fishery will never push the population below MSY, the profit-maximizing owner of a bred fishery will never allow the population to grow up to the point of MSY. Can you explain why?

The advantage of the wild mode is that the base population is larger, containing more biodiversity, providing more ecological fund services, and feeding is free. The advantage of the bred mode is that the smaller base population takes up less ecological space, leaving more room for other life, wild or human. Of course the food for the captive population requires some ecological space as well as collection. Controlled breeding and genetic engineering would seem to require the breeding mode. Burgeoning human populations and the technical thrust toward genetic engineering push the shift toward bred genetically engineered populations (e.g., the famous 200-pound salmon). Also, the quest for higher biological growth rates to keep up with the interest rate favors the bred stock, since a smaller base population yielding the same annual increment obviously implies a higher rate of growth. But smaller populations growing more rapidly, and ever more dependent on human management for feeding, reproduction, and disease control, surely will increase the instability, brittleness, and vulnerability to uncertainty of the whole biotic system.

BOX 12-2 | GEO-ENGINEERING OR COSMIC PROTECTIONISM

"We are capable of shutting off the sun and the stars because they do not pay a dividend."—John Maynard Keynes, 1933

As evident from Chapter 17 we are not free traders. But we do recognize limits to protectionism, especially when applied to protect rival goods against competition from non-rival goods.

Frederic Bastiat's classic satire, "Petition of the Candlemakers Against the Sun", has been given new relevance. Written in 1845 in defense of free trade and against national protectionism in France, it can now be applied to the cosmic protectionists who want to protect the global fossil fuel-based growth economy against "unfair" competition from sunlight —a free good. The free flow of solar radiation that powers life on earth should perhaps be diminished, suggest some, including American Enterprise Institute's S. Thernstrom (*Washington Post* 6/13/09, p. A15), because it threatens the growth of our candle-making economy that requires filling the atmosphere with heat-trapping gasses. The protectionist "solution" of partially turning off the sun (by albedo-increasing particulate pollution of the atmosphere) will indeed make thermal room for more carbon-burning candles. Although this will likely increase GDP and employment, it is attended by the inconvenient fact that all life is pre-adapted by millions of years of evolution to the existing flow of solar energy. Reducing that flow cancels these adaptations wholesale— just as global warming cancels myriad existing adaptations to temperature. For reasons explained in the first chapters of this book, artificially reducing our most basic and abundant source of low entropy in order to burn up our scarcer terrestrial source more rapidly is contrary to the interests both of our species and of life in general. Add to that the fact that "candles", and many other components of GDP, are at the margin increasingly unneeded and expensive, requiring aggressive advertising and Ponzi-style debt financing in order to be sold, and one must conclude that "geo-engineering" the world for more candles and less photosynthesis is an even worse idea than credit default swaps. Why then do some important people advocate geo-engineering? As the lesser evil compared to absolutely catastrophic and imminent climate disaster, they say. If American Enterprise Institute has now stopped offering scientists money to write papers disputing global warming, and in fact has come around to the view that climate change is bad, then why have they not advocated carbon taxes or cap and trade limits? Because they think the technical geo-fix is cheap and will allow us to buy time and growth to better solve the problem in the future. Just one more double whiskey to help us get our courage up enough to really face our addiction....

Profit-Maximizing Harvest When Profits Can Be Invested: Net Present Value

Returning to the higher-cost scenario (TC), suppose we make sure that there is a single owner, not open access. Are we then sure that we will end up at N_1? Unfortunately not, because of the troublesome issue we swept under the rug initially: What happens to the fish that are part of the stock reduction rather than the annual recruitment? They are not thrown away, and their number is large relative to the annual growth. Those fish are sold. Stock reduction fish have the advantage of being available now; you don't have to wait for them to be hatched and grow. But the more you reduce the stock of fish today, the fewer fish you will have tomorrow, and the more difficult it will be to catch those fish. The population of fish is like the proverbial goose that lays golden eggs in perpetuity. Surely no rational capitalist would kill such a productive goose.

Or would she?

If the capitalist wanted to maximize the sum of golden eggs from now until the end of time, then obviously she would not kill the goose. But the goose also has a liquidation value as a cooked goose. Suppose the capitalist could kill the goose, cook it, and sell it for a sum of money, which when put in the bank at the going interest rate would yield an annual sum greater than the value of the golden eggs? Then it's goodbye goose, hello bank! The population growth rate of the goose (its egg-producing fecundity) is in direct competition with the interest rate, the "fecundity" of money. Neoclassical economists argue that money itself may have no reproductive organs, but it is a surrogate for many other things that can reproduce, and on average those other things can reproduce faster than the goose. So the goose-killing, reinvesting capitalist has converted a slow-growing asset into a fast-growing one, and therefore we are all better off. According to economists, cooking the goose in this case maximizes **net present value (NPV)**: the value to us today of all cost and benefit streams from now into the future. Economists calculate NPV by using a discount factor to give less weight to costs and benefits the farther in the future they occur (see Chapter 10).

Let's take the story a bit further in a thought experiment. Suppose an economy consists only of renewable resources. The interest rate is equal to some weighted average of the growth rates of all renewable resource populations. Everything that grows more slowly than the average (the interest rate) is a candidate for extinction (unless at some stock its growth rate rises above the interest rate). But something is always below average. When the below average is eliminated, what happens to the average in the next period? It goes up, of course. The tendency, it seems, would be to end up with only the fastest-growing species. Biodiversity would entirely disappear. In a world in which everything is fungible,[10] that would not matter. We could all eat algae, if that were the fastest-growing species.

But we have forgotten prices. Surely prices would rise as particular slow-growing species became scarce, and the rising price would compensate for a low biological growth rate, so that the value of the species would grow at a rate equal to the rate of interest before it became extinct. Yes, but remember that when the price goes up, the price of the existing stock rises as well as the price of the flow of recruits. As the price increases, the incentive to liquidate the now more valuable remaining stock rises, along with the incentive to reduce current offtake to allow an increase in the more valuable new recruits. If demand for the species is inelastic, then total revenues increase as harvests fall. If harvest costs increase only slowly, price increases are an added incentive to fish a species to extinction.

The bluefin tuna is an excellent example of this argument. In 2001, a single 444-pound bluefin tuna sold in Japan for nearly $175,000, or about $395/lb. Although this was an anomaly, restaurants in Japan regularly pay up to $110 per pound for bluefin tuna.[11] Admittedly this occurs under a regime of imperfect property rights,[12] but how confident are you that private ownership of the bluefin would solve the problem? The higher price means higher liquidation value, as well as higher future revenue, from new tuna.

How do we decide whether to harvest a (marginal) ton of fish that if left in the water reproduces (giving us a flow of golden eggs) and if harvested (giving us a cooked goose) yields interest on the profit?

The neoclassical approach is to ask: What are all the opportunity costs of harvesting the fish today? Obviously, if we harvest the fish today, that same ton of fish is not there to harvest tomorrow. Unlike oil or iron, how-

---

[10]Something is fungible if one unit of it substitutes indifferently for another unit. For example, two buckets of water from the same well are fungible (you can't tell any difference between them). But two buckets of water from two different wells may not be fungible because of qualitative differences such as hardness and taste. Money is fungible; we cannot tell if the money the government spends on foreign aid came from my tax dollars or from yours. Things convertible into money—goods, services, even biological species—acquire a kind of artificial or abstract fungibility, even though physically they are not at all fungible. This makes it easy to commit the fallacy of misplaced concreteness (see Chapter 2).

[11]G. Schaeffer, Tuna Sells for Record $175,000, Associated Press International, January 5, 2001.

[12]Fishing of bluefin tuna in the Atlantic at least is regulated, with quotas for the Eastern Atlantic held by European countries and quotas in the Western Atlantic held by the U.S., Canada, and Japan. But evidence indicates that these quotas are too high and that the two populations are not even distinct. Thus, regulations designed to assign property rights to a formerly open access resource may fail in their objective of preserving the species. See T. Bestor, How Sushi Went Global, *Foreign Policy* 121 (November/December 2000).

ever, we lose not only the opportunity to harvest that ton of fish tomorrow but also the offspring those fish would have had, as well as the increase in biomass they would have experienced, if left in the water. In addition, as the economy grows, people demand more fish, and growing human populations further increase the demand. We lose the additional profit from tomorrow's higher prices if we harvest today. Moreover, if we leave the fish in the water to reproduce, the greater population means that it will be cheaper and hence more profitable to catch a ton of fish next period than this period.

In contrast, the benefit of catching the fish this period rather than next is that the profits from their harvest can be invested; that is, the opportunity cost of not catching the fish is the money forgone from not being able to invest the profits from that ton of fish between this period and the next.

The economist will therefore favor harvesting as long as the diminishing marginal benefits of catching the next ton of fish are greater than the rising marginal costs and will stop when they are equal. The tricky thing about the decision is that when we consume more stock today in exchange for less stock and less yield tomorrow, it is not just one tomorrow, but tomorrow and tomorrow and tomorrow in perpetuity. We must compare a one-time benefit with a perpetual loss. As we mentioned above, economists address this problem rather unsatisfactorily by the financial convention of discounting and present value maximization. They argue that money in the bank is as real an asset as fish in the ocean, and because it grows faster, it is a more profitable one. What does all of this mean in practical terms?

Compared to our static analysis when we ignored the stock reduction needed to reach the annual profit-maximizing equilibrium, the opportunity to invest profits from stock reduction will lead to a lower stock of fish (or any other renewable resource). If we depicted this on Figure 12.2, the profit-maximizing harvest would be to the left of N_1, and the higher the interest rate, the further to the left it would be.

Take the case of bluefin tuna, where the cost of capture of one fish may be a negligible portion of its sales price. Imagine current harvests were sustainable and in the vicinity of MSY, and there was a single resource owner intent on maximizing profits. The reduction in sustainable yield from MSY to a somewhat lower stock may be small, while the stock liquidation needed to get there is still quite large. If the interest payments on the profits from the sale of that liquidated stock are greater than the value of the lost annual yield, then profit maximization favors sustainable harvest at a stock lower than MSY and closer to the minimum viable population.

What are the implications of this scenario in an extreme but not at all unrealistic case where harvest costs ($P_E E$) are negligible compared to harvest revenue ($P_Y Y$), even for very low-resource stocks? Timber is a good example. Say that a forest of redwoods not yet biologically mature[13] increases in size and thus value by 3.5% per year.[14] In contrast, the average real growth rate of money on the U.S. stock market over the last 70 years was about 7%.[15] Clearly, harvesting the resource now and investing the profits in the bank will maximize private monetary gains. In fact, for any species that is inexpensive to harvest and grows more slowly than alternative investments, harvesting the species to extinction maximizes profits. In general, averaged over the time it takes to reach harvest size, many valuable species grow quite slowly relative to alternative investments, and technology tends to reduce unit harvest costs over time. And for such resources it is, once again, goodbye golden eggs, hello bank.

In summary, the advantage of the catch-per-unit-effort curve analysis is that it builds in from the beginning the stock reduction effect and in that way is more realistic. The advantage of the TR–TC diagram is that it shows that annual profit maximization can be sustainable and efficient in the absence of open access, and with some limits on the biologically blind financial logic of discounting and present value maximization.

■ RENEWABLE RESOURCE FUNDS AND SERVICES

The analysis of optimal harvests of renewable resources so far has only treated them as stocks and flows of raw materials. But, as we discussed in Chapter 6, renewable resources are also funds that provide ecosystem services, and we cannot ignore one when deciding how best to allocate the other. While natural resource stocks and flows have some characteristics of market goods, the services generated by funds typically do not. Such services are generally nonexcludable, and for many, no feasible institutions or technologies could make them excludable. Thus, free markets will not produce them. They are also nonrival, and selling them in an otherwise perfect market would not still equate marginal costs with marginal benefits.

Treating the destruction of ecosystem services as a negative externality

---

[13]Biological maturity occurs when growth rates for the forest taper off toward zero, i.e., new growth is just matched by rates of decay.

[14]This is actually an unrealistically high rate of growth. Data exist on a 1-acre plot of redwoods that has been monitored for over 70 years. Though the rate of growth on this plot is so high that it is widely known as the "wonder plot," the most rapid 10-year mean annual increment in total stand volume was 3.5%. The mean annual increment in growth from 1923 to 1995 is well under 1%. These figures were calculated by the authors from data provided in G. Allen, J. Lindquist, J. Melo, and J. Stuart, Seventy-Two Years Growth on a Redwood Sample Plot: The Wonder Plot Revisited (no date). Online: http://www.cnr.berkeley.edu/~jleblanc/WWW/Redwood/rdwd-Seventy-.html.

[15]S. Johnson, Are Seven Percent Returns Realistic? Online: http://www.sscommonsense.org/page04.html. Common Sense on Social Security.

of aggregate economic production offers some insights into "optimal" harvest levels.[16] Returning to the analysis of renewable resource stock-flows, we would need to add all external costs to total private harvest costs. Marginal external costs are likely to increase at increasing rate, especially near an ecological threshold, such as the minimum viable population, which will inevitably be reached when harvest effort is too great to be sustained. In fact, as we near ecological thresholds, marginal activities lead to largely unpredictable nonmarginal outcomes, and marginal analysis is no longer appropriate. Figure 12.4 is similar to Figure 12.2 but relabels the TC curve as the total private cost (TPC) curve and includes a total social cost (TSC) curve that adds external costs to the TPC. The TSC curve approaches the vertical (i.e., unacceptably high marginal social costs) as we approach the minimum viable population. The optimal harvest is where marginal social costs equal marginal revenue, labeled N_4 on the graph. Whenever the renewable resource contributes to the provision of ecosystem services, the optimal harvest from an ecological economic perspective will always be at a higher stock with lower private costs than in the annual profit-maximizing equilibrium.

Of course, optimality would require the micro-level internalization into prices of all ecosystem services. Yet human impacts on these services are characterized at best by uncertainty (we know the possible outcomes of damage to ecosystem funds on ecosystem services but don't know the probabilities) and more often than not by ignorance (we don't even know the range of possible outcomes). In fact, we almost certainly do not know the full extent of the ecosystem services from which we benefit. In addition, the value of all externalities would need to be worked out by economists, ecologists, and others, and incorporated into the prices of the

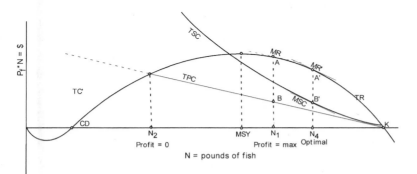

Figure 12.4 • Optimal harvest of renewable resources when accounting for ecosystem services.

---

[16]To remind yourself of what we mean by "optimal," you might want to review Box 2.1.

goods that generate the externalities. And, of course, the marginal value of an ecosystem service changes along with the quantity of the ecosystem service supplied, so the value of externalities would be constantly changing. As we have pointed out, all economic production incurs externalities. The notion of calculating the constantly changing values of all externalities for all goods would be a Promethean task. Once achieved, it would still require some institution to incorporate the fees into market prices. And we must remember that the magic of the market is precisely its unplanned, decentralized nature, and its ability to use "knowledge not given to anyone in its totality."

Effectively internalizing externalities, in contrast, requires precisely the opposite: centralized planning by individuals provided with knowledge in its totality. While an optimal allocation of everything is not a feasible goal, in Part VI (Chapters 21–24), we will explore approaches to achieving a satisfactory allocation.

Note also that we have again left out the potential for investing the profits from reducing the stock. This is intentional. No potential return could substitute for either life-sustaining ecosystem services or the raw materials essential for all economic activities. In addition, many investments are profitable precisely because they do not account for the opportunity costs of resource depletion (MUC) or the social costs of the ecosystem services inevitably degraded or destroyed through resource extraction.

Nonmarket ecosystem fund-services simply cannot be converted to money and invested, as we can do with a cooked goose. Also, natural resources are growing physically scarcer. Technology seems to be developing new uses for most natural resources faster than it develops substitutes, which increases future demand. Increasing demand and decreasing supply imply greater value for natural resources in the future, not less. Instead of intertemporal profit maximization, we concur with Geoffrey Heal and other environmental economists that we should seek to maximize well-being from renewable natural resources for the current generation without diminishing the capacity of future generations to benefit from those resources. Heal and others have called this principle the Green Golden Rule.[17] Applied to the stock-flow alone, the Green Golden Rule corresponds to our analysis of maximizing sustainable annual profit, as opposed to present value maximization.

In summary, the more we have of an ecological fund, all else being equal, the more services we can expect it to provide. If we are concerned only with the service provided by a fund, the optimal amount of the fund

[17]G. Heal, *Valuing the Future: Economic Theory and Sustainability,* New York: Columbia University Press, 1998. Note that Heal proposes a number of objective functions in addition to the Green Golden Rule.

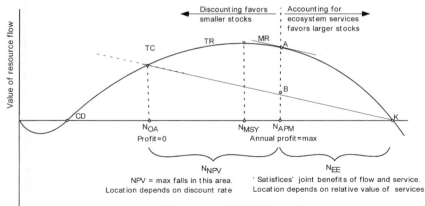

Figure 12.5 • Optimal harvest levels for renewable resources with respect to different objectives and management regimes. From left to right, N_{OA} is the open access equilibrium, at which profit is zero. If total costs decrease over time, the N_{OA} may become unsustainable. N_{NPV} is the stock at which net present value is maximized. At very high discount rates, this will be the same as the open access equilibrium and at a zero discount rate will be equal to N_{APM}. N_{APM} is the annual profit-maximizing stock. N_{EE} is the objective of ecological economists and strives for "satisficing" (seeking a sufficient, rather than the maximum, amount) the joint benefits of both flow and service.

is the carrying capacity, as measured on the *x*-axis in Figures 12.1 and 12.4. In contrast, optimal harvest of stocks is solely a function of flow (the *y*-axis in the same figures). Unless we recognize the values of both the fund-service and the stock-flow, resource extraction rates will not be optimal. Figure 12.5 summarizes the discussion of optimal stocks and harvests from renewable resources.

The Natural Dividend from Renewable Resources

Unearned income is the amount above and beyond what is needed to bring a resource to market. In the case of nonrenewables, the unearned income is called scarcity rent. In the case of renewables, there can also be an unearned income deriving from nature's reproductive capacity, which we propose to call the **natural dividend**. Extraction and harvesting impose real costs, and those engaged in these activities are earning a legitimate income. This is included in the TC curve as "normal profit," which is the opportunity cost of the owner's labor, capital, and perhaps entrepreneurial ability. Profit above normal profit (e.g., AB) is called "pure economic profit." Since it is beyond the opportunity cost of the owner, we think of it as an unearned growth dividend from the reproductive power of nature. We have already seen that in an open access equilibrium, total costs are just equal to total revenues, and profits are zero (at stock N_{OA}

The *natural dividend* is the unearned income from the harvest of renewable resources. As nature and not human industry produces renewable resources, all profits above "normal" profit (included in TC) are unearned, and the natural dividend is equivalent to TR–TC.

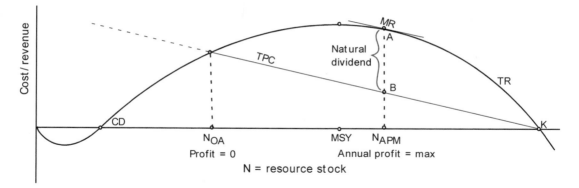

Figure 12.6 • The natural dividend from renewable resource harvests.

in Figure 12.6). However, if there is a single owner of the resource, or an outside regulator such as a government agency, it's possible to limit harvest to the profit-maximizing level, stock N_1. The pure profits in this case (AB in Figure 12.6) are the natural dividend and arise not from special abilities of the particular owner but from the reproductive powers of nature. The natural dividend can also be thought of as the value of the resource in the ground (or sea). While the natural dividend typically accrues to the owner of the resource, it is a purely unearned profit from production. To whom should the dividend belong? That is a political decision, more in the realm of fair distribution than efficient allocation.

■ Waste Absorption Capacity

Waste absorption capacity is really just another ecosystem service. We treat it separately here because it is extremely important and because it has different characteristics from most other ecosystem services. Waste absorption capacity is the ability of the ecosystem to absorb and process pollution, and the economics of pollution is the predominant focus of neoclassical environmental economics. As we pointed out previously, waste absorption is a rival good. If I dispose of my sewage in a wetland, there is less capacity subsequently for that wetland to process someone else's wastes. As we also pointed out, many countries are trying to create institutions that make waste absorption capacity an excludable good. These can range from regulations that directly limit industrial emissions to mandatory catalytic converters in cars to tradable emission permits for sulfur oxides. Tradable permits and quotas for pollution essentially make waste absorption capacity a private good. These mechanisms will be discussed at length in Chapter 21.

However, we must bear in mind that pollution is pure externality and a **public bad**, which is something that is nonrival, nonexcludable, and undesirable. Therefore, even when a market exists in air pollution, for example, this is not the same as saying that a market exists for clean air. Nonetheless, pollution permits are one of several mechanisms that can help achieve a socially optimal level of pollution (see Chapters 20–22).

There is, of course, no direct social benefit to pollution per se, but as we have repeatedly stated, virtually all productive processes generate some pollution, and if we prohibit all pollution, we virtually prohibit production. This is why economists use the apparent oxymoron of "optimal pollution." By *optimal*, economists simply mean potentially Pareto efficient. A reasonable estimate of the benefits of pollution is therefore the marginal net private benefits of production (MNPB) associated with a unit of pollution. The problem is, of course, that our knowledge of external costs of pollution is characterized predominantly by ignorance and uncertainty, to a lesser extent by risk, and to a minimal extent by certainty. Since we do not know the full social costs of pollution, it is exceedingly difficult to balance costs with benefits. Policy makers are also not well informed concerning the MNPB of pollution to polluters.

We have to recognize that waste absorption capacity is a dynamic process, and we must define carefully what we mean by it. We define **waste absorption capacity** as the ability of an ecosystem to assimilate a given flow of byproducts of economic activity that have direct or indirect negative impacts on human well-being. If the waste flow exceeds the waste absorption capacity, then waste will accumulate. Ecosystems are highly adaptive and may evolve mechanisms for processing greater waste loads, though perhaps at the expense of important ecosystem services. For example, marine systems around the world absorb excessive nutrient loads through algal blooms, which create hypoxic conditions and dead zones when the algae decay. Alternatively, positive feedback loops are possible, in which excessive waste loads induce changes that reduce absorption capacity, leading to even more rapid accumulation. For example, warmer oceans may absorb less CO_2 (think of a bottle of soda giving off CO_2 bubbles as it warms up). Regardless of ecosystem response, as long as waste flows exceed absorption capacity, waste will build up indefinitely. The resulting loss of critical ecosystem services may well be irreversible.

What is the marginal external cost of the increase in flow rate that takes the system beyond the point of no return? It is the value of the lost services from that ecosystem for all time. If the ecosystem provides vital functions, the marginal external cost is basically infinite. However, if the waste flow is halted before crossing critical thresholds, the system can slowly process the waste and restore itself. This appears to be what is happening in Lake Erie on the U.S.–Canada border. If the ecosystem in question is simply a local ecosystem with similar ecosystems nearby, even after collapse, stopping the waste flow can lead to recovery. Wastes will be absorbed, dissipate, and settle out of the system, new organisms will colonize the system, and the restoration process will begin.

It is worth illustrating these points graphically. Figure 12.7 is an appropriately modified version of the analysis of externalities from Figure 10.2. Note that economic output and waste output are measured on the same axis, in recognition of the laws of thermodynamics. In reality, the relationship is not as fixed as Figure 12.7 indicates; many technologies are available for producing different goods, some of which generate less pollution than others, though for any given technology, the relationship will be fixed.

We assume here that the economic output in this example is not essential; either substitute products and processes are readily available, or the good itself is simply not that important to human quality of life. Chlorine-bleached paper from wood pulp is a good example. Unbleached papers from kenaf or hemp are excellent substitutes, and paper itself is important but not essential to life. Among the many wastes emitted by paper mills are organochlorines, and paper mills are the largest emitter of organochlorines into the water supply in the U.S. and numerous other countries as well.

Organochlorines resist biodegradation, and therefore the waste absorption capacity for these substances is quite small; it is indicated by the perpendicular at point Q_A, W_A in Figure 12.7. Organochlorines include some of the most toxic substances known, such as dioxins, which readily accumulate in the environment and in animals, including humans. Health problems associated with dioxins include cancer, immune system disorders, and developmental problems in children and fetuses. We assume without too much exaggeration that surpassing the waste absorption capacity for organochlorines for extended periods could make the affected areas essentially uninhabitable for humans, and on our crowded planet, this is an unacceptably high cost. This is indicated on the graph by the marginal external cost (MEC) curve, which approaches vertical as we near the waste absorption capacity for these toxins (Q_A, W_A). These nearly infinite marginal costs occur only if the accumulation of pulp mill waste continues unabated for some time.

On the other hand, the extreme levels of uncertainty and ignorance concerning waste absorption capacity and the impacts of paper mill pollutants are not reflected in this graph. In reality, both the MEC curve and the line depicting waste absorption capacity should be thick smears instead of the fine lines depicted. We have labeled the optimal level of output or pollution as Q^*,W^* but again caution that with our given state of knowledge, this is a broad range, not a precise point. The reasoning behind the location of Q^*,W^* was described in Chapter 10.

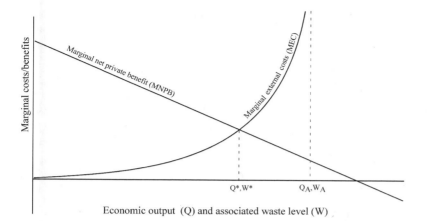

Figure 12.7 • Waste absorption capacity: marginal costs and benefits of pollution.

We cannot stress enough the importance of looking at pollution flows and waste absorption capacities as dynamic. Some economists have argued that pollution causes zero damage before reaching point Q_A, W_A, because the ecosystem is capable of assimilating the waste. But there clearly are substantial costs to pollution even when the ecosystem is capable of assimilating them. MEC may approach infinity at Q_A,W_A.

Depending on the population level and the level of economic activity (i.e., based on scale), MNPB may become zero before or after reaching the waste absorption capacity of the environment. The more full our planet becomes—the larger our scale—the more likely that MNPB will still be positive when we reach the waste absorption capacity and MEC approaches the vertical. Why? Goods and services are characterized by diminishing marginal utility. This means that for a given amount of goods, more people imply lower per-capita consumption and hence higher per-capita utility from each unit consumed. This shifts the MNPB curve upward.

Finally, we must stress once again here that even if policy makers could measure the full marginal costs and benefits of pollution and charge polluters accordingly, pollution markets would still fail to generate all the wonderful outcomes associated with the free market. Different individuals obviously have different preferences (utilities) with respect to polluted environments. Markets are widely extolled because they allow the individual to choose what she produces and consumes so that her marginal benefits from either are exactly equal to her marginal costs. Pollution, however, affects public goods, and all individuals must consume the same amount. It would be impossibly complex to create a system in which each individual was paid by the polluter according to his or her own dislike of pollution.

This does not mean we are opposed to markets in waste absorption capacity, but it does mean that we should not associate with them all the market virtues associated with the buying and selling of market goods.

■ Biotic and Abiotic Resources: The Whole System

To move toward a more a sustainable, just, and efficient economic system, we clearly must understand the nature of the resources upon which that system depends. We need to understand the role these resources play in meeting the needs of humans and other species on this planet and the characteristics that affect their allocation within and between generations via market and nonmarket mechanisms. It would, of course, be impossible to analyze every individual resource. Instead, we introduced the important concepts of rivalness, excludability, externalities, ignorance and uncertainty, and stock-flow and fund-services and applied these concepts to specific categories of natural resources. To facilitate this, we created a rough taxonomy of biotic and abiotic resources subdivided into eight categories and applied the above concepts to each of these categories.

Our first goal with this approach was to help you understand precisely why markets fail to efficiently allocate each individual resource and start you thinking about what types of institutions and mechanisms might work better. We began with abiotic resources, which are fairly simple to understand. We then moved on to the stock-flows provided by nature, the raw materials on which the economy depends. We began to see the emergence of complexity: unpredictable ecological thresholds beyond which a population will collapse, impacts from outside variables such as climate change and habitat degradation. The analysis grew a bit more complicated. Once we turned to ecosystem services and waste absorption capacity, it became obvious that these were elements of a whole system and could not be understood apart from that system. Ecosystem fund-services, including waste absorption capacity, are provided by the complex interaction of ecosystem stock-flows and are necessary to sustain those stock-flows. We can't think of allocating the stock-flow independently of the fund-service. Both fund-service and stock-flow are seriously affected by the waste flows from nonrenewable abiotic resources. So what does this mean?

Our second goal with this approach was, paradoxically, to guide you toward the conclusion that the first goal (described in the preceding paragraph) is insufficient. While it helps to understand the particular characteristics of each individual natural resource, it is more important by far to recognize that these resources are so intimately intertwined that we cannot allocate any one resource without considering the impact it will have on others. The reductionist approach (breaking down low-entropy resources into narrow categories) offers some useful insights but is inadequate on a complex, living planet. Seemingly efficient allocation of each resource individually will not necessarily lead to the efficient allocation of all resources together. Ecological economics is concerned with integrated systems, not individual commodities, and with complex societies, not atomistic individuals. Breaking a system down to better understand its individual components is a useful analytic tool, but it can seriously mislead us unless we subsequently synthesize these components into an integrated understanding of the whole.

BIG IDEAS to remember

- Stock-flow versus fund-service resources
- Sustainable yield, maximum sustainable yield
- Absorptive capacity
- Per-unit effort curve
- Stable and unstable equilibrium

- Maximizing annual profit versus maximizing net present value
- Exploitation of wild versus bred populations
- Natural dividend

Human Behavior and Economics

Economists agree that all the world lacks is,
A suitable system of effluent taxes,
They forget that if people pollute with impunity,
This must be a sign of lack of community.

—Kenneth Boulding, "New Goals for Society."

Conventional economic theory, as described in Chapters 8 and 9, assumes that certain economic behaviors are innate. That is, they are highly predictable across time and cultures. Embedded in this assumption is a core belief about human nature, which turns out to have profound implications for the development of economic theory. It is crystallized in the concept of *Homo economicus*, which emerges from the discipline's foundations in utilitarianism and incorporates the following traits:

1. Insatiability. What we really want is more stuff. Put another way, more is always better, and consumption is the major source of utility (i.e., well-being).

2. Perfect rationality. Individuals have stable, exogenously determined preferences (i.e., preferences are not affected by advertising, by the preferences of others, by the number of choices available, etc.) and make choices that best satisfy these preferences in the face of given constraints of time, income, and so on.

3. Perfect self-interest. Individuals (or at least families) do not care how their choices affect others and are not affected by the "utility"

others experience. Social interactions matter only to the extent that they affect one's own consumption, leisure, and wealth.

Broader economic behavior is simply the aggregation of decisions by rational, self-interested individuals. Though most economists recognize that the assumptions of *Homo economicus* are somewhat of a caricature of real human behavior, these assumptions nonetheless form one of the central pillars of conventional microeconomic theory. Competitive free markets in theory take advantage of our self-interest to create a system through which competitive, selfish behavior generates the greatest good for the greatest number. Since the market works its magic through the price mechanism, market prices reflect our values and desires.

THINK ABOUT IT!

As we explore the behavioral basis for economics, you might find it useful to think about how you would answer these questions: (1) Does H. economicus *really describe us?; (2) If not, then what does?; and (3) What do you think this means for the study of economics?*

Box 13-1 | **ECOLOGICAL ECONOMICS AND THE SELF-INTEREST ASSUMPTION**

Ecological economics has inherited from both of its parents the idea that individual selfishness and competitive struggle lead to the greater collective good. From economics, beginning with Adam Smith, comes the "invisible hand." From biology, via ecology, comes Darwin's natural selection of the best-adapted individuals in the face of competition for the limited means of subsistence forced by Malthusian population pressure. In part, these are two assumptions about how the world works rather than the affirmation of self-interest as a moral value. Competition is taken as a fact. But in both cases the assumption is blessed by its purported consequences: market efficiency and evolutionary progress.

There are other traditions in both economics and biology that contradict the assumption of selfishness. Adam Smith himself, in *The Theory of Moral Sentiments*, emphasized cooperation and community as the overall context in which competition could be trusted. Darwin recognized that group selection favored the evolution of moral values and cooperation,[a] and Kropotkin[b] emphasized mutual aid as a factor in evolution. Nevertheless, in both disciplines the selfishness tradition has been quite dominant, and we should be aware in ecological economics that we have received a double dose of this inheritance, for better or worse.

---

[a]*C. Darwin*, Descent of Man, *1871. Online: http://www.infidels.org/library/historical/ charles_darwin/descent_of_man/.*

[b]*P. Kropotkin*, Mutual Aid: A Factor of Evolution, *1902. Online: http://www.calresco .org/texts/mutaid.htm.*

This chapter explores what we know about human behavior, whether or not the conventional model of that behavior, *Homo economicus*, is adequate, and the implications for ecological economics. In so doing, we will explore research that addresses human desires and the roles of rationality and emotionality, of selfishness and altruism, of competitiveness and cooperativeness. We will also touch on what the field of evolution tells us about our behavior, as well as the difficult question of cultural evolution and the extent to which it may be possible to change human behavior.

Box 13-2   HOW WOULD YOU WANT PEOPLE TO BEHAVE?

Before going any further with this chapter, make up a list of five personality traits that you associate with good people and five personality traits that you associate with evil people. Physical traits such as strength, intelligence, athletic ability, looks, and so on are irrelevant.

Once you've completed your list, look back at the model of *H. economicus*. Do you think the explicit and implicit behaviors of *H. economicus* most closely resemble a good or evil person?

Now perform the following thought experiments. If you were to place one good person and one evil person together on a desert island, both with the same physical traits, who do you think would be most likely to thrive? Why? Imagine that a society started out with equal numbers of good and evil people, but those who thrive are able to leave more descendants, and they pass on their own characteristics to those descendants (either through genetics or culture). What will happen to the composition of society over time?

If you placed 10 good people on one island and 10 evil people on another island, which population would be most likely to thrive? Why? Imagine that those who thrive will increase in number and populate other islands with descendants sharing their characteristics. What will happen to the global society over time?

Do you think most people are good, evil, or somewhere in between? Do you think the human race would be more likely to thrive if we behaved like good people or evil people?

These questions are adapted from a book by D. S. Wilson, *Evolution for Everyone* (New York: Delacorte Press, 2007). He has repeatedly asked these questions of his students and found that "traits associated with 'good' cause groups to function well as units, while traits associated with 'evil' favor the individual at the expense of the group" (125). Is this true for the traits you chose?

■ CONSUMPTION AND WELL-BEING

We began this book by arguing that we must have a picture of the desirable ends of economic activity before deciding what and how to allocate.

Economists often talk about maximizing utility, which in the original util-itarian philosophy was equated with happiness. While happiness was once considered too subjective to evaluate, in recent years the study of happiness has turned into a respectable academic pursuit. We do not claim that that the ultimate end of economics is simply to make people happy, but psychic satisfaction is certainly an important goal of economic activity. It is therefore well worth examining what makes people happy and unhappy.

Is Consumption the Path to Happiness?

The Easterlin paradox is the evidence that within a country, wealthier people tend to be happier than the less wealthy. However, beyond a certain threshold, citizens of wealthier countries do not seem to be much happier than citizens of less wealthy ones, and overall happiness within countries beyond this threshold does not seem to increase with increasing income.

Most economists (and most of society, for that matter) seem to believe that ever-increasing consumption is the ultimate desirable end. In Chapter 1, we argued that humans are not by nature insatiable (a topic we return to in Chapter 14). For most of human history, we were hunter-gatherers who depleted resources in a small area and then moved on, often traveling over 20 miles a day. If we accumulated more than we could carry, we could not keep up with our food supplies and starved. Accumulation meant death.

But what we desire is clearly influenced by culture as well as our evo-lutionary history. In the modern world, does more income correlate with greater happiness? There are at least four ways to look at this question. First, are individuals in wealthier nations happier than individuals in poorer nations? Second, are wealthier individuals within a society happier than poorer individuals? Third, do nations grow happier over time with increases in income? And fourth, does wealth correlate with happiness within a single individual's lifetime?

Researching such questions in the 1970s, Richard Easterlin came upon a puzzling paradox. As economists would predict, wealthier individuals within a nation reported greater happiness[1] than less wealthy ones. How-ever, once countries had sufficient wealth to meet the basic needs of their citizens, reported levels of happiness across nations showed little correla-tion with national income. Furthermore, reported levels of happiness within a country did not increase even with dramatic increases in national income over time[2]. These basic results have been replicated numerous times in subsequent years, and also hold true for satisfaction with life as a

---

[1]Many of you might be skeptical about the accuracy of self-reported happiness levels, but neu-roeconomists have shown that these correlate with the level of activity in certain parts of the brain (e.g., H. Plassmann et al. Marketing Actions Can Modulate Neural Representations of Experienced Pleasantness, *PNAS* 105:1050–1054 [2008]).

[2]R. Easterlin, Does Economic Growth Improve the Human Lot? In P. David and M. Rede, eds., *Nations and Households in Economic Growth: Essays in Honor of Moses Abramovitz*, New York: Aca-demic Press, 1974.

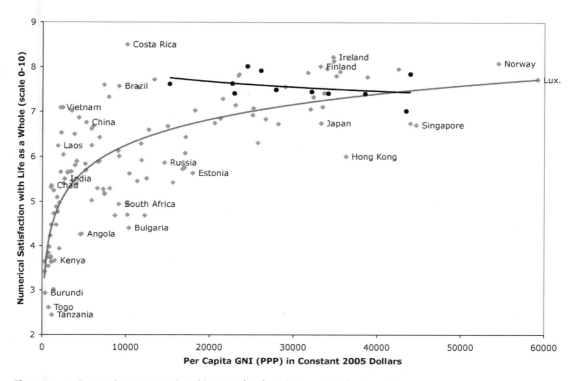

Figure 13.1 • Per capita gross national income (GNI) and mean satisfaction with life as a whole. The gray diamonds represent a cross section of 121 countries around the world, and the gray line a logarithmic trend curve. The black squares represent a time series for the United States, and the black line a trend curve. (Sources: Weenhoven, R., *World Database of Happiness, Distributional Findings in Nations*, Erasmus University Rotterdam. Available at: http://worlddatabaseofhappiness.eur.nl, 2010; Bureau of National Economic Accounts. Current-dollar and "real" GDP. US Department of Commerce. Available at: http://www.bea.gov/national/index.htm, 2007; World Bank Group. Development Indicators. Available at: http://devdata.worldbank.org/data-query/, 2010.

* For each country, we used the most recent survey results from 2004–2008 for the question "All things considered, how satisfied or dissatisfied are you with your life as-a-whole these days?" on a scale of 0–10, or on a scale 1–10, adjusted to 0–10, if the former was not available. No surveys were available for the missing countries. The US data consists of all years available for the same question. To standardize all income measurements into the same unit, we created a conversion factor: CFt = (US "real" GDP per capita (in 2005 dollars))t / (US GNI per capita PPP current international dollars)t, for t = 2004–2008. The x-axis is in units of CF* GNI per capita PPP.

whole (see Figure 13.1).[3] A recent, widely publicized study claims to have disproved Easterlin's paradox, finding evidence of a strong correlation between GDP and happiness both across time and across countries.[4] However, a rebuttal by Easterlin shows that the time series correlations between GDP and happiness exist only for short time periods and disappear over the long term. Surprisingly, this holds true for developed, developing, and transition (e.g., former Soviet Union) countries.[5]

One explanation of this paradox is that while absolute levels of income do not matter much, relative income does matter. For example, more detailed studies of income disparities within countries show that middle-class people in a rich neighborhood typically have lower life satisfaction than middle-class people in middle-class or poor neighborhoods.[6] Furthermore, individuals compare their own income with their own past. They quickly adapt to income gains, taking them as the new normal, but less quickly to income losses (a topic we'll discuss further). This explains the presence of a short-run correlation between GDP and happiness, which occurs when countries enter a recession or when they recover, and the absence of a long-run correlation.[7] But if only relative income matters, then increasing income, especially in the wealthy countries, may do little to increase happiness. But surely, having more opportunities, more choices, must make us better off? At the very least, it would seem that more choices could not make us any worse off.

Unfortunately, a number of empirical studies identify numerous ways in which too many choices make the act of choosing more difficult, stressful, and unpleasant. Choice can create conflict. In many cases, people simply respond by not choosing (which may mean choosing the default option) or by making a worse choice.

For example, one study asked people to suppose they were considering buying a CD player from a store that had a one-day clearance sale on a popular Sony model for $99. Given the choice of buying the player or

---

[3]E.g., R. Layard. *Happiness: Lessons from a New Science* (New York: Penguin Press, 2005); M. Max-Neef, "Economic Growth and Quality of Life: A Threshold Hypothesis," *Ecological Economics* 15(1995):115–118; R. Lane, *The Loss of Happiness in Market Economies* (New Haven: Yale University Press, 2000); M. Shields and S. Wheatley Price. "Exploring the Economic and Social Determinants of Psychological Well-being and Perceived Social Support in England," *Journal of the Royal Statistical Society Series A*, 168(2005):513–538.

[4]B. Stevenson and J. Wolfers. 2008. Economic Growth and Subjective Well-Being: Reassessing the Easterlin Paradox. IZA Discussion Paper No. 3654, Institute for the Study of Labor.

[5]R. Easterlin and L. Angelescu. 2009. Happiness and Growth the World Over: Time Series Evidence on the Happiness-Income Paradox. IZA discussion paper No. 4060, Institute for the Study of Labor.

[6]E. Luttmer. "Neighbors as Negatives: Relative Earnings and Well-Being," *Quarterly Journal of Economics* 120(2005):963–1002 .

[7]R. Easterlin and L. Angelescu, Happiness and Growth the World Over.

waiting to learn more about the various models in this low-conflict situation (i.e., only two choices), the majority preferred to buy the Sony. A separate group of people was given a similar scenario, with the difference that
the store now had, in addition to the Sony, a top-of-the-line Aiwa model
on sale for $169 (i.e., a high-conflict situation, with more choices that are
difficult to compare). In this situation, the majority of the participants
chose to await more information. Adding a third choice should not affect
the preference ordering of the first two choices, yet it does. Furthermore,
people like to avoid conflict, even in their own minds, and increasing the
number of choices increases potential conflict.[8] Numerous studies show
that too many choices may not only make it more difficult to make a good
choice but can also be an unpleasant experience on its own.[9]

Increasing choice also has cost implications. If there were only a few
dozen varieties of cars, computers, or bicycles, for example, it would be
very easy for stores to stock spare parts and very easy to salvage spare
parts from broken items (as evidenced by the popularity of auto junkyards
as sources for spare parts in the 1950s, though perhaps few of you will remember that). In modern society, firms launch a suite of new products
every year, and most have very little standardization. With so many
choices available, it becomes very difficult to stock or even manufacture
replacement parts. This makes it more difficult to repair broken items
(which in effect reduces consumer choice) and increases the flow of waste
back into the environment.

The Sources of Happiness

If income and choice are not what makes us happy, then what does?
Dozens of studies have identified mental health, satisfying and secure
work, a secure and loving private life, strong social networks, freedom,
and moral values. Behavioral economists now discuss **procedural utility**,
which is essentially the pleasure you get from doing something, not from
just having things.[10] Other researchers argue that "because identity is fundamental to behavior, choice of identity may be the most important 'economic' decision people make."[11] In other words, *being* may be more

---

[8]A. Tversky and E. Shafir, "Choice Under Conflict: The Dynamics of Deferred Decision," *Psychological Science* 3(6)(1992):358–361.

[9]D. Ariely, *Predictably Irrational: The Hidden Forces That Shape Our Decisions* (New York: Harper
Collins, 2008); B. Schwartz, *The Paradox of Choice: Why More Is Less* (New York: Harper Perennial, 2004).

[10]M. Benz, "The Relevance of Procedural Utility for Economics." In B. Frey and A. Stutzer,
eds., *Economics and Psychology: A Promising New Cross-Disciplinary Field*, pp. 199–228. Cambridge,
MA: MIT Press, 2007.

[11]G. Akerlof and R. Kranton. Economics and Identity. *Quarterly Journal of Economics* 115:717
(2000).

important than *having*. Considering that most academic economists could earn much more in the business sector than they do in academia, it's surprising how much emphasis they give to the utility from consumption while treating production as a source of disutility and largely ignoring self-identity. Socializing and interacting with family, friends, community, and religious groups also increases self-reported quality of life. When asked in an interview about what he had learned from a 72-year longitudinal study of a group of Harvard men (admittedly not a random sample of society), the lead researcher, George Vaillant, replied that "the only thing that really matters in life are your relationships to other people."[12]

Conversely, comparing yourself to others and yearning for money, possessions, image, and fame appears not only to reduce vitality and increase depression but also to increase physical symptoms such as aches and pains.[13] Similarly, earning less than one's neighbors or less than one's own aspirations makes people unhappy. While an increase in one's income correlates with greater happiness (at least temporarily, as discussed later), it also imposes a negative externality by making others less happy. It is interesting to note that the term *keeping up with the Joneses,* which means buying whatever your neighbors, the Joneses, have acquired, in order to maintain your status, has been transformed into the verb *Jonesing* by the drug culture—drug abusers always need more just to stay happy. While one's own income correlates positively (though only slightly) with happiness, the income of others (and hence per-capita GNP) has a negative correlation because the wealthier other people are, the less pleasure one gets from a given salary.[14]

Getting a raise certainly does make people feel good. However, abundant research has shown that we adapt to simply feeling good (e.g., to the happiness we get from a raise, from winning the lottery, or from buying a new big-screen TV), leading to a hedonic treadmill in which we always want more. People often return to a given "set point" (i.e., their typical level of happiness) after winning the lottery or suffering a major accident, leading some researchers to believe that the pursuit of happiness is fruitless.[15] However, positive psychologists have learned that while we may adapt to feeling good, doing good, "devoting resources to others, rather than indulging a materialist desire," leads to a lasting sensation of well-

[12]J. Wolfshenk, What Makes Us Happy? *The Atlantic.* Online: http://www.theatlantic.com/doc/200906/happiness.

[13]T. Kasser, *The High Price of Materialism,* Cambridge, MA: MIT Press, 2002.

[14]R. Layard, *Happiness: Lessons from a New Science,* New York: Penguin Press, 2005.

[15]P. Brickman and D. Campbell, "Hedonic Relativism and Planning the Good Society." In M. Appley, ed., *Adaptation Level Theory: A Symposium,* pp. 287–302. New York: Academic Press, 1971

being.[16] Other research shows that those who devote resources to others have higher self-esteem, better health, less stress, and more energy than those who do not.[17]

■ RATIONALITY

At its most extreme, economics assumes that individuals understand the full impacts of all their decisions, from now into the future, and make rational choices that maximize their utility. However, the real world is too complex and people too imperfect to make fully rational decisions.[18] In fact, we think that the absurdities of rigid assumptions of rational behavior and perfect information are so obvious that we needn't waste our time illustrating them.

It is worth asking whether people are rational when making simple decisions with adequate information and, if not, what are the implications for economic systems. We have already explained how increasing choice can lead to a reversal of preference orderings (e.g., given a choice of A and B, I choose A, but given the choice of A, B, and C, I choose B). While such contradictions of rationality are interesting and informative, Nobel laureates Tversky and Kahneman have conducted a number of studies with far more important implications. One study asked people to imagine that the United States has to prepare for an outbreak of an unusual Asian disease—let's say avian flu. If nothing is done, 600 people will die. Half the subjects were told to choose between Program A, which would save 200 of these lives, and Program B, which would have a 1/3 probability of saving all 600 lives and a 2/3 probability of saving no one. Of these subjects, 72% chose A. The other half of the subjects were told to choose between Program C, which would result in 400 deaths, and Program D, which would have a 1/3 probability of no one dying and a 2/3 probability of 600 deaths. Of these subjects, 78% chose program D. *Objectively,* Program A is identical to Program C, and B is identical to D. Furthermore, the expected outcome (which is the probability of an outcome multiplied by its value)

---

[16]D. Keltner, interviewed by D. DiSalvio, Forget Survival of the Fittest: It Is Kindness That Counts. *Scientific American.* Online: http://www.iterasi.net/openviewer.aspx?sqrlitid=jc7toxobhk simmar_seqqg.

[17]D. Wilson, *Evolution for Everyone: How Darwin's Theory Can Change the Way We Think About Our Lives,* New York: Delacorte Press, 2007.

[18]H. Simon, *Reason in Human Affairs,* Stanford, CA: Stanford University Press, 1983. As chairman of an admissions committee, one of us was reminded by a colleague from history that we must be careful to avoid unintended consequences in our decisions. Granted, but what made him say that? He explained, "Remember that what Hitler most wanted was to be an artist, but the Vienna Art Academy turned him down." Consequences are hard to predict!

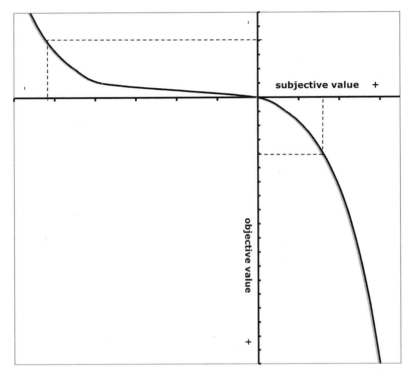

Figure 13.2 • A hypothetical value function. The x-axis depicts the actual out-come of an event, and the y-axis how people value it. Both losses and gains show diminishing marginal value, and people weigh losses more than they do equivalent gains.

of choices B and D is equal to the certain outcome in choices A and C. The significant difference in preferences was entirely due to the description of outcomes and our attitudes toward risk: the choice between A and B looked at gains (i.e., saving lives) and between C and D looked at losses (people dying). It turns out that *subjectively*, people have different ways of looking at gains and losses.

Using this and other studies, Tversky and Kahneman teased out a number of important results. Interestingly, it appears that most people experience diminishing marginal utility with respect to lives as well as money and material possessions (some of their studies used money rather than lives)—for example, their research found that people do not view 600 saved lives as being three times as good as 200 saved lives, and therefore view a sure bet of saving 200 lives as better than a 1/3 chance of saving 600. Furthermore, it turns out that people in general are more risk averse with respect to gains than with losses: given the choice of a $50 sure loss or a 50% chance of losing $100, most people take the gamble, but given the choice of a $50 sure gain or a 50% chance of gaining $100, we take the sure thing. This is shown in the hypothetical value function depicted in Figure 13.2. They also found that people weigh events with low probability more heavily than events with moderate or high probability, relative to their expected outcomes. One result of this is that problems can be framed so that people are likely to choose a *dominated alternative*—a clearly inferior choice, which is to say one that is worse than other options in some situations and better in none.[19]

Preferences are heavily influenced by how a choice is framed: people prefer a surgery that offers an 80% survival rate to one that offers a 20% mortality rate.[20] Preferences are heavily influenced by default choices. There is an enormous disparity across countries in the number of people who agree to donate their organs to others in the event of death, a decision people are typically asked to make when they get a driver's license. It turns out that disparity is explained almost entirely by whether checking a box opts you in to donating organs or opts you out. In European countries, where one has to opt in to donating organs, donation rates range from 4.25% to 27.5%, while in countries where one has to opt out, rates range from 86% to 100%.[21] Similarly, if people have to opt in to retirement savings, they are much less likely to save than if they have to opt out. Preferences are heavily influenced by whether we think about the benefits of policy first or the costs (we're more likely to favor a policy if we think about benefits first) and if we make decisions on our own or in a group (being in a group may lead us to give more weight to the future, for example).[22]

■ SELF-INTEREST

When choosing between different options that affect only their personal well-being, nonrational behavior can lead people to make the wrong choice. When choosing between options that benefit either the individual or society, both nonrational and selfish behavior can undermine social well-being. Most of us know from simple introspection that we are not purely self-interested, and we constantly hear of people making significant sacrifices for others. At the same time, however, evidence of purely selfish

---

[19]A. Tversky and D. Kahneman, The Framing of Decisions and the Psychology of Choice, *Science* 211:453–458 (1981). We refer you to this article for examples of dominated alternatives.

[20]C. Sunstein and R. Thaler, Libertarian Paternalism Is Not an Oxymoron, *University of Chicago Law Review* 70:1159–1202 (2003).

[21]E. Johnson and D. Goldstein, Medicine: Do Defaults Save Lives? *Science* 302:1338–1339 (2003).

[22]J. Gertner, Why Isn't the Brain Green? *New York Times Magazine*, April 16, 2009.

behavior abounds and explains the degradation and underinvestment in open-access resources and public goods worldwide. We review here the current state of knowledge on human behavior as it relates to self-interest, other-interest, competition, cooperation, and fairness.

■ Experimental Evidence

One of the simplest studies of human behavior with the most obvious results is the dictator game, in which one experimental subject is given a sum of money and the option to give as much as he wishes to an anonymous stranger. The rational self-interested person would obviously keep all the money. In experiments with college students, this is indeed the most common choice, but only about 20% of the subjects make it, while the rest give at least some of the money away. The mean offer is 20% of the pot.[23] Perhaps more interesting are studies done with three widely differing tribal groups in Africa and South America, one pastoral, one horticultural, and one foraging. In these experiments, using about the equivalent of a day's income, almost no one offered zero, and the mean offers ranged from 20% to 32%.[24] The only possible explanations for such behavior would appear to be that a significant percentage of humans care about fairness, community, or the well-being of others—that we are in fact social animals.

The ultimatum game is slightly more complex but even more revealing of human behavior. In this game, one player (the proposer) is given some money or other good and told to propose a split with another player (the decider), who typically remains anonymous. The decider then has the option of accepting the split, in which case both players keep their share, or rejecting it, in which case neither player gets anything. A rational and self-interested player would prefer some money over none, whatever the division. A rational self-interested proposer assuming that the decider is also rational should therefore make a minimal offer of, say, 1%. In studies with college students, however, most people proposed much more equal divisions, contradicting the assumptions of rational self-interest. Furthermore, deciders typically reject offers that they deem unfair (typically anything less than 30% in the United States), in effect sacrificing their own welfare to punish the proposer for selfish behavior. Such punishment presumably deters selfish behavior in the future and as a result has been called **altruistic punishment.** We'll explore its significance shortly.

The ultimatum game has also been played among different cultures

---

[23]C. Camerer, *Behavioral Game Theory,* Princeton, NJ: Princeton University Press, 2003.

[24]J. Henrich et al., "Economic Man" in Cross-Cultural Perspective: Behavioral Experiments in 15 Small-Scale Societies, *Behavioral and Brain Sciences* 28:795–855 (2005).

around the world. This revealed greater variation than found among college students but still negligible support for rational self-interest: among 15 different cultures, mean offers ranged from 26% to 58%, and modal offers from 15% to 50%.[25] The size of the stakes involved seems to have minimal effect on the percentage of the stake offered or on the percentage thresholds for rejection.[26] Cultural variations are important and appear to be closely tied to the nature of the economic system. For example, in cultures where cooperation is important (e.g., among whale hunters) offers were quite high, whereas in cultures where cooperation is less important (e.g., among relatively independent horticulturalists) offers were low, and even low offers were not rejected. Curiously, in cultures based on reciprocal gifting (i.e., gift giving is common but obliges the recipient to reciprocate at some time in the future), offers of over 50% were common and in many cases were rejected, even though both proposers and deciders were anonymous.[27]

A game with more obvious analogues in real life is known as the **prisoner's dilemma,** whose structure is nicely summarized on Wikipedia:

> *Two suspects are arrested by the police. The police have insufficient evidence for a conviction, and, having separated both prisoners, visit each of them to offer the same deal. If one testifies (defects) for the prosecution against the other and the other remains silent, the betrayer goes free and the silent accomplice receives the full 10-year sentence. If both remain silent, both prisoners are sentenced to only six months in jail for a minor charge. If each betrays the other, each receives a five-year sentence. Each prisoner must choose to betray the other or to remain silent. Each one is assured that the other would not know about the betrayal before the end of the investigation. How should the prisoners act?*

If the suspects are rational and self-interested, then they prefer less time in jail to more and do not care what happens to the other suspect. In this case, if A defects, B spends less time in jail by also defecting. If A is silent, B still spends less time in jail by defecting: no matter what suspect A does, suspect B is better off defecting, and vice versa. The dominant strategy therefore is for both players to defect, leading both to spend 5 years in jail, when through cooperation they could have gotten away with 6 months.

In real life, of course, people develop a reputation as someone who cooperates or someone who defects, and others will react accordingly. People will refuse to engage in prisoner's dilemma–type situations with

---

[25]Ibid.

[26]L. Cameron, Raising the Stakes in the Ultimatum Game: Experimental Evidence from Indonesia, *Economic Inquiry* 37(1): 47–59 (1999).

[27]Henrich et al., op. cit.

defectors, who will therefore lose the opportunity for gains through cooperation, while anyone will be happy to engage in such situations with cooperators. In other words, in a world in which people can and do cooperate, cooperation is a more rational strategy than defection.

A more realistic situation is one in which there are many people engaged in a prisoner's dilemma. Overuse of common-pool resources (rival resources in open-access regimes) and underprovision of public goods (nonrival, nonexcludable resources) are both good examples. In a fishery that is open to everyone who wants to fish (an open-access regime), cooperation involves protecting breeding stock and reducing harvests enough so that stocks and reproduction rates remain healthy and resilient. If all fishermen did this, the fishery could generate high output at low cost, as described in Chapters 10 and 12. If no fishermen cooperate, the fishery is likely to be severely depleted, driving up the costs of harvest, lowering profits, and even risking economic or biological extinction. However, if some players cooperate and others defect, the defectors get a bigger harvest than ever in the short run since they are able to take the cooperators' share as well as their own. The real-life result is the serious depletion of resources in open-access regimes, ranging from oceanic fisheries to the waste absorption capacity for CO_2 and other pollutants.

The provision of public goods is quite similar, as explained in Chapter 10, and similarly prone to **free-riding**. Restoring the wetlands surrounding New Orleans, for example, would reduce storm surges in the event of another hurricane. Assume the expected benefits of restoration outweigh the expected costs. If everyone cooperated to restore the wetlands, the city could be spared considerable damage in the event of another hurricane, and everyone comes out ahead. However, each individual can contribute only a small share to the restoration process and bears the full cost of this contribution. According to conventional theory, most people will defect— they will free-ride on the efforts of others, gaining almost as much benefit but incurring no costs. The result is that the wetlands will not be adequately restored. However, experimental games and innumerable real life studies show that this is not always the case, and these exceptions have much to teach us.

Two experimental games closely approximate the problems of common-pool and public-good resources. In the common-pool game, participants can withdraw any amount up to some fixed limit from a common pot. What remains in the pot then "grows" by some prespecified proportion, say 50%, and is redistributed equally to all, regardless of how much each person withdrew. In the public-good game, participants start with a fixed sum and are allowed to donate as much as they want to a fixed pool. This money is then doubled (or increased by some other prespecified amount) and redistributed equally to all, regardless of how much each person contributed. If people act in their rational self-interest, then they will withdraw as much as possible and contribute as little as possible in the two games, even though minimum withdrawal and maximum contribution generate the greatest wealth for the group as a whole.

Once again, experimental evidence fails to support conventional economics' assumption that people act only out of pure self-interest. Most people in the voluntary contribution game contribute something to the common pool. University students tend to contribute 40%–60% of the total amount they are given, on average, with one mode at zero contribution and a typically smaller one at full contribution. However, in repeated games either among the same group or with different group members (i.e., each person plays the game multiple times but with different people), contribution rates fall. It appears that those who initially cooperate engage in a tit for tat strategy: the most generous individuals decrease their contributions to the mean contribution, which further drives down the mean.[28]

Is there a way to avoid this suboptimal outcome? In one variation of the game, participants learn after each round who contributed and how much, and they are allowed to punish those who did not contribute. Punishment is costly; for example, the punisher may have to give up 1/3 unit of reward to punish defectors by 1 unit. Yet when punishment is allowed, the rates of cooperation go up with repeated rounds, not down. This is another example of altruistic punishment and helps explain the significance of the term: individuals sacrifice their own welfare to make defection a losing strategy, encouraging cooperation even from people who are purely selfish, and even when they make up a significant percentage of the group. In other words, altruistic punishment can make cooperation the dominant strategy in prisoner's dilemma-type situations, even for selfish individuals. One could argue that in a repeated public-good game, the punisher is ultimately rewarded by increased cooperation in future rounds, but in the ultimatum game described earlier players are not rewarded for altruistic behavior. Clearly, altruism plays a role in both. In fact, it's interesting to note that **neuroeconomic** studies, which measure brain activity, find that the same areas of the brain are stimulated by altruistic punishment as are stimulated by receiving money.[29]

Altruistic punishment is not the only way to achieve cooperation, however. If participants in experimental games are allowed to talk about their

---

[28]"Tit for tat" simply means acting toward your partners as they acted toward you in the previous round. In a famous experiment, tit for tat was found to be the most successful overall strategy in repeated prisoner dilemma games (R. Axelrod, *The Evolution of Cooperation*, New York: Basic Books, 1984).

[29]C. Camerer, M. Bhatt, and M. Hsu, "Neuroeconomics: Illustrated by the Study of Ambiguity-Aversion." In B. S. Frey and A. Stutzer, eds., *Economics and Psychology: A Promising New Cross-Disciplinary Field*, Cambridge, MA: MIT Press, 2007.

strategies ahead of time, they are much more likely to cooperate. This is true even for "cheap talk," which means that the decisions participants ultimately make are not revealed to others, and there is no way to create binding contracts.[30]

Nor are such results confined to the laboratory. A number of studies have shown that real-life behavior corresponds closely with what is seen in laboratories.[31] For example, Elinor Ostrom and her colleagues have done extensive studies of the management of common-pool resources in real life. They have found that while in many cases such resources are indeed overexploited, in numerous other circumstances institutions emerge that lead to sustainable, just, and efficient management. One key to making such institutions work is that community members own the resources in common, while non-community members are not allowed to use them; they are common goods when viewed from within the community but private goods from the perspective of other communities. It also helps when community members have broad input into management strategies, can effectively monitor resource use, sanction those who fail to respect community rules, and have access to mechanisms for cheaply and easily mediating any conflicts.[32]

The Role of Money and Incentives in Cooperative Behavior

Much of our behavior seems to be guided by intrinsic motivations rather than extrinsic ones. Many people volunteer free time, refrain from stealing even when they know they could not possibly be caught, and help others when they know that there is no chance of reciprocation in the future. Not everyone behaves this way, of course, and economists conventionally argue that we can create **extrinsic incentives** to promote desirable behavior. Unfortunately, there is increasing evidence that, for those guided by intrinsic motivations, extrinsic incentives may actually "crowd out" such motivations. A much-cited example comes from an experiment conducted in Israel, in which a day care, suffering from too many parents arriving late to pick up their children, began charging fines penalizing tardy parents. Rather than decreasing, in the undesired behavior actually increased. Apparently, parents who were unwilling to arrive late when they felt they had a social obligation to be on time had no problem doing so when it became a market issue (though presumably if the fines were

---

[30]E. Ostrom, J. Walker, and R. Gardner, Covenants With and Without a Sword: Self-Governance Is Possible, *American Political Science Review* 86:404–416 (1992).

[31]S. Gachter, "Conditional Cooperation: Behavioral Regularities from the Lab and the Field and Their Policy Implications." In B. S. Frey and A. Stutzer, eds., *Economics and Psychology: A Promising New Cross-Disciplinary Field*, Cambridge, MA: MIT Press, 2007.

[32]E. Ostrom, *Governing the Commons: The Evolution of Institutions for Collective Action*, Cambridge: Cambridge University Press, 1990.

high enough, they would ultimately deter the behavior). Another study found that students performed worse on an IQ test when paid for each correct answer relative to a control group that was not paid. While raising the payment for each correct answer did increase scores, it did not raise them above the no-payment control.[33]

Equally interesting from a policy perspective, other studies have examined the impact of monetary cues on social behavior. In them, subjects were exposed to monetary cues ranging from posters of money, screen savers displaying money, linguistic puzzles referring to money, or simply being given play money, while others were exposed to nonmonetary cues. Social behaviors included helping pick up spilled pencils, helping someone to understand directions, asking for help in solving a problem, placing a chair near an unacquainted participant's chair, choosing individual or social leisure activities, and so on. In each case, participants exposed to monetary cues subsequently proved less cooperative and less social, helping to pick up fewer pencils, giving less help with directions, working longer at a problem before asking for help, placing their chairs farther away from other participants, and choosing individual leisure activities over those with friends.[34]

Such studies raise serious questions about the role of market mechanisms in addressing so-called market failures. For example, if we want to reduce pollution, effluent taxes might not only prove less effective than promoting community ties but could also lead to increased pollution where community ties are already strong.

Box 13-3	**THE CO-EVOLUTION OF ECONOMICS AND EVOLUTIONARY BIOLOGY**

It's interesting to note that the academic disciplines of economics and evolutionary biology seem to have evolved together. There is considerable evidence that Darwin was influence by Adam Smith; in the words of Stephen Jay Gould, "Darwin grafted Adam Smith upon nature to establish his theory of natural selection."[a] While Darwin clearly recognized the survival advantages of cooperation, Herbert Spencer's notion of natural selection as "survival of the fittest"—clearly a situation of competition—seemed to have more influence on economists for many years. While the Great Depression illustrated the advantages of cooperation in economic systems, it was really not until the 1960s that John Maynard Keynes' call for a government role in the economy (a form of cooperation) became ac-

[33]U. Gneezy and A. Rustichini, Pay Enough or Don't Pay at All, *The Quarterly Journal of Economics* 115:791–810 (2000).

[34]K. Vohs, N. Mead, and M. Goode, The Psychological Consequences of Money, *Science* 314: 1154–1156 (2006).

cepted wisdom among economists. At the same time, the evolution of altruism through group selection was a hot topic in evolutionary biology. The stagflation of the 1970s, the writings of Milton Friedman and other "Chicago boys," and the political rise of Margaret Thatcher in England and conservative ideology in the United States led to a resurgence in the belief in unregulated competition in both economic theory and practice. At the same time, group selection and the notion of true altruism was being rejected in evolutionary biology. Richard Dawkins, for example, popularized the notion of the selfish gene and claimed that apparent altruism was merely the result of genes maximizing their own fitness in a purely selfish way, just as rational self-interest in a market setting generates benevolent outcomes.[b] In the words of another evolutionary biologist, "The economy of nature is competitive from beginning to end. . . . Where it is in his own interest, every organism may reasonably be expected to aid his fellows. . . . Yet given a full chance to act in his own interest, nothing but expediency will restrain him from brutalizing, from maiming, from murdering—his brother, his mate, his parent, or his child. Scratch an 'altruist' and watch a 'hypocrite' bleed."[c] Economics and evolutionary biology even shared the same tools, using game theory constructs such as the prisoner's dilemma to show the difficulties with the evolution of altruism. In the 1980s, behavioral economists began to challenge some of the assumptions about rational self-interest in humans, and evolutionary biologists began to reassert the role of group selection in the evolution of altruism. We now see a growing emphasis in both economics and evolutionary biology on cooperation and altruism, although the selfishness model remains dominant in economics.

[a]S. Gould, Ever Since Darwin: Reflections in Natural History. New York: Norton, 1977, p. 100.

[b]R. Dawkins, The Selfish Gene, 2nd ed., New York: Oxford University Press, 1990.

[c]M. Ghiselin, The Economy of Nature and the Evolution of Sex, Berkeley: University of California Press, 1974, p. 274, cited in E. Sober and D. Wilson, Unto Others: The Evolution and Psychology of Unselfish Behavior, Cambridge, MA: Harvard University Press, 1998.

■ THE SPECTRUM OF HUMAN BEHAVIOR

Prosocial behavior is behavior motivated by the desire to help someone else, without concern for private gains.

Perhaps the most important insight from research on human behavior is that humans are highly heterogeneous. Studies from behavioral economics suggest that about 20%–30% of people are purely selfish by nature, like *H. economicus;* about 50% are **conditional cooperators** (*H. reciprocans*); and about 20%–30% are very prosocial (*H. communicus*).[35] A rigorous study in-

[35]S. Meier, "A Survey of Economic Theories and Field Evidence on Pro-Social Behavior." In B. Frey and A. Stutzer, eds., *Economics and Psychology: A Promising New Cross-Disciplinary Field*, Cambridge, MA: MIT Press, 2007.

volving thousands of participants found that the actual distribution of **prosocial behavior** in a typical population approximates a normal distribution, with tails of extremely selfish and extremely selfless on either end.[36]

One interesting question is the extent to which prosocial behavior is the result of nature or nurture. Convincing studies comparing monozygotic (identical) and dizygotic (fraternal) twins suggest that only 10%–20% of prosocial behavior is genetic.[37] While this suggests that human behavior is malleable, it is also true that there are powerful biophysical forces influencing it. Researchers have found that administering aerosolized oxytocin, an important neurotransmitter, can increase the level of cooperation shown in trust games.[38] Cooperation can also increase oxytocin levels, thus leading to further cooperation. However, we do not recommend putting aerosolized oxytocin in our air conditioners. Fortunately, there appear to be more appropriate ways to stimulate cooperative behavior that are a lot more relevant to policy.

BOX 13-4 OXYTOCIN, TRUST, AND COOPERATION

Oxytocin is a neurotransmitter and hormone found in species ranging from fish to humans, with a number of different functions. For example, oxytocin induces labor contractions in pregnant mammals and is induced by sexual stimulation. More relevant to the topic at hand, however, oxytocin is involved in pair-bonding between mother and child, sexual partners, and even friends and community. Oxytocin induces labor, then reinforces the bond between mother and child when the baby is born. If a ewe is separated from its lamb for more than six hours after birth, the ewe may fail to nurture the lamb. However, shepherds discovered (don't ask us for details on how!) that if the ewe were subsequently sexually stimulated, she would then bond with the lamb. Oxytocin is also induced by breastfeeding.[a]

As it turns out, intentional signals of trust from a stranger will also increase oxytocin levels, and high oxytocin levels correlate with trustworthy behavior.[b]

---

[a]N. Angier, *A Potent Peptide Prompts an Urge to Cuddle*, New York Times, *January 22, 1991*; P. Zak, *The Neurobiology of Trust*, Scientific American, *June, 2008:62–67*.

[b]P. Zak and A. Fakhar, *Neuroactive Hormones and Interpersonal Trust: International Evidence*, Economics & Human Biology *4:412–429 (2006)*.

---

[36]D. Wilson and M. Csikszentmihalyi, "Health and the Ecology of Altruism." In S. Post, ed., *The Science of Altruism and Health.* Oxford: Oxford University Press, 2006.

[37]D. Cesarini, C. Dawes, J. Fowler, M. Johannesson, P. Lichtenstein, and B. Wallace, Heritability of Cooperative Behavior in the Trust Game. *Proceedings of the National Academy of Sciences* 105:3721–3726 (2008).

[38]M. Kosfeld et al., Oxytocin Increases Trust in Humans. *Nature* 435:673–676 (2005).

This suggests that cooperation has evolutionary origins, and indeed numerous studies from the field of evolutionary biology support this notion. For example, if you throw a bunch of *Pseudomonas fluorescens* bacteria in a beaker, mutations emerge that exhibit cooperative behavior (see Box 13.5). Such cooperative behavior is also evident in higher-level organisms and lends the species exhibiting it a selective advantage (see Box 13.6).

Among humans, evolution takes place at the cultural level as well, which is far more relevant to policy. Human culture has a profound impact on human behavior, as can be surmised from the numerous experiments and studies described above, and cultures are constantly evolving. Different cultures evolve different economic institutions, and when those institutions reinforce adaptive behavior, those cultures are more likely to persist. If a culture consists of independent family groups with little social or economic interaction and few gains from cooperation, then selfish behavior may be quite adaptive. If a culture consists of larger social units and an economic system that enjoys gains from cooperation, such as a whale hunting society, then cooperative behavior may be most appropriate.

Box 13-5	THE EVOLUTION OF COOPERATION

Throw a bunch of *Pseudomonas fluorescens* bacteria in a beaker, and they will rapidly reproduce until they become starved for oxygen. At this point, the survival advantage shifts to a mutant type known as the "wrinkly spreader," which can create a film that binds them together into a floating colony with access to oxygen from above and nutrients from below. Cooperation allows the group to thrive. However, within this cooperative colony there may be some defectors; they produce none of the sustaining film, but instead free-ride on that produced by others. With the energy they save by not producing the film, they are able to have more offspring than the cooperative *Pseudomonas*. Competitive individuals (i.e., defectors) within the group outcompete cooperative ones. However, if there are too many defectors, the colony can no longer stay afloat and plunges to the depths of the beaker, losing its relative fitness. Colonies with fewer defectors will continue to thrive and leave more descendants than others.[a] What we see is two distinct types of evolutionary pressure, at the individual and group level. The basic rule is that "Selfishness beats altruism within single groups. Altruistic groups beat selfish groups."[b]

[a] D. Wilson, Evolution for Everyone: How Darwin's Theory Can Change the Way We Think About Our Lives, *New York: Delacorte Press, 2007.*

[b] D. Wilson and E. Wilson, *Rethinking the Theoretical Foundations of Sociobiology,*

Quarterly Review of Biology *82:327–348 (2007), esp. p. 345.*

BOX 13-6   COOPERATION IN OTHER SPECIES

Numerous other species also cooperate and punish defectors, presumably conferring a survival advantage. If a rhesus monkey finds a tree laden with fruit, it will call out to the rest of the tribe to share in the harvest, thus reducing its own share, a seeming act of pure altruism. However, if the monkey fails to call out and is discovered by others in the tribe gorging alone, it is subject to a severe beating—punishment for defecting.[a] Tamarin monkeys are reciprocal cooperators, but if one member of a cooperating pair defects, the other member will generally begin cooperating only after two unexpected acts of cooperation from the defector—two tits for a tat.[b]

[a]N. Angier, *Taxing, a Ritual to Save the Species,* New York Times, *April 14, 2009.*

[b]M. Chen and M. Hauser, *Modeling Reciprocation and Cooperation in Primates: Evidence for a Punishing Strategy,* Journal of Theoretical Biology *(2005):5–12.*

Historically, cultures with economic systems that require cooperation have developed low-cost mechanisms for punishing defectors, which can increase the returns to cooperation and induce even self-interested indi-

---

[39]E. Fehr and U. Fischbacher, Why Social Preferences Matter: The Impact of Non-selfish Motives on Competition, Cooperation and Incentives, *Economic Journal* 112:C1–C33 (2002).

viduals to cooperate.[39] In small hunter-gatherer societies, institutions for inducing cooperation may be as simple as the widespread practice of ostracizing those who refuse to share food or who simply eat alone, while in more complex societies, inducements might range from imprisonment and fines to restrictions on marriage and childbearing.[40]

Free market economies obviously stress competition and self-interest. Rather than ostracizing those who take the most for themselves, modern society tends to idolize them. For rival and excludable resources with minimal externalities in production and consumption and mutual gains to voluntary exchange, pursuit of self-interest may lead to adequate outcomes. However, if the most important resources are common pool or public in nature, then sustainable, just, and efficient allocation may require cooperation.

■ A NEW MODEL OF HUMAN BEHAVIOR

Conventional economics assumes that people are always rational, competitive and self-interested. The alternative assumption of a heterogeneous population that includes *H. economicus, H. reciprocans,* and *H. communicus* has much greater explanatory power. With some types of resources and some types of institutions, a heterogeneous population will sometimes act like everyone is self-interested and at other times like everyone is prosocial. It explains empirical results from the dictator, ultimatum, public-good, and common-pool resource games, as well as the outcomes from real-life institutions that promote cooperation and others that promote competition.[41]

THINK ABOUT IT!

The modern scientific method is based on the notion of falsifying hypotheses. One can never conclusively prove something is true, only that it has proven true so far. However, it is possible to prove a theory or assumption false. Once a theory has been proven false, true scientists then seek a more robust model that better explains all available data. Do economists utilize the scientific method? What other basic economic assumptions or theories need testing?

The results we have presented clearly falsify the neoclassical assumption that people always act in their rational self-interest.

This is an important insight, because the nature of economic problems

---

[40]E. Sober and D. Wilson, *Unto Others: The Evolution and Psychology of Unselfish Behavior.* Cambridge, MA: Harvard University Press, 2002.

[41]E. Fehrand and K. Schmidt, A Theory of Fairness, Competition, and Cooperation, *Quarterly Journal of Economics* 114:817–868 (1999).

is changing, rendering conventional economic theories less and less adequate to explain and guide the full range of economic activity. As we have moved from an empty planet to a full planet, natural capital has become scarcer than manmade capital. As natural capital has dwindled, knowledge, a purely nonrival resource, plays an increasingly critical role in economic production and will be needed to address the most serious problems society now faces.

Not only have the scarce resources changed in recent decades, but so have their physical characteristics. In times past, the scarcest resources were rival and excludable, but now the resources most essential to our sustainable well-being are neither. To understand this, let's take a look at two of the most difficult economic problems we now face.

Peak Oil. Energy plays a central role in economic production, usually in the form of fossil fuels, which are quintessential market goods, both rival and excludable. Competition for scarce fossil fuel supplies is inevitable. Indeed, as we point out in Chapter 5, the market economy emerged at the very same time as the fossil fuel economy, and if we ignore externalities (which we too often do), the two seem tailor made for each other. But many analysts conclude that we have passed the global peak in fossil fuel production and must find alternatives.

Box 13-7	COOPERATION OR COMPETITION? THE CHICKEN AND THE EGG

Chicken breeders did an interesting experiment that sheds some light on the cooperation versus competition question. The goal of the chicken breeders was to increase egg production in chickens. They used two approaches, each beginning with nine cages full of hens. In the first approach, the breeders selected the most productive hen from each of the nine cages, then used these hens to produce enough chickens to fill another nine cages. In the second approach, the breeders selected the cage that produced the most eggs, and used these hens to produce enough chickens to fill another nine cages. They continued the experiment for six generations.

Which approach resulted in the greatest increase in egg production? As it turned out, the experiment was truncated after six generations because the treatment using the most productive hen from each cage could no longer produce enough hens to fill nine more cages. Many of the individual hens were the most productive because they bullied the other hens into underproduction. The breeders were selecting for the hen version of psychopathic bullies. The cooperative hens, in the meantime, had doubled egg production.[a]

ᵃD. Wilson, Evolution for Everyone: How Darwin's Theory Can Change the Way We Think About Our Lives, *New York: Delacorte Press, 2007.*

Ultimately, the only sustainable and widely available replacement is solar energy. Though photons are technically rival, no matter how many photons we capture in one country (with the possible exception of highly advanced space-based technologies) it will have no impact on the number of photons striking another. The current constraint on capturing solar energy is information—we need to develop more efficient technologies that do not rely on exceedingly rare elements. Information of course is purely nonrival, or even additive,[42] in that it improves through use. While we may still compete for rare elements required by solar technology, more information may help us overcome these constraints as well.

We have the option of providing information cooperatively and making it a public good or providing it competitively and making it a market good. If private firms compete to develop information, it may take longer to develop (as explained in Chapter 10), and price rationing will create artificial scarcity (see Box 9.2). If information provision is cooperative, we face the problem of public-good provision. One solution in this case is to make information a club good—institutions (e.g., countries or corporations) that contribute a fair amount to developing alternative energy technologies (members of the club) will be allowed to use them freely, while those who fail to contribute will be charged a fair contribution to costs of development or denied access as nonmembers. If payments are then dedicated to further technological improvements, the result is no different from cooperation. This approach solves the problem of free-riding, though failure to allow free use remains inefficient. However, the greater the number of institutions contributing to produce a given amount of information, the lower the cost per institution, and the more worthwhile it becomes to join the club. Unlike a country club, an information club can never become congested.

Climate Change. Global climate change can be defined as underprovision of the public good of climate stability or excessive use of the common-pool resource of waste absorption capacity. At least part of the solution will undoubtedly involve the new carbon-neutral technologies needed to solve the peak oil problem. From the perspective of climate change, though, there is no free-rider problem in the deployment of these technologies. In the absence of climate change, one institution has nothing to gain from others using alternatives to fossil fuels. In the presence of climate change, such technologies become additive: the more people use them, even with-

---

[42]We consider using the word *antirival* to emphasize the contrast with rival goods. However, rival goods are also known as subtractive.

out paying, the better off the inventor becomes, since she, too, benefits from a more stable climate. The countries best able to fund research into carbon-neutral energy sources are precisely those countries that have made the most significant contributions to climate change. This means that cooperative provision of such technologies by those countries would promote ecological sustainability, just distribution, and allocative efficiency. Private, competitive provision would undermine all of these goals.

Climate is one of the most important services provided by Earth's ecosystems. In Chapter 10 we explained that most ecosystem services are public goods or common-pool resources, both of which seem to require cooperative provision. The good news is that sound science amply confirms our most basic common sense that cooperative, other-regarding behavior is widespread. Economic analysis of the most serious problems currently faced by society reveals that cooperation will be needed to solve them. It would therefore be extremely foolish to blindly follow an economic model that fosters competition and claims that true cooperation is well nigh impossible. This is not to say that competitive market forces have no role in our economy but rather that we cannot rely on the market for the sustainable, just, and efficient allocation of all resources.

It would be just as foolish to argue for an economy based purely on cooperation as it is to argue for one based purely on competition, however. Allocative mechanisms must be tailored both to specific desirable

BIG IDEAS to remember

- *Homo economicus, H. reciprocans, H. communicus*
- Behavioral economics
- Neuroeconomics
- Prosocial behavior
- Easterlin paradox
- Choice under conflict
- Extrinsic and intrinsic incentives
- Dictator game
- Ultimatum game
- Prisoner's dilemma
- Public-good and common-pool resource game
- Altruistic punishment
- Conditional cooperation

Conclusions to Part III

In Chapters 8 and 9, we provided a brief introduction to market micro-economic theory, showing how spontaneous order can arise naturally from the rational, uncoordinated actions of millions of individuals. We then described how this spontaneous order can lead producers to efficiently allocate raw materials, labor, and capital toward their most profitable uses while allowing consumers to efficiently allocate their resources toward consumption in a manner that maximizes well-being. However, in Chapter 10 we extended this analysis to show how the occurrence of these optimal outcomes depends entirely on the specific characteristics of the goods and services we wish to allocate. In Chapters 11 and 12, we examined the natural resources upon which all economic production depends to see whether they met the criteria for market allocation. Unfortunately, none of the goods and services provided by nature had all of the characteristics needed for efficient market allocation. Finally in Chapter 13 we compare the conventional economic assumptions concerning human behavior to the more nuanced understanding emerging from the fields of behavioral economics, neuroscience, and evolution. It turns out that humans are not the insatiable, egotistic, and perfectly rational automatons of neoclassical theory but rather are capable of a broad range of behaviors which are influenced by economic institutions and culture in general. Specifically, with the proper institutions in place, we are capable of the cooperative behavior required to address the problems with market allocation described in Chapters 10–12. In subsequent chapters, we will build on this analysis to propose policies and institutions better suited to allocating the economic means provided by nature toward the desirable end of a high quality of life for this and future generations. Before we turn to this task, however, we will first examine the macroeconomy, Part IV of our text.

PART IV

Macroeconomics

Macroeconomic Concepts: GNP and Welfare

Microeconomics focuses on how markets function. It is useful for analyzing markets, with the aim of ensuring that they operate efficiently. Beyond that, however, it steers clear of policy recommendations. In essence, it assumes that the best policy is to let the market do its thing without interference. Macroeconomics looks at the economy as a whole, at the national or global level. In contrast to microeconomics, macroeconomics more often recognizes the importance of policy interventions, especially fiscal policy (government spending and taxation) and monetary policy (money supply and interest rates).[1] These policy interventions are important. However, the making of policy implies a goal. The traditional goal for macroeconomic policy is stable market-driven economic growth without limit, and to a lesser extent full employment. But unlimited economic growth is impossible. Many of the scarcest resources are nonmarket goods and services, and many of the most serious problems we now face extend beyond the borders of the nation-state.

In ecological economics, optimal scale replaces growth as a goal, followed by fair distribution. Traditional macroeconomics generally leaves allocation to market forces at the microeconomic level. Ecological economics more often recognizes that markets are inadequate for allocating

---

[1]There are many schools of macroeconomics, some of which do not call for policy interventions. For example, new classical macroeconomics, also known as rational expectations theory, argues that policy interventions are ineffective. The monetarists argue that policy interventions can be counterproductive. Both schools are quite conservative, favoring small, weak government. In practice, however, policy makers do use macroeconomic policy in efforts to attain policy objectives.

many scarce resources, and policy interventions are necessary to supply adequate quantities of these nonmarket goods. These different goals of ecological economics will favor different uses of traditional policies and also suggest an array of alternative policy interventions.

In Chapters 14–17 we provide a brief introduction to some of the concepts, issues, and policy tools of mainstream macroeconomics and apply them to the policy goals of ecological economics. Remembering the circular flow diagram, we recall that macroeconomics deals with the aggregate flows of national product and income (the real sector). It also deals with the aggregate money supply and with interest rates (the monetary sector). Following a discussion of the relationship of macroeconomics to microeconomics, this chapter will first look at the aggregate measure of the real sector, gross national product. Then we will have a look at money and the aggregate monetary sector, followed by a consideration of welfare indices other than GNP.

In Chapter 15, we will look at the medium by which wealth is measured in conventional economics: money. We then focus on distribution in Chapter 16. In Chapter 17, we will present the basic macroeconomic model for combining the two sectors into a simple general equilibrium model of the economy. This model (the IS-LM model) shows how the behavior of savers and investors in the real sector interacts with the behavior of the monetary authority (usually a national central bank, such as the Federal Reserve Bank in the U.S.) and the money-holding public to determine the interest rate and the level of national income and employment. We will show how the goals of ecological economics lead to different policy recommendations than those supported by the mainstream. We will then discuss the possibilities for extending the IS-LM model to incorporate ecological constraints.

■ A Troubled Marriage

Microeconomics developed historically prior to macroeconomics. Indeed, if we accept the behavior of the decision-making units—firms and households—and how competitive markets work, then we will come to accept Adam Smith's "invisible hand," that individuals seeking only their own benefit will automatically serve the common good. The microeconomic search for maximum private benefit will automatically result in the large-scale consequence of the greatest public welfare for all, or so it was thought. There was no need for special consideration of the macroeconomic picture, since the invisible hand would guarantee that if the microeconomics are right, then the macroeconomic picture will be right. Monopoly can ruin this nice result (as we showed in Chapter 8), so markets must be kept competitive, but that is about the only collective action needed.[2]

In Chapter 2, we met Say's Law (supply creates its own demand), which asserts that production always generates sufficient aggregate income to purchase aggregate production. Therefore, there can be no general glut of all products—at worst an imbalance in the mix of products, a misallocation, too much of something, too little of something else. That misallocation will soon be corrected by relative price changes in competitive markets. The same applies to the labor market—if there is unemployment (a surplus of labor), it simply means real wages are too high, not that there is a problem on a larger scale. If unemployment persists, you just need to let wages fall some more.

This view lasted well into the Great Depression of the 1930s and still has its adherents today. But under the leadership of John Maynard Keynes, economists began to think that prolonged unemployment, though theoretically impossible, was sufficiently real to warrant rethinking the theory. This rethinking led to the discovery of the leakages and injections from the circular flow and the problem of making sure that total injections equal total leakages. We considered this in Chapter 2.

In addition, economists remembered the **fallacy of composition**, the false belief that whatever is true for the part must be true for the whole, or vice versa. For example, one spectator in a football stadium can get a better view by standing up. But all spectators cannot. If all stand on tiptoe, then no one has a better view than when everyone was comfortably seated. Similarly, one country can have a surplus or deficit in its balance of payments. But for the world as a whole, neither surplus nor deficit is possible because the sum of all exports must identically equal the sum of all imports. One worker may gain employment by being willing to work for a lower wage, but all workers probably could not, because lower wages for everyone means less income for the majority of the people, which means less spending on goods and services and less demand for labor even at the lower wage. Reduced spending leads to reduced investment, which further lowers aggregate demand. In addition, any individual can easily convert his money holdings into real assets, but the community as a whole cannot, because when everyone tries to exchange money for real assets, someone has to end up holding the money.

---

[2]Many economists even question the need to prevent monopolies. For example, Alan Greenspan, the former chair of the U.S. Federal Reserve Bank, has argued that only government-protected monopolies are harmful. When a firm in a free market develops a monopoly position in an industry, it is a just reward for its efforts and only promotes social welfare. A. Greenspan. "Antitrust" in Ayn Rand, *Capitalism: The Unknown Ideal.*

There were, in sum, ample reasons to begin to develop a "macroeconomics" to deal with aggregate phenomena, especially unemployment and inflation. But contrary to what one might expect, the new macroeconomics did not build on the foundations of microeconomics. The macroeconomy is the aggregate of all the micro units, but macroeconomics is not just microeconomics aggregated. If it were, then we would be back to the invisible hand and the conclusion that macroeconomics was not necessary. The entire economy described in microeconomic terms is the general equilibrium model. In it, all supply-and-demand relations in all markets are presented as one great interdependent system of simultaneous equations—say, a million equations in a million unknowns. This system of equations is solved by the market, a giant social computer that works by trial-and-error iteration. Economists have devoted much effort to counting equations and unknowns and making sure they were equal so that the general system, at least theoretically, could be solved. While the general equilibrium model is enlightening and conceptually satisfying, it is not very helpful from a policy perspective simply to know that everything depends on everything else. Policy needs a few leverage points at which to influence the gross behavior of the big system in its most important aspects. That is what macroeconomics has sought: simple models of the economy in terms of key aggregate variables, such as the money supply, aggregate price level, the interest rate, aggregate consumption and investment, exports, and imports. And, of course, the biggest goal and leverage point of all: the rate of growth of GNP.

Ecological economics challenges today's standard emphasis on growth. Growth, yesterday's panacea, is rapidly becoming today's pandemic. Growth was a panacea because it was thought to be the solution to the macroeconomic problems of overpopulation, inequitable distribution, and involuntary unemployment. Microeconomists do not have much to say about growth, although not many would oppose it. Microeconomics is dominated by the concept of optimum and its associated "when to stop rule." As we argued in Chapter 2, if the macroeconomy is a part rather than the whole, then the logic of microeconomic optimizing applies, and at some point people trained in microeconomics will have to ask the macroeconomist, What is the optimal scale beyond which this economic subsystem should not grow? And when growth becomes uneconomic, as it must once we are at the optimum, then how are we going to deal with overpopulation, inequitable distribution, and involuntary unemployment?

It is the job of ecological economists to think about that: What happens after we reach the optimal scale, and how do we return there if we accidentally surpass it?

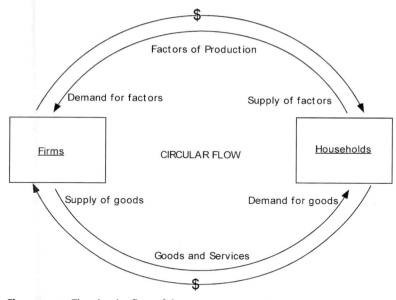

Figure 14.1 • The circular flow of the economy.

The whole is something more than the sum of its parts. In recognizing this, ecological economics bridges micro- and macroeconomics, though the exact relation between the two remains a bit mysterious. For our purposes, the relation between macro and micro is that shown in the circular flow diagram, Figure 2.4, repeated here as Figure 14.1.

The firms and households are our focus of attention in microeconomics. The firm as producing unit decides a supply plan for goods and a demand plan for factors. The household as consuming unit decides a demand plan for goods and a supply plan for factors. Microeconomics deals with these supply-and-demand decisions and their interactions in markets to determine the price and quantity of goods and factors exchanged in the markets. Because of its focus on prices, microeconomics is often called "price theory." Macroeconomics deals with the total volume of aggregate goods and services flowing through the goods market (national product) and the total volume of factors flowing through the factors market (national income). Because of its focus on aggregate income, macroeconomics is sometimes called "income theory."

Although we will speak of production and consumption because these terms are well established, it is important to remember (from earlier chapters) that in a *physical* sense, there is neither production nor consumption—only transformation. Raw materials are transformed into useful things (and waste) by "production." Useful things are transformed into waste by "consumption." What we are producing and consuming are "utilities"—useful temporary arrangements of matter and energy that serve our purposes. The throughput remains fundamental in both micro- and macroeconomics, even though it is not explicit in the accounts of firms and households or in the aggregate accounts of nations. And the throughput is governed by the First and Second Laws of Thermodynamics, not by circular flow accounting conventions.

■ GROSS NATIONAL PRODUCT

Because economic growth is the paramount goal of nations, it is important to know just how it is measured. Growth in *what*, exactly? Economic growth is measured as growth in gross national product (or GDP, gross domestic product).[3]

Gross national product (GNP) is the market value of final goods and services purchased by households, by government, and by foreigners (net of what we purchase from them) in the current year.

As previously discussed in terms of the circular flow diagram, we have two measures of the aggregate circular flow that give the same number: national product and national income. Sometimes they are called national product at consumer goods prices (lower loop in Figure 14.1) and national product at factor prices (upper loop in Figure 14.1), or national income. Let's focus first on the lower loop, national product at consumer goods prices.

In this measure, **gross national product (GNP)** is the market value of final goods and services purchased by households, by government, and by foreigners (net of what we purchase from them) in the current year. With a few exceptions, anything not purchased this year is not counted.[4] Household production for the household itself is not sold and thus not counted; cooking, cleaning, childcare, and so on are omitted unless done by a paid domestic helper. Intermediate transactions between firms are not counted. Only the sale of the final product to the household is counted. The wheat sold by the farmer to the miller is not counted, the flour sold by the miller to the baker is not counted; only the bread sold by the baker to the household for final consumption is counted. The value of the bread is the sum of the values added by the farmer, by the miller, and by the baker. Values added to what? To the basic natural resource: the wheat seed, the soil, the rain, the sunlight, and so on. The basic natural resources in most cases are considered to be free. Therefore, GNP is the sum of value added. It does not include any attribution of value to that to

---

[3]The difference, not significant for our purposes, is that GNP counts production by all U.S. citizens whether at home or abroad. GDP counts all production within the geographic borders of the U.S., whether by citizens or by foreigners.

[4]E.g., annual rent is imputed to measure the current service of owner-occupied houses. The owner is thought of as renting his house from himself in the current year. Yet the owners of automobiles are not thought of as renting their cars to themselves.

which the value was added. What is it that adds value to free natural resources? The transforming services of labor and capital funds.

Note that these accounting conventions are consistent with the neoclassical production function discussed in Chapter 9—namely, that production is a function of labor and capital only.[5] The exchange of existing assets is not counted because it is not current-year production. The value of a used car bought this year is not counted because it is a transfer of an existing asset. But the commission of the used car salesman will be counted as a service rendered this year. And of course the total value of a new car will be counted this year. The same holds for trading stocks on the stock market.

Total GNP is often divided by the population and stated as per-capita GNP. This is a simple mean and tells us nothing about the distribution of per-capita GNP of individuals about the mean. The mean may or may not reflect a representative central tendency in the distribution. Often modal or median per-capita income is a better measure of central tendency.[6]

GNP is measured in units of "dollar's worth." Dollar's worth of what? Of final goods and services traded in the market in the current year. It is the quantity of all such goods and services, times their price, all summed up. Changes in GNP over time can reflect price changes or quantity changes. To eliminate the effect of price level changes (inflation or deflation), economists correct the dollar figure by converting current dollars into dollars of constant purchasing power. This conversion is done by dividing nominal GNP by a price index that measures the rate of inflation. Suppose that there has been 20% inflation between 1990 and 2000. To convert year 2000 nominal GNP into real GNP, measured in dollars of 1990 purchasing power, we divide GNP in 2000 by 1.20; this is the price index that in the base year of 1990 would have been 1.00 but because of 20% inflation rose to 1.20 in 2000. This gives "real GNP," or rather GNP measured in dollars of constant purchasing power as of a base year.

---

[5]One might object that natural resources are not really free. A ton of coal does cost money on the market, but the money price is equal to the labor and capital cost of finding and extracting the coal. Coal in the ground, or in situ, as the resource economists say, is considered a free gift of nature. A particularly rich and accessible coal mine will require less labor and capital per ton of coal than a marginal mine. Will its coal sell for less than that of the marginal mine? No, and this gives rise to producer surplus or differential rent. The more accessible mine earns a rent, which results from saved labor and capital relative to the marginal mine. Coal in situ is still a free gift of nature, but some free gifts are nicer than others, and differential rent takes that into account. The rent is attributed to the value of labor and capital saved in extraction, not to any original value of the coal in the ground.

[6]The mode is the income category that has the most members. The median is the per-capita income number for which there are as many members above as below. As students of statistics will know, for a normal distribution, the mean, median, and mode will coincide, all giving the same measure of central tendency.

Changes in real GNP are due to changes in quantities, not price levels. So real GNP, although measured in value units, is an index of quantities of something physical and is therefore considered a better measure of economic growth than nominal GNP. Just as a dollar's worth of gasoline corresponds to a definite physical quantity of gasoline, so a dollar's worth of real GNP corresponds to some aggregate of physical goods and services. But because different goods and services have differing material and energy intensities, there is not a tight one-to-one relationship between real GNP and physical throughput, as there is in the case of dollar's worth of gasoline and the throughput it represents.[7]

The point to emphasize is that although GNP is measured in value terms and cannot be reduced to a simple physical magnitude, it is nevertheless an index of an aggregate of things that all have irreducible physical dimensions. The relationship between real GNP and throughput is not fixed, nor is its variability unlimited. And to the extent that one believes that GNP growth can be uncoupled from throughput growth, one must be willing to accept limits on throughput growth. If the environmental protection achieved by limiting throughput costs little or nothing in terms of reduced GNP growth, then no one should oppose it. If GNP could grow forever with a constant throughput, then ecological economists would have no objection.

GNP and Total Welfare

GNP is a measure of economic activity, not a measure of welfare. It tells us how fast the wheels are turning, not where the car is going. Economists all say that. Yet in the absence of a true measure of welfare, most policy makers look to the GNP as a trustworthy index of the general direction of change of welfare, based on the following:

Total welfare = economic welfare + noneconomic welfare

The faith-based assumption is that economic welfare and total welfare move in the same direction. But the increase in economic welfare could induce a more than offsetting decline in noneconomic welfare. For example, GNP goes up as labor becomes more mobile. But the welfare of being close to family and friends gets sacrificed as people have to move. Also, the extra income and job satisfaction of two-earner households raise eco-

---

[7]But even here, economists try to keep the aggregate mix constant in calculating the price index. They assume a given basket of goods and given relative prices of goods in the basket in order to calculate a weighted average price of the basket and its change over time. This average price is not supposed to reflect either changes in relative composition of the basket of goods or changes in *relative* prices of the goods in the basket. Since relative prices inexorably do change over time, as does the composition of the representative basket of goods consumed, price level indexes inevitably "wear out" over time and have to be recalculated. Therefore, real GNP figures lose comparability over longer time periods.

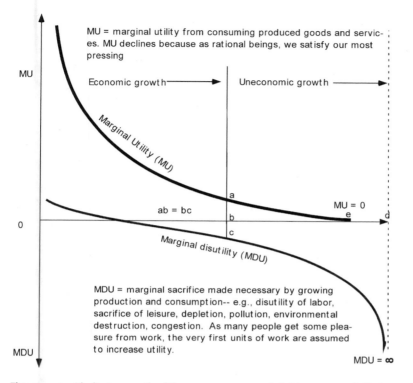

MU = marginal utility from consuming produced goods and services. MU declines because as rational beings, we satisfy our most
pressing

Economic growth⎯⎯⎯⎯▶   Uneconomic growth ⎯⎯⎯▶

Marginal Utility (MU)

ab = bc   MU = 0

Marginal disutility (MDU)

MDU = marginal sacrifice made necessary by growing
production and consumption-- e.g., disutility of labor,
sacrifice of leisure, depletion, pollution, environmental
destruction, congestion. As many people get some pleasure from work, the very first units of work are assumed
to increase utility.

MDU = ∞

Figure 14.2 • Limits to growth of the macroeconomy. Point b = economic limit or optimal scale, where marginal utility (MU) = marginal disutility (MDU) (maximum net positive utility); e = futility limit, where MU = zero (consumer satiation), d = catastrophe limit, where MDU = infinity. At point d, we have gone beyond sustainable scale.

nomic welfare, but the stress of lost leisure and the extra financial burden and lost satisfactions resulting from external childcare reduce noneconomic welfare. Pollution-induced illnesses constitute an enormous loss of noneconomic welfare. Because the category "noneconomic welfare" is unmeasured while economic welfare has a numerical measure, we tend to overestimate the importance of the latter and underestimate the importance of the former. In Figure 14.2, the MDU curve, traditionally missing in economic analysis, represents the loss of "noneconomic welfare."

It's worth pointing out that much of the marginal disutility from growth is caused by negative impacts on global public goods, including critical ecosystem life support functions. This means that a country whose economy is growing gains most of the utility from growth but shares the costs with the rest of the world. Many of these costs, such as waste emissions, habitat degradation, and resource depletion, are cumulative, which means that the marginal costs of growth are likely to increase as we move from an empty world to a full world. In support of this conclusion, recent studies have found that the marginal costs of growth outweigh the benefits in China and Thailand, and benefits just barely outweigh costs in India and Vietnam, all countries exhibiting phenomenal rates of growth.[8] Furthermore, while Figure 14.1 suggests that economic growth in the U.S. is futile, as measured by increases in overall happiness, other studies have found that happiness levels in China actually exhibited a mild (not statistically significant) decline in recent decades.[9]

Defensive Expenditures and the Depletion of Natural Capital

Two other categories are problematic in national income and product accounts: regrettably necessary defensive expenditures and the depletion of natural capital. Let's have a look at each.

Regrettably necessary defensive expenditures, or defensive expenditures for short, are those that we have to make to protect ourselves from the unwanted consequences of the production and consumption of other goods by other people—for example, extra thick walls and windows to block out the sound of living near an airport or busy street or medical services resulting from pollution-induced asthma. In the sense of just measuring activity, these are freely chosen expenditures that people make in order to be better off in their concrete circumstances, and therefore they should be counted—they are if not "goods," at least "anti-bads." In another sense, they are really involuntary intermediate costs of production that should not count as welfare to the final consumer or as final consumption. This category could be broadly or narrowly defined. The examples just given reflect a narrow definition. Some would include all costs of global warming and the extra legal and law enforcement costs resulting from a general breakdown in trust and increases in complexity attributed to economic growth. Exactly where to draw the line is a matter of judgment.

The depletion of natural capital is a more clear-cut category. GNP is *gross* national product. It is gross of depreciation of capital. If we deduct depreciation of manmade capital, we get net national product (NNP), which is a closer approximation to what we can consume without eventual impoverishment. But even in calculating NNP, there is no deduction for the depreciation and depletion of natural capital. Even NNP is gross of natural capital consumption (as well as gross of defensive expenditures). What's more, manmade capital is not a perfect substitute for natural cap-

> Regrettably necessary defensive expenditures, or defensive expenditures, are those that we have to make to protect ourselves from the unwanted consequences of the production and consumption of other goods by other people.

---

[8] P. Lawn and M. Clarke, *Sustainable Welfare in the Asia-Pacific: Case Studies Using the Genuine Progress Indicator*, Cheltenham, UK: Edward Elgar, 2008.

[9] R. A. Easterlin and L. Angelescu, 2009, Happiness and Growth the World Over: Time Series Evidence on the Happiness–Income Paradox. IZA Discussion Paper No. 4060, Institute for the Study of Labor.

ital for the simple reason that the former cannot exist without the latter. The two are complements. Putting a dollar value on the depreciation of both manmade capital and natural capital implicitly assumes that both types of capital are perfect substitutes and that we can accept the loss of natural capital as long as manmade capital grows by a compensating amount. In reality, less natural capital makes our manmade capital less valuable as well. Of what use is a car if there is no gas to put in it?

■ SUSTAINABLE INCOME

The true definition of **income**, implicitly stated above, is the maximum that a community can consume in a given time period without causing itself to have to consume less in future time periods.[10] In other words, income is the maximum you can consume this year without reducing your capacity to produce and consume the same amount next year, and the year after—without reducing future productive capacity, that is, without consuming capital. Strictly speaking, it is redundant to say "sustainable income" because income by definition is sustainable. Yet this feature of income has been so overlooked that a bit of redundancy for the sake of emphasis seems useful. If it's not sustainable it is, at least in part, capital consumption, not income.

The whole idea of income accounting is the prudent concern to avoid inadvertent impoverishment by consuming capital. Of course, there are times when we may choose to consume capital—for example, using a nest egg during retirement or liquidating the inventory of a store going out of business. Most of us, however, prefer not to run our national economy and ecosystem as if it were a business in liquidation. Certainly you may choose to consume capital and voluntarily become impoverished. The income accountant's job is to make sure you know what you're doing, not to tell you what to do. But if the accountant does not deduct the consumption of natural capital in calculating income, then she has failed at her professional duty.

To be concrete, if you cut only this year's net growth of a forest, that's income because you can do the same thing again next year. If you cut down the whole forest, you cannot do it again next year, and the value of the cut forest is mostly capital consumption, not income. Yet in GNP, we count the whole amount as this year's income. The same is true for over-exploited fisheries, waste sinks and croplands, and depleted mines, wells, and aquifers.[11] Some neoclassical economists have come to realize that

---

[10]J. Hicks, *Value and Capital*, 2nd ed., Oxford, England: Clarendon, 1948.

[11]The running down of renewable stocks or funds of natural capital is *depreciation*, analogous to the depreciation of a machine. The running down of nonrenewable natural capital is *liquidation*, analogous to the liquidation of an inventory. Both represent capital consumption.

nature's services are a huge infrastructure to the economy, and we are failing to maintain that infrastructure.

Why do our national accountants fail to subtract natural capital consumption in calculating income? Neoclassical economics does not count natural capital consumption as a cost because in its preanalytic vision of the world, nature is not scarce. The reason natural funds and resource flows are absent from the usual neoclassical production function is also the reason there is no deduction for natural capital consumption in national income accounting.

GNP as Cost

Years ago, Kenneth Boulding suggested that GNP be relabeled GNC, for gross national cost. While Boulding's plea may have been tongue-in-cheek, it bears close examination. GNP is a measure of the final goods and services a society produces multiplied by the price at which they sell on the market. But demand for the most important resources such as food, energy, and life-saving medicines is inelastic. As you'll recall from Chapter 9, this means that large changes in price have little impact on how much people want to consume, and conversely, that a small change in quantity will lead to a large change in price. Imagine that one year the food and oil industries decided to work less and reduced output by 20% over previous years. Because people would not want to reduce their consumption of food and energy, they would bid up the prices for these commodities dramatically. In fact, something like this really did happen, in 2008, when a small drop in grain supplies relative to annual consumption led to a 200% increase in prices, and a drop in the rate of increase in oil production led to a similar increase in oil prices. If we multiplied 80% of 2007's output by 300% of 2007's price, GNP would show a 140% increase in economic activity in these sectors instead of a 20% decrease. Real GNP would be lower, due to inflation, but the share of these commodities in GNP would nonetheless soar.

Even when GNP reflects economic activity, it may not reflect well-being. For example, compared to the other developed countries, the United States ranks last on a wide variety of health care measures, ranging from infant mortality to life expectancy. It also has by far the highest percentage of uninsured individuals. By such measures, the U.S. health care system provides fewer benefits than the systems in other developed nations. However, in 2008 the United States spent 50% more per capita on health care than any other nation,[12] and these expenditures were ris-

---

[12]OECD Health Data 2009: Frequently Requested Data. Online: http://www.oecd.org/docu ment/16/0,3343,en_2649_34631_2085200_1_1_1_1,00.html.

ing rapidly. Aside from those who reap income from health care, no one claims this is a good thing. Yet if we measure well-being by the market value of health care goods and services, the United States has by far the best health care system in the world.

The fact is that one person's income is another person's expenditure, so GNP is also an explicit measure of costs. As long as costs and benefits are closely correlated, this does not matter, but we can't take such a correlation for granted. Striving to maximize expenditures on health care, food, energy, or anything else would obviously be crazy.

What should be done about GNP? One approach would be to disaggregate GNP into two separate accounts: a national benefits account and a national costs account (we'll explore the challenges to this below). As the scale of the economy grows, both benefits and costs will increase. We could compare those benefit and cost increases at the margin to find the optimal scale (see Figure 14.2).[13] It makes absolutely no sense to add them together.

Another option is to move beyond consumption-based measures of well-being altogether, as we discuss below. If the aim of economic activity is to maximize human well-being, then health, nutrition, literacy, family, friends, social networks, and so on are probably the most important indicators, perhaps best measured by overall levels of happiness and satisfaction with life (see Box 14.1).

Nonetheless, absent more rational measures of well-being, we can't help feeling a certain nostalgia for the good old days when newscasters regaled us with quarterly changes in the GNP. Now we are subjected to hammer-banging, gong-clanging reports of hourly changes in the Dow Jones and Nasdaq stock price indices—numbers that are an order of magnitude further removed from either welfare or income than GNP is. For example, in 2008, global stock markets lost trillions of dollars in value with virtually no change in real productive assets. This is because stock market values are forward-looking, based on expectations of future earnings (even on speculators' estimates of the expectations of others). By contrast, GNP is backward-looking, a historical record of what has already happened. Since the past is better known than the future, GNP is inherently a more trustworthy number than stock market values.

---

[12]For an effort in this direction for Australia, see P. A. Lawn, *Toward Sustainable Development: An Ecological Economics Approach*, Boca Raton, FL: Lewis Publishers, 2001.

GROSS NATIONAL HAPPINESS AND THE HAPPY PLANET INDEX

Box 14-1

In the late 1980s, the country of Bhutan declared that it would strive to increase Gross National Happiness (GNH) rather than GNP, in an approach that "stresses not material rewards, but individual development, sanctity of life, compassion for others, respect for nature, social harmony and the importance of compromise."[a] Rather than attempting to measure happiness itself, Bhutan seeks to measure and improve the factors that contribute to happiness. The first global study on GNH included multicriteria measures of economic, environmental, physical, mental, social, workplace, and political wellness.[b] While initially seen as a quixotic goal, GNH is much less of a departure from economists' historical conceptions of utility than is GNP, and the idea has taken off, along with the study of happiness. A related measure is the Happy Planet Index, which divides a country's happy life years (life expectancy adjusted by subjective well-being) by its ecological footprint as an estimate of ecological economic efficiency or sustainable happiness. By this measure, Costa Rica is the world leader in sustainable development.[c]

[a]*Bhutan Planning Commsion,* Bhutan 2020: A Vision of Peace, Prosperity, and Happiness, *Thimphu: Royal Government of Bhutan Planning Commission, 1999, p. 19.*

[b]*International Institute of Management, Gross National Happiness (GNH) Survey. Online: http://www.iim-edu.org/polls/GrossNationalHappinessSurvey.htm.*

[c]*http://www.happyplanetindex.org.*

■ ALTERNATIVE MEASURES OF WELFARE: MEW, ISEW, AND GPI

In the early 1970s, there was considerable criticism of GNP growth as an adequate national goal—so much so that economists felt obliged to reply. The best reply came from William Nordhaus and James Tobin.[14] They questioned whether growth was obsolete as a measure of welfare and thus as a proper guiding objective of policy. To answer their question, they developed a direct index of welfare, called Measured Economic Welfare (MEW), and tested its correlation with GNP over the period 1929–1965. They found that, for the period as a whole, GNP and MEW were indeed positively correlated; for every six units of increase in GNP, there was, on average, a four-unit increase in MEW. Economists breathed a sigh of relief,

---

[14]W. Hordhaus and J. Tobin, "Is Growth Obsolete?" In *Economic Growth*, National Bureau of Economic Research, New York: Columbia University Press, 1972.

forgot about MEW, and concentrated again on GNP. Although GNP was not designed as a measure of welfare, it was, and still is, thought to be sufficiently well correlated with welfare to serve as a practical guide for policy.

Some 20 years later, Daly and Cobb revisited the issue and began to develop an Index of Sustainable Economic Welfare (ISEW) with a review of the Nordhaus and Tobin MEW. They discovered that if one takes only the latter half of the Nordhaus-Tobin time series (i.e., the 18 years from 1947 to 1965), the positive correlation between GNP and MEW *falls* dramatically. In this most recent half of the total period—surely the more relevant half for projections into the future—a six-unit increase in GNP yielded on average only a one-unit increase in MEW. This suggests that GNP growth at this stage in U.S. history may be quite an inefficient way of improving economic welfare—certainly less efficient than in the past.

The ISEW was then developed to replace MEW, since the latter omitted any correction for environmental costs, did not correct for distributional changes, and included leisure, which both dominated the MEW and introduced many arbitrary valuations.[15] The Genuine Progress Indicator (GPI) is a widely used, updated version of the ISEW that does account for the loss of leisure time. The ISEW and GPI, like the MEW, though less so, were positively correlated with GNP up to a point (around 1980), beyond which the correlation turned slightly negative.[16] Figure 14.3 shows estimates of GNP and ISEW for seven different countries.

Measures of welfare are difficult and subject to many arbitrary judgments, so sweeping conclusions should be resisted. However, it seems fair to say that for the United States since 1947, the empirical evidence that GNP growth has increased welfare is weak and since 1980 probably nonexistent (see also Figure 14.1 for further support of this claim). Consequently, any impact on welfare via policies that increase GNP growth

---

[15]The concept of leisure is an important part of welfare, but the problems of valuing leisure are difficult. Is the leisure chosen or unchosen? Should sleep time count as leisure? Is commuting time leisure or "time cost of working"? Should we use the wage rate? The minimum wage? Should the "leisure" of mom taking care of children be valued at her opportunity cost if she's a doctor, or at the cost of avoided daycare? Such difficult choices have a big effect on the index.

[16]Neither the MEW nor the ISEW considered the effect of individual country GNP growth on the *global* environment, and consequently on welfare at geographic levels other than the nation. Nor was there any deduction for legal harmful products, such as tobacco or alcohol, or illegal harmful products, such as drugs. No deduction was made for overall diminishing marginal utility of income resulting from GNP growth over time (although a distributional correction for lower marginal utility of extra income to the rich was included). Such considerations would further weaken the correlation between GNP and welfare. Also, GNP, MEW, GPI, and ISEW all begin with personal consumption. Since all four measures have in common their largest single category, there is a significant autocorrelation bias, which makes the poor correlations between GNP and the three welfare measures all the more surprising.

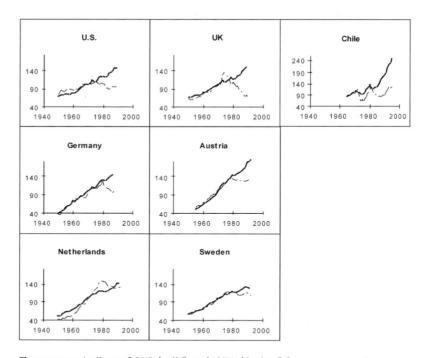

Figure 14.3 • Indices of GNP (solid) and ISEW (dashed) for seven countries. 1970 = 100 in all cases. (Source: R. Costanza, J. Farley, and P. Templet, "Quality of Life and the Distribution of Wealth and Resources." In R. Costanza and S. E. Jørgensen, eds., *Understanding and Solving Environmental Problems in the 21st Century: Toward a New, Integrated Hard Problem Science*, Amsterdam: Elsevier, 2002.)

would also be weak or nonexistent. In other words, the "great benefit," habitually used to justify sacrifices of the environment, community standards, and industrial peace, appears, on closer inspection, not even likely to exist.[17] Certainly if economic growth is to be the number-one goal of nations and the central organizing principle of society, then citizens have a right to expect that the index by which we measure growth, GNP, would reflect general welfare more accurately than it does. Continued use of GNP as a proxy for welfare reminds us of the quote often attributed to Yogi Berra: "We may be lost, but we're making great time."

The objective, accurate scientific measurement of national costs and national benefits is not a realistic goal. Both costs and benefits of economic growth are spread out over time, and how we treat costs and benefits that affect future generations is an ethical issue, not a scientific one.

---

[17]For further evidence from other countries, see M. Max-Neef, Economic Growth and Quality of Life: A Threshold Hypothesis, *Ecological Economics* 15:115–118 (1995).

The use of a particular discount rate to address intertemporal distribution, for example, is clearly a value-laden decision. Ecosystem change and evolution are not predictable, and how we treat the resulting uncertainty is also an ethical issue. Even using monetary measures of market goods is not objective; markets will yield different monetary values depending on the initial distribution of the wealth, and what constitutes a desirable initial distribution is an ethical judgment. Monetary values for a given resource also vary depending on the amount of the resource society is using; for example, the price of oil depends primarily on current rates of extraction of oil. Oil is such an important input into so many economic processes that all prices are affected by how much oil we are using. Using prices determined by resource use in this period to decide the appropriate amount of a resource to use is therefore a case of circular reasoning; you can't do it on a computer spreadsheet, and you can't do it in real life. Efforts to put monetary values on nonmarket goods such as ecosystem services not only compound these ethical issues with serious methodological problems but also imply that natural capital and manmade capital are perfect substitutes, a position that most ecological economists strongly reject.

■ BEYOND CONSUMPTION-BASED INDICATORS OF WELFARE

Personal consumption is not an end in itself but merely one means toward achieving the end of enhancing human welfare. GNP is inadequate as a proxy for income, and income is only one element among many that provide human welfare. For example, the ecosystem services that increasing GNP inevitably encroaches upon are at least as important as GNP in providing welfare.[18]

Human Needs and Welfare

Do other factors not yet discussed contribute to our welfare? It is reasonable to assume that welfare is determined by the ability to satisfy one's needs and wants. What are our needs? Absolute needs are those required for survival and are biologically determined. Some 1.4 billion individuals globally and 26% of the population in the Third World currently live in extreme poverty (less than $1.25 per day), and 2.6 billion earn less than

---

[18]See R. Costanza et al., The Value of the World's Ecosystem Services and Natural Capital, *Nature* 6630:253–260 (1997), in which the value of global ecosystem goods and services is found to outweigh global GNP. While this article does put monetary values on natural capital for purposes of comparison with manmade capital, it also explicitly discusses many of the problems with this approach.

$2.00 per day. These people have difficulty meeting even these absolute needs.[19] For this group, greater consumption is probably very closely correlated to greater welfare.

Once absolute needs have been met, as is the case for the remaining three-fifths of the world's population, then welfare is determined by the satisfaction of a whole suite of primary human needs. Numerous researchers have proposed a variety of human needs, typically claiming that they are pursued in hierarchical order, with Maslow's hierarchy (1954) (in which consumption is the lowest rung on the needs ladder) being the most famous. The hierarchical ordering, though generally not seen as rigid by these researchers, still leaves something to be desired. Even the 1.2 billion people living in absolute poverty seek to fulfill needs other than mere subsistence.

Manfred Max-Neef[20] has summarized and organized human needs into nonhierarchical axiological[21] and existential categories (Table 14.1). In this **matrix of human needs**, needs are interrelated and interactive—many needs are complementary, and different needs can be pursued simultaneously. This is a better reflection of reality than a strict hierarchy in which we pursue higher needs only after lower ones have been fulfilled. Also important in Max-Neef's conception, needs are both few and finite. This stands in stark contrast to the assumption of infinite wants, or the nonsatiety axiom in standard economics.

If we are to evaluate the success of economic policies both now and in the future (assuming that providing a high level of welfare for humans for the indefinite future is our economic goal), then we must develop measurable indicators that serve as suitable proxies for needs fulfillment and welfare.

To state the obvious, we cannot precisely measure welfare, which in the present context is equivalent to quality of life (QOL). In the words of Clifford Cobb,[22]

> *The most important fact to understand about QOL indicators is that all measures of quality are proxies—indirect measures of the true condition we are seeking to judge. If quality could be quantified, it would cease to be quality. Instead, it would be quantity. Quantitative measures should not be judged as true or false, but only in terms of their adequacy in bringing us closer to an unattainable goal. They can never directly ascertain quality. (p. 5)*

---

[19]M. Ravallion and S. Chen, "The Developing World Is Poorer Than We Thought but No Less Successful in the Fight Against Poverty." Policy Research Working Paper Series 4211, The World Bank. 2008.

[20]M. Max-Neef, "Development and Human Needs." In P. Ekins and M. Max-Neef, *Real-Life Economics: Understanding Wealth Creation*, London: Routledge, 1992, pp. 197–213.

[21]Axiology is the study of the nature of values and value judgments.

[22]C. W. Cobb, *Measurement Tools and the Quality of Life: Redefining Progress*, Oakland, CA. Online: http://www.rprogress.org/pubs/pdf/measure_qol.pdf.

Table 14.1

Max-Neef's Matrix of Human Needs

Axiological Categories	Existential Categories			
	Being	**Having**	**Doing**	**Interacting**
Subsistence	Physical health, mental health, equilibrium, sense of humor, adaptability	Food, shelter, work	Feed, procreate, rest, work	Living environment, social setting
Protection	Care, adaptability, autonomy, equilibrium, solidarity	Insurance systems, savings, social security, health systems, rights, family, work	Cooperate, prevent, plan take care of, cure, help	Living space, social environment, dwelling
Affection	Self-esteem, solidarity, respect, tolerance, generosity, receptiveness, passion, determination, sensuality, sense of humor	Friendships, family, partnerships with nature	Make love, caress, express, emotions, share, take care of, cultivate, appreciate	Privacy, intimacy, home, space of togetherness
Understanding	Critical conscience, receptiveness, curiosity, astonishment, discipline, intuition, rationality	Literature, teachers, method, educational policies, communication policies	Investigate, study, experiment, educate, analyze, meditate	Settings of formative interaction, schools, universities, academies, groups, communities, family
Participation	Adaptability, receptiveness, solidarity, willingness, determination, dedication, respect, passion, sense of humor	Rights, responsibilities, duties, privileges, work	Become affiliated, cooperate, propose, share, dissent, obey, interact, agree on, express opinions	Setting of participative interaction, parties, associations, churches, communities, neighborhoods, family
Idleness	Curiosity, receptiveness, imagination, recklessness, sense of humor, tranquility, sensuality	Games, spectacles, clubs, parties, peace of mind	Daydream, brood, dream, recall old times, give way to fantasies, remember, relax, have fun, play	Privacy, intimacy, space of closeness, free time, surroundings, landscapes
Creation	Passion, determination, intuition, imagination, boldness, rationality, autonomy, inventiveness, curiosity	Abilities, skills, method, work	Work, invent, build, design, interpret	Productive and feedback settings, workshops, cultural groups, audiences, spaces for expressions, temporal freedom

Continued

■ Table 14.1

Max-Neef's Matrix of Human Needs (Continued)

Axiological Categories	Existential Categories			
	Being	**Having**	**Doing**	**Interacting**
Identity	Sense of belonging, consistency, differentiation, self-esteem, assertiveness	Symbols, language, religion, habits, customs, reference groups, sexuality, values, norms, historical memory, work	Commit oneself, integrate oneself, confront, decide on, get to know oneself, recognize oneself, actualize oneself, grow	Social rhythms, everyday settings, settings in which one belongs, maturation stages
Freedom	Autonomy, self-esteem, determination, passion, assertiveness, open-mindedness, boldness, rebelliousness, tolerance	Equal rights	Dissent, choose, be different, run risks, develop awareness, commit oneself, disobey	Ability to come in contact with different people at different times in different places

The column of Being registers attributes, personal or collective, that are expressed as nouns. The column of Having registers institutions, norms, mechanisms, tools (not in material sense), laws, etc. that can be expressed in one or few words. The column of Doing registers locations and milieus (as time and spaces). It stands for the Spanish estar or the German befinden, in the sense of time and space. As there is no corresponding word in English, Interacting was chosen for lack of something better.

Source: M. Max-Neef, "Development and Human Needs." In P. Ekins and M. Max-Neef, Real-Life Economics: Understanding Wealth Creation. London: Routledge, 1992, pp. 197–213.

Objective Measures

Numerous efforts have been made to objectively measure welfare. The problem is that these studies have found only weak relationships between objective measures of welfare and the subjective assessments of the same by the subjects concerned.[23] However, both these studies and the various types of national accounts seem to include a narrow range of objective indicators, often placing what we consider to be an excessive emphasis on consumption. Quite possibly the problem is that welfare is too rich a gumbo for us to recapture its flavor with so few ingredients. An important research agenda in economics is to develop a methodology for measuring access to "satisfiers" (the means by which we satisfy a given need) for Max-Neef's axiological and existential categories of human needs as indicators of welfare. With sufficient ingredients, we can produce something reasonably close to the flavor of welfare.

---

[23]B. Haas, A Multidisciplinary Concept Analysis of Quality of Life, *Western Journal of Nursing Research* 21(6):728–743 (1999).

Max-Neef's human needs matrix as the basis of a welfare measure is a dramatic departure from existing national accounts, as well as from most of the proposed alternatives, differing even in its theoretical underpinnings. Neoclassical economics and GNP are explicitly utilitarian. Within utilitarian philosophy, individual welfare is determined by the degree to which individuals can satisfy their desires, and it is generally accepted that the goal of society is to provide the maximum amount of utility for its citizens. As utilitarian philosophy has been operationalized by NCE, citizens are the best able to determine what provides utility. Because it is extremely difficult to measure utility directly, economists have taken to using revealed preferences as a proxy. Preferences are revealed by people's objectively measurable choices in the market. In the market economy, preferences are revealed through market decisions, and market decisions can be made only with money. Under this conception of utilitarianism, the philosophy values only end-states and requires only "having" such things as possessions and experiences. Sustainable income accounting and measurements of economic welfare are basically just extensions of this philosophy, and they similarly value only having.[24]

In Max-Neef's framework, having things is important, but it is just one of the elements required to meet our needs. Thus, a benevolent dictator with the resources to provide us with all the physical things we need for happiness would fail to meet our existential needs for being, doing, and interacting, as well as our axiological needs for creation, participation, and freedom. Also, within Max-Neef's conception, people are not always best able to determine what contributes to their quality of life; for example, advertising may falsely convince people that consumption satisfies their need for affection, freedom, or participation.

This approach, which values human actions independently of their outcomes, has been dubbed the "human development" approach to welfare. Its main proponents include Nobel Prize–winning economist Amartya Sen and philosopher Martha Nussbaum. In a similar tone to Max-Neef, they argue that "capabilities" and "functionings" are critical to welfare.[25] Roughly speaking, "functionings" correspond to human needs, while "capabilities" include both states of being and opportunities for doing and are therefore analogous to access to satisfiers for these needs in Max-Neef's matrix (see Table 14.1). In utilitarian theory, we might have several different options, of which we choose one. If all options but that one were

---

[24]C. W. Cobb, *Measurement Tools and the Quality of Life* (San Francisco, CA: Redefining Progress, 2000). Online: http://www.rprogress.org/pubs/pdf/measure_qol.pdf.

[25]Ibid.; M. Nussbaum, "Aristotelian Social Democracy," in R. B. Douglass, G. M. Mara, and H. S. Richardson, eds., *Liberalism and the Good*, New York: Routledge, 1990, pp. 203–252; R. Sugden, "Welfare, Resources, and Capabilities: A Review of *Inequality Reexamined* by Amartya Sen," *Journal of Economic Literature* 31 (December, 1993): 1947–1962.

eliminated, it would not affect our welfare. In the human development approach, losing options restricts our capabilities and would therefore affect our welfare. The human development approach is less concerned with the actual choices that people make than with the options they are free to choose from, and the marketplace is only one of many spheres in which choice is important.

Operationalizing Human Needs Assessment as a Measure of Welfare

Measuring the extent to which human needs are satisfied is, of course, an exceptionally difficult task and highly subjective. Following the lead of Sen and Nussbaum, it would be most useful to measure capabilities, that is, the extent to which individuals have access to satisfiers. However, as noted by Max-Neef, specific satisfiers may vary by culture, and the difference in satisfiers required to meet a human need may indeed be one of the key elements that defines a culture. This means that objective "welfare accounts" must be very culture-specific. Second, some satisfiers might help fulfill several human needs, while other needs require several satisfiers. Further complicating matters, satisfiers may change through time. And humans are social creatures who inhabit a complex environment; needs are satisfied not only in regard to the individual but also in regard to the social group and environment.[26] Furthermore, while needs are different and distinct, they are also interactive and may complement each other, and therefore may not be additive. Abundant access to satisfiers for one set of needs does not compensate for a lack of satisfiers for another set of needs. This suggests that separate "accounts" should be kept for access to satisfiers to different needs.

In developing welfare accounts based on **human needs assessment (HNA)**, it would be useful to test measurements of satisfiers empirically in studies comparing these objective measures against subjective assessments of welfare to determine their effectiveness. These empirical tests, as well as efforts to operationalize HNA accounts, must involve people in dialogues to confirm or refute the validity of the needs Max-Neef specifies, as well as the validity of the satisfiers we use to assess the degree to which needs are met. Such dialogues would almost certainly elicit additions and alternatives to the generic satisfiers, the entries in the columns of Table 14.1.[27] While the average person may not always know exactly what sat-

---

[26]Max-Neef, op. cit., 1992.

[27]E.g., food and shelter are specific dimensions of "having" that are satisfiers of the need for "subsistence." How we actually meet our needs for food and shelter are culture-specific. A traditional Inuit might be satisfied with walrus blubber and an igloo, while a New Yorker would require hamburgers and a high-rise apartment.

isfiers will best meet their needs, interactive discussion with people is nonetheless essential to select and test appropriate indicators. We would also need to develop group-based methodologies to determine the effectiveness of our indicators in a social setting.

It is clear that Max-Neef's approach is very difficult to operationalize, even if his concept is theoretically more compelling than GNP or even ISEW. The debate over which approach to take to national accounting—theoretically sound measures or ease of accounting—is old. As Irving Fisher argued back in 1906, the appropriate measure, even of income, is one that captures the psychic flux of service (i.e., satisfaction of needs and wants) and not simply the final costs of goods and services.[28] And at the time Fisher wrote, the absence of suitable data for calculating either psychic flux of service or final costs no doubt led many to ignore the debate as entirely academic. The widespread use of GNP indicates that in practice, Fisher lost this earlier debate. However, measures such as the ISEW suggest that the GNP is becoming increasingly incapable of measuring economic welfare, much less general human welfare. Even if we can never quantify access to satisfiers as precisely as we currently quantify GNP, as Sen suggests, perhaps it is better to be vaguely right than precisely wrong.[29]

Accepting Max-Neef's human needs matrix as a framework for the specific elements of human welfare, and access to satisfiers as potentially the best objective indicator of welfare, has profound implications with respect to scale, distribution, and allocation. First, most of the possible indicators suggested by Max-Neef require few, if any, material resources beyond those needed to sustain human life and hence are less subject to physical exhaustion. Thus, for most elements of human welfare, increases for one person or one generation do not leave less for others. Second, explicitly accepting that there is a limit to material needs implies that we can limit consumption greatly with little, if any, sacrifice of welfare. This result is critical, because the laws of thermodynamics make it impossible to uncouple physical consumption from resource use and waste production. Abundant evidence suggests that current levels of consumption could not be sustainably met with renewable resources alone, and we must therefore limit consumption or else threaten the welfare of future generations.

The difficulty of operationalizing Max-Neef's framework may actually be a point in its favor. Why do we want to measure welfare in the first place? It's not just to track its rise or fall but to help us create policies to

---

[28]H. Daly and J. Cobb, *For the Common Good: Redirecting the Economy Toward Community, the Environment, and a Sustainable Future*, Boston: Beacon Press, 1989.

[29]D. Crocker, "Functioning and Capability: The Foundations of Sen's and Nussbaum's Development Ethic, Part 2." In M. Nussbaum and J. Glober, eds., *Women, Culture, and Development: A Study in Human Capabilities*, Oxford, England: Oxford University Press, 1995.

improve it. Simply providing statistical data on welfare doesn't help us achieve this end. However, applying Max-Neef's framework would require extensive surveys asking people to think deeply about what their needs really are and how they can satisfy them. Ultimately, improving welfare falls to decisions by political, cultural, and religious groups about what they want and how they want to achieve their goals, and making the correct decisions will require people to think deeply about what it is they ultimately desire.

BIG IDEAS to remember

- Fallacy of composition
- General equilibrium model versus aggregate macroeconomics
- Optimal scale of macroeconomy
- Gross national (or domestic) product
- Total welfare = economic welfare + noneconomic welfare

- Defensive expenditures or "anti-bads"
- Natural capital consumption
- Sustainable income
- MEW and ISEW
- Gross national cost
- Relative wealth and welfare
- Human needs and welfare
- Matrix of human needs (Max-Neef)
- Human needs assessment (HNA)

15

Money

Money ranks with the wheel and fire as ancient inventions without which the modern world could not function. Probably more people today are run over and burned by out-of-control money than by out-of-control wheels and fires. Money is mysterious. Unlike matter and energy, it *can* be created and destroyed, evading the laws of thermodynamics. Private citizens (counterfeiters) are sent to jail for making even small amounts of it, yet private commercial banks make almost all of it, and we pay them for it! Sometimes money is a costly commodity (gold) and sometimes a costless token (paper notes). It is easily transferable into real assets by individuals, but the community as a whole cannot exchange its money into real assets at all, since someone in the community ends up holding the money. Some economists think the money supply should be determined by fixed rules, others think it should be manipulated by public authorities. And some people think the love of money is the root of all evil! Anyone who is not confused by money probably hasn't thought about it very much.

Money functions as a medium of exchange, a unit of account, and a store of value. The functions are interrelated but worth considering separately. To measure exchange value, we need a unit—call it a dollar, a peso, a franc, or a yen. If the unit is stable over time (no inflation or deflation), then money automatically serves as a store of exchange value. To function as a medium of exchange and let us escape the inconvenience of barter, money must hold its value at least long enough to effect both sides of the transaction, which in barter, of course, are simultaneous. A moment's reflection shows how tremendously inefficient barter is and consequently how efficient money is. In barter there must be a coincidence of wants. It is not enough that I want what you have to trade; you also have to want whatever it is that I have to trade, and we have to find each other. Money provides a common denominator that everyone wants simply because everyone else is willing to accept it. It is a standard, well-defined commodity (or later a token) that breaks the two sides of a difficult barter arrangement into two separate and easy transactions.

Karl Marx analyzed transactions as follows. First we have simple barter, which he denoted as

$$C—C^*$$

Commodity C is exchanged for commodity C^*. You have C and prefer C^*; I have C^* and prefer C. We are both better off after the transaction. We both increase the use value of what we own. Exchange value is not separated from use value. No money is needed, but we were lucky to have found each other.

Next for Marx comes simple commodity production:

$$C—M—C^*$$

Now we have money functioning as a medium of exchange. Exchange value, the sum of money, M, is entirely instrumental to bringing about an increase in use values by facilitating the exchange. The process begins and ends with commodity use values. The goal is to increase use value, not exchange value.

For Marx the critical change comes in the historical shift from simple commodity production to capitalist circulation, which he symbolized as

$$M—C—M^*$$

The capitalist starts with a sum of money capital, M, uses it to make commodity C, and then sells C for the amount M^*, presumably greater than M. Thus:

$$M^*—M = \Delta M$$

ΔM is profit, or surplus value in Marxist terms. For us the important thing is not Marx's notion of surplus value, which is tied up with his very problematic labor theory of value, but the simple observation that in moving from $C—M—C^*$ to $M—C—M^*$ the driving motive has shifted from increasing use value to increasing exchange value.

Use value arises from the actual use of commodities; it is concrete and physically embodied. **Exchange value** is abstract and inheres in money. It has no necessary physical embodiment.[1] Real wealth—commodities—obeys the laws of thermodynamics. Money, a mere symbolic unit of account, can be created out of nothing and destroyed into nothing. There is a physical limit to the accumulation of use values. There is no obvious

---

[1]Though, of course, exchange value is real only if something exists for which money can actually be exchanged.

limit to the accumulation of exchange value. Fifty hammers are not much better than two (one and a spare) as far as use values are concerned. But in terms of exchange value, fifty hammers are much better than two, and better yet in the form of fifty hammers' worth of fungible money that can be spent on anything, anywhere, and at any time.

Box 15-1   DIAMONDS-WATER PARADOX

The distinction between use value and exchange value goes back to Aristotle and was used to "resolve" the diamonds-water paradox—the paradox that although water is a necessity it has a low price, while diamonds are practically useless but have a high price. Economists dealt with this conundrum by declaring that there are two basic kinds of value, use value and exchange value, and one has nothing to do with the other. In the late 1800s the marginalist revolution in economic thinking resolved the paradox as follows: Exchange value is determined by marginal utility, and use value is determined by total utility; that is, exchange value equals marginal use value. Water has enormous total utility, but it is so plentiful that at the margin we use it for trivial satisfactions. This marginal utility determines exchange value. How do we know that? If you want to buy a gallon of water from me, what determines how much you will have to give me in exchange? If I give you a gallon of water, I won't stop drinking and go thirsty, nor will I stop bathing and be dirty. I'll probably water my petunias less often. The petunias are my least important use value, my marginal utility of water, my opportunity cost for a gallon of water. Since the *marginal* utility of water is what I will sacrifice by trading away a gallon, that's what determines the exchange value of water. Exchange value is determined by the least important use value, the value sacrificed. Water is abundant, so its marginal utility is very small; diamonds are scarce, so their marginal utility is still high.

A hoard of hammers takes up space and is subject to rust, termites, fire, and theft. Fifty hammers' worth of money is not subject to rust, rot, and entropy, and far from costing a storage fee will earn interest from whoever gains the privilege of "storing" it for you. Production for use value is self-limiting. Production for the sake of exchange value is not self-limiting. Since there is no limit to the accumulation of abstract exchange value, and since abstract exchange value is convertible into concrete use value, we seem to have concluded that there must not be any limit to concrete use values either. This has perhaps led to the notion that exponential growth, the law of money growing in the bank at compound interest, is also the law of growth of the real, or material, economy.

■ VIRTUAL WEALTH

Frederick Soddy summarized all this by carefully distinguishing wealth from debt.[2] He noted that "a weight, although it is measured by what it will pull up, is nevertheless a pull down. The whole idea of balancing one thing against another in order to measure its quantity involves equating the quantity measured against an equal and opposite quantity. Wealth is the positive quantity to be measured and money as the claim to wealth is a debt" (p. 103).[3] Monetary debt, the measure of wealth, is negative wealth, say minus two pigs. It obeys the laws of mathematics but not of physics. Wealth, on the other hand, plus two pigs, obeys the laws of thermodynamics as well as mathematics. Positive pigs die, have to be fed, and cannot reproduce faster than their gestation period allows. Negative pigs are hyper-fecund and can multiply mathematically without limit. As Soddy put it, "You cannot permanently pit an absurd human convention, such as the spontaneous increment of debt (compound interest), against the natural law of the spontaneous decrement of wealth (entropy)" (p. 30).

The holding of token money by the public to avoid the inconvenience of barter gives rise to the curious phenomenon that Soddy called **virtual wealth**, which he defined as the aggregate value of the real assets that the community voluntarily abstains from holding in order to hold money instead. Individuals can always convert their money holdings into real assets, but they choose not to in order to avoid the inconvenience of barter. This raises the question of whether money should be counted as a part of the real wealth of the community. Yes, if money is a commodity like gold that circulates at its commodity value. No, if it is token money like a dollar bill whose commodity value is nil but whose exchange value is significant. Even though each person can at an instant convert his money into real assets, it is impossible for the community as a whole to do this, as we have previously noted.

Money, therefore, represents not real wealth but, in Soddy's term, virtual wealth. More exactly, it is the magnitude of virtual wealth that determines the value of money. What happens if the government puts into circulation more money than people currently want to hold? People will exchange money for real assets and drive up the price of real assets. As the price of real assets rises, the real value of money falls until it again coincides with the virtual wealth of the community. If there is too little money, people will exchange real assets for money, thereby driving down the price

---

[2] F. Soddy, *Wealth, Virtual Wealth, and Debt*, London: George Allen & Unwin, 1926.

[3] When banks create money by providing someone with a loan (see below), they actually create a debt as the first step. On the asset side of the accounting books, the banker enters a debt for the amount of money borrowed (to be paid off with interest). This borrowed money is then placed in a bank account, which is listed in the bank's books as a liability.

of real assets. As the prices of real assets fall, the value of money increases until it again equals the virtual wealth of the community. The value of a dollar, then, is the virtual wealth of the community divided by however many dollars are in circulation. It follows that the value of a unit of token money is determined not by the total wealth of a community, nor by its annual GNP, but by its virtual wealth relative to the money supply.

Box 15-2   VIRTUAL WEALTH AND FIDUCIARY ISSUE

Nobel laureate economist James Tobin comes very close to Soddy's concept of virtual wealth in his explanation of the "fiduciary issue":

The community's wealth now has two components: the real goods accumulated through past real investment and fiduciary or paper "goods" manufactured by the government from thin air. Of course, the nonhuman wealth of such a nation "really" consists only of its tangible capital. But as viewed by the inhabitants of the nation individually, wealth exceeds the tangible capital stock by the size of what we might term the fiduciary issue. This is an illusion, but only one of the many fallacies of composition which are basic to any economy or society. The illusion can be maintained unimpaired as long as society does not actually try to convert all its paper wealth into goods.[a]

[a]*J. Tobin, "Money and Economic Growth,"* Econometrica *(October 1965), p. 676.*

■ SEIGNIORAGE

Who owns the virtual wealth? Since it does not really exist, we might say that no one owns it. It is a collective illusion. Yet individuals voluntarily hold money instead of real assets, and they behave as if money were a real part of their individual wealth, even if they understand that collectively it is only "virtual" or illusory. Every member of the community who holds money had to give up a real asset to get it—except for the issuer of money. The one who creates the money and is the first to spend it gets a real asset in exchange for a paper token. The difference between the monetary value and the negligible commodity value of the token, the profit to the issuer of money, is called **seigniorage**, in recognition of the lordly nature of this privilege. Who is this fortunate person? Historically it was the feudal lord, or the king, the sovereign, who issued money within his domain. One might expect that this privilege would have been passed on to the sovereign's legitimate heir, the democratic state. To some extent this is the case, because only governments can issue currency or legal tender. However, over 90% of our money supply today is not currency but demand deposits created by the private commercial banking system.[4] They

---

[4]Demand deposits are ordinary checking accounts from which money is payable on demand to the bearer of your check.

are created out of nothing and loaned into existence by the private commercial banks under rules set up by the government. Who gets the seigniorage? Seigniorage from currency goes to the government. Seigniorage from demand deposits goes to the private sector, initially to commercial banks. To the extent there is competition between banks for savings, they will redistribute some of the seigniorage to depositors. Sectors of society too poor to save will receive nothing.

What does money consist of in our economy? A further mystery of money is that it has several definitions. The most restrictive is "currency plus demand deposits in the hands of the nonbank public." More expansive definitions include savings deposits and even credit card debt. Most of our money supply bears interest as a condition of its existence. Whoever borrowed it into existence must pay back what he borrowed plus interest. Thus, a requirement for growth (or else inflation) is built into the very existence of our money supply. Moreover, the money supply, *ceteris paribus*, expands during boom times when everyone wants to borrow and invest, and contracts during recessions when loans are foreclosed, thereby aggravating cyclical instabilities.

On learning for the first time that private banks create money out of nothing and lend it at interest, many people find it hard to believe. Indeed, according to Joseph Schumpeter, as late as the 1920s, 99 out of 100 economists believed that banks could no more create money than cloakrooms could create coats. Yet now every economics textbook explains how banks create money. We will explain how it works in a minute, but first we'll let the strangeness of it sink in. Of course, this is not the only way to create money. Nonetheless, most economists today accept this situation as normal. But the leading economists of the early twentieth century, Irving Fisher and Frank Knight, thought it was an abomination. And so did Frederick Soddy.

| **Box 15-3** | **LOCAL CURRENCIES AND LOCAL EXCHANGE TRADING SYSTEMS** |

Currencies are not created exclusively by governments. A variety of non-government legal currencies exist in countries throughout the world, and a closer look at local currencies can provide important insights into money. There are three ways to design a currency system. Most national currencies are created by fiat. There is nothing to back up fiat money but faith that someone else will accept it in exchange for goods ("In God We Trust," or as the Ithaca HOUR says, "In Ithaca We Trust"). Second, a currency can be valued in terms of a commodity and may or may not be redeemable in terms of that commodity. For example, the Constant, one of the earliest alternative currencies and a forerunner of today's local currencies, was introduced in the 1970s on an experimental basis in Exeter, New Hampshire. The Constant was designed to maintain a constant value against a basket of 30 different commodities. Finally, a currency can be backed by a commodity, which means it can freely be exchanged for that commodity. Such was the case for U.S. currency in the nineteenth century, when money holders could theoretically exchange gold-backed dollars for gold at any time, and the necessary gold reserves were physically available to do this.[a]

The city of Ithaca, New York, has one of the best-developed local currency systems in the world. The currency is known as Ithaca HOURs. An individual can participate in the HOURs system simply by agreeing to accept HOURs in exchange for the goods or services she produces. New money must be issued to chase this greater supply of goods and services. Where does this new money come from?

Published backers of the HOUR directory, which is considered a service provided to Ithaca HOURs, are paid 2 HOURs (the equivalent of approximately $20 US) on first participation and again when they renew their commitment. Technically speaking, the participant is being paid for publicly backing HOURs, but one could also say that in the HOURs system, the person who agrees to generate new goods and services earns the right to seigniorage. While at first glance it may seem strange that one would be entitled to money for simply agreeing to accept money, new money must clearly come from somewhere, and it's reasonable for part of it to go to the person responsible for creating the new wealth.

Theoretically, the amount of new money created times the velocity with which the money circulates should equal the amount of new goods and services being offered. So far it seems that new participants have on average offered more than enough goods and services to use up their 2 new HOURs. Several mechanisms are used to increase the money supply and prevent deflation. Residents of Ithaca may request interest-free loans of HOURs, organizations may request grants of HOURs, employees of member businesses can accept HOURs as a regular part of their pay, and people may purchase HOURs into circulation with dollars, from the HOUR bank. Additional money is created to finance administrative costs of the system. The circulation committee of Ithaca HOURs is responsible for deciding how many HOURs to create. So far, Ithaca HOURs are holding their own against the U.S. dollar, and they continue to trade at a ratio of 1 HOUR to 10 U.S. dollars.[b]

---

[a]R. Swann and S. Witt, Local Currencies: Catalysts for Sustainable Regional Economies. Revised 1988 Schumacher lecture, 1995/2001. Online: http://www.schumachersociety. org/currencypiece.html (E. F. Schumacher Society).

[b]See also http://www.ithacahours.com. Paul Glover, the founder of Ithaca HOURs, was also very helpful in providing information.

■ THE FRACTIONAL RESERVE SYSTEM

What allows banks to create money is our fractional reserve system. If banks had to keep 100% reserves against the demand deposits they create, then there would be no creation of money. Therefore, the reform called for by Soddy, Fisher, Knight, and others was for a 100% reserve requirement on demand deposits. Banks would still provide the convenience of checks and safekeeping, and they would charge for these services. They could still lend other people's money for them and make a profit, but those people would have access to that money only after it was repaid. Banks could not create money any longer.

Exactly how does the fractional reserve system enable banks to create money? Suppose the law required banks to keep 10% reserves against their demand deposits (actually, it is much less). Reserves are either cash or deposits with the Federal Reserve Bank owned by the commercial bank. The bank needs reserves only to settle the difference between daily deposits and withdrawals, which nearly always balance to within a few percent. Therefore, the bank feels that keeping 100% reserves is excessively cautious. It can keep only 10% reserves and meet all imbalances that are statistically likely to ever happen. The "excess reserves" can be loaned at interest, thereby increasing the bank's profits. The government has concurred in this practice and made it legal; it is known as **fractional reserve banking**. It works as long as all depositors do not demand their money at once, as happens in a bank panic (when depositors doubt the solvency of the bank and all rush to get their money out at the same time). To avoid panics, the government set up the Federal Deposit Insurance Corporation (FDIC). If depositors are insured against loss when a bank fails, then they will be less likely to panic and cause the very failure they feared. (They will also be less likely to demand prudence from their bank, but that's another story we leave for later.)

How do banks actually create money? Let's first consider a monopoly commercial bank. Because it is the only bank, it knows that any check drawn against it in one branch will be deposited with it in another branch. When it clears its own check, there is no transfer of money, of reserves, to another bank. Therefore, if it has a new cash deposit of $100 that counts as reserves, and the reserve requirement is 10%, it can lend out in newly created demand deposits an amount of $900. People and businesses borrow only what they intend to spend, so it is certain that this $900 will be spent. Its total additional demand deposits are $100 in exchange for the new cash deposit, plus $900 in new loans, giving $1000 in new demand deposits backed by $100 in new reserves, thus satisfying the 10% reserve requirement. Net addition to the money supply is $900 worth of demand deposits.

Now let's consider a competitive banking system rather than a single monopoly bank. Suppose Bank A receives a new cash deposit of $100. Unlike the monopoly bank, Bank A cannot lend out $900 because nearly all of the checks written on that amount of new demand deposits will be deposited in other banks, not Bank A. Clearing will necessitate a transfer of reserves to other banks. If it had lent out $900, it would surely soon have to transfer almost that amount to other banks. But it only has $100 in new reserves and thus will not be able to meet its legal reserve requirement of 10%.

So how much can Bank A lend as a result of a new cash deposit of $100? If it safely assumes that all checks drawn on its loans will be deposited in other banks, it can only lend out $90. Therefore, it still creates money—$90 in new demand deposits above the $100 demand deposit in exchange for the $100 cash. But the process does not stop here. The $90 of excess reserves safely lent by Bank A end up being transferred to Bank B, which can now safely lend 90% of that, or 0.9 ($90) = $81. So now the money supply has gone up by $90 + $81 = $171. But then the new $81 excess reserves of Bank B get transferred to Bank C, which can create new deposits of 0.9 ($81) = $72.90. And the process continues in an infinite series, the sum of which turns out to be—can you guess it? Exactly $900 of new money, as with the monopoly bank, or $1000 of new demand deposits, remembering the exchange of $100 cash for a $100 demand deposit that started the whole process.[5] The whole process works in reverse when someone withdraws cash (reserves) from the bank.

Just as money is created when banks loan it into existence, money is destroyed when it is paid back. Interest-bearing loans require that more be paid back than was initially borrowed, however, demanding a constant increase in the money supply. The net result of simultaneous processes of money creation and destruction determines the net growth of the money supply.

■ MONEY AS A PUBLIC GOOD

Money is a collective phenomenon, not a privately owned resource. In a peculiar but very real way, money is a true public good. You might think that if you own money, you can exclude others from using it, but if you did so completely, your money would have no value whatsoever. Money has value only if everyone can use it. And money is certainly nonrival, in that my spending a dollar in no way decreases the value of that dollar for

---

[5] If r is the required reserve ratio, then the demand deposit multiplier is this infinite series:

$$1 + (1 - r) + (1 - r)^2 + (1 - r)^3 + \cdots + (1 - r)^n = 1/r$$

the next person. Since money is a public good, one would expect seigniorage to be public revenue, not private. The virtual wealth of the community could be treated as a publicly owned resource, like the atmosphere or electromagnetic spectrum. But that is not the case. The money supply is privately loaned into existence at interest. The fact that most of our money was loaned into existence and must be paid back at interest imparts a strong growth bias, as well as cyclical instability,[6] to our economy. There is no economic reason why the monetary system must be linked with the private commercial activity of lending and borrowing.

What are the alternatives? Soddy offered three reforms. His first proposal was to gradually raise the reserve requirement to 100%. That would put the private banks out of the money creation business and back into the business of borrowing and lending other people's real money, providing checking services, and so on. Control of the money supply would then belong to the government. How, then, would the government regulate the money supply? Soddy's second policy suggested an automatic rule, based on a price level index. If the price level index is falling, the government should finance its own activities by simply printing new money and spending it into existence. If the price level is rising, the government should cease printing money and tax more than it spends, that is, run a surplus. This would suffice for a closed economy, but for an open economy, one that engages in international trade, the domestic money supply can be increased or decreased by international payments balances. Soddy's third proposal (back in 1926, under the gold standard) was freely fluctuating exchange rates. Currencies would trade freely and directly against each other; an equilibrium exchange rate would eliminate any surplus or shortage (deficit) in the balance of payments and consequently any international effect on the domestic money supply. Remember our discussion of surplus and shortage in Chapter 9.

Of course, this is not what we have now.

The gold standard has been abandoned, and fixed exchange rate regimes have given way to flexible exchange rates, but not to freely floating exchange rates, which are thought (rightly or wrongly) to be too volatile and disruptive of international trade. (We return to the topic of exchange rates in Chapter 20.) The money supply is determined largely by the commercial banking system, subject to some manipulation, but not control, by the Federal Reserve (the Fed). The **Federal Reserve System**

---

[6]We will explain this instability in more detail in Chapter 17. In the meantime, it is enough to understand that banks are eager to loan new money into existence during economic booms, increasing the money supply and favoring more economic growth. During recessions, banks prefer to collect more in old loans than they loan out in new ones, reducing the money supply and aggravating the recession.

is a coordinated system of district central banks in the U.S. that influences interest rates and money supply.

The Federal Reserve has three tools for manipulating the money supply. First, the Fed can set the reserve requirements, within limits prescribed by law, and thus reduce or expand the supply of money created by banks, as explained above. This tool is used infrequently, because it has large impacts on the financial sector. Second, the Fed can change the interest rate it charges to lend reserves to the commercial banks (known as the discount rate), thus making it more or less profitable for the commercial banks to lend to their customers, and in doing so expand (or limit the expansion of) the money supply. Third, the Fed can conduct open market operations, directly increasing or decreasing the money supply by buying and selling government securities in the open market. When the Fed buys government securities, it does so by crediting the bank account (at Reserve Banks) of securities dealers. This directly increases the available supply of money by the amount of the purchase. The deposit also increases the bank's reserves, allowing the bank to make more loans and create even more money. When the Fed sells government securities, the money supply contracts.

■ MONEY AND THERMODYNAMICS

Frederick Soddy was a Nobel Prize winner in chemistry and a great believer that science should be used to benefit humankind. He doubted that this would happen, however, and even predicted back in 1926 the development of the atomic bomb. Why are the fruits of science often badly used? Because, thought Soddy, we have a flawed and irrational economic system. Unless we reform that system, scientific progress will only help us destroy the world faster. Soddy spent the second half of his 80-year life studying the economic system. He understood thermodynamics and entropy and the biophysical basis of economics, and he forcefully called attention to this interdependence. But he focused his attention mainly on money. Why? Because money was the one thing that did not obey the laws of thermodynamics; it could be created and destroyed. And yet this undisciplined, imaginary magnitude was used as a symbol and counter for real wealth, which has an irreducible physical dimension and cannot be created or annihilated. Money is the problem precisely because it leads us to think that wealth behaves like its symbol, money; that because it is possible for a few people to live on interest, it is possible for all to do so; and that because money can be used to buy land and land can yield a permanent revenue, money can yield a permanent revenue.

Because of this fallacy, M. King Hubbert recently had to remind us that **exponential growth**—growth at a constant percentage rate—is a transient phase in human history.[7] The classic example of the power of exponential growth is the story about putting a grain of wheat on the first square of a chessboard, two grains on the second, four on the third, and so on. At the next-to-last, or 63rd, square the board contains 2^{63} grains of wheat, far more than the world's whole wheat crop, and the last, or 64th, square will by itself contain that much again. Hubbert's conclusion was that the world cannot sustain 64 doublings of even a grain of wheat. In our world, many populations are simultaneously doubling—populations of people, livestock, cars, houses—things much bigger than a grain of wheat. How many times can each of these populations double? How many times can they all double together? A few tens at most, was Hubbert's answer. Our financial conventions, on the other hand, assume that this doubling will go on forever.

This expectation gets played out in reverse when we discount future values to an equivalent present value. We simply run the exponential calculation backward, asking: How much would we have to deposit in the bank today at today's interest (discount) rate in order to have the given future amount at a given future date? This discounting procedure is, as we have seen, at the heart of the financial model of present value maximization, which has displaced the more traditional economic model of profit maximization. The error that bothered Soddy is deeply ingrained in present economic thinking. We have already encountered it in our discussion of why renewable resources are driven to extinction.

It is convenient to dismiss Soddy as a "monetary crank" and to remark what a pity it was that such a brilliant chemist wasted so much of his time on a topic that he was unqualified to think about. This is exactly the treatment that Soddy was given. It was harder to dismiss Irving Fisher and Frank Knight, who also called for 100% reserve requirements, because they were the leading economists of their generation. But their ideas on money were simply classed separately from the rest of their economics, treated as a peccadillo, and were ignored.

Our previous statement—that money does not obey the laws of thermodynamics—needs some qualification. Exchange value is hardly a value if there is nothing for which it can be exchanged. If money is issued without real wealth to back it up, spending that money simply drives up the prices of goods and services, causing inflation and bringing "real money" back closer into line with real wealth (more on inflation later).

What about virtual wealth? Are there limits to the amount of real wealth people are willing to forgo in order to hold money? If not, then the amount of real money in circulation can continue to grow independently

---

[7]M. King Hubbert, "Exponential Growth as a Transient Phenomenon in Human History." In H. Daly and K. Townsend, eds., *Valuing the Earth*, Cambridge, MA: MIT Press, 1993.

of the production of real goods and services. Financial assets are neither money nor real wealth, but they are bought and sold in the market, and people will hold more money to be able to meet their demand for transactions in these assets. In addition, people trade in money itself, using one national currency to buy another, and this similarly increases the demand for money. Both currency speculation and growth in financial assets have increased dramatically in recent years.

The M—C—M* equation previously showed how money has become less a means for facilitating exchange, more an end in itself. In reality, the M—C—M* equation has itself been dwarfed by pure currency speculation and trading in financial paper. John Maynard Keynes warned back in the 1930s, "Speculators may do no harm as bubbles on a steady stream of enterprise. But the position is serious when enterprise becomes the bubble on a whirlpool of speculation. When the capital development of a country becomes a by-product of the activities of a casino, the job is likely to be ill-done."[8] While global production of marketed goods and services is roughly on the order of $30 trillion per year,[9] the trade in paper purchasing paper (or, more accurately these days, electrons purchasing electrons) with no intervening commodity is almost $2 trillion *per day*.[10] This means that the buying and selling of paper assets and currencies, M—M*, is more than 20 times greater than exchanges in the real economy! Real enterprise has indeed become a bubble on the whirlpool of speculation. As no productive activity intervenes in these speculative purchases, the sole result seems to be a magical growth in money. But is such growth actually possible indefinitely?

Growth in money is meaningless unless there is a corresponding increase in real wealth, so now we must ask: Does financial speculation lead to growth in real wealth? Some paper-paper purchases are purchases of new stock offerings that do provide financial capital, which can mobilize physical factors of production, but this is only an estimated 4% of stock purchases. Speculation in currency, in which millions of dollars are traded back and forth for very small margins over short time scales, clearly produce nothing. Indeed, such transactions almost certainly contributed to the crises in several Southeast Asian economies in 1997–1998 as speculators sold off regional currencies, and these crises meant dramatic *decreases* in production from those economies. Yet such speculation would not be

---

[8]J. M. Keynes, *The General Theory of Employment, Interest and Money,* Orlando, FL: Harcourt Brace, 1991, p. 159.

[9]Official estimates based on purchasing power parity (PPP) are on the order of $40 trillion per year; the numbers for speculation are in nominal dollars, not PPP.

[10]D. Korten, *The Post-Corporate World: Life After Capitalism,* San Francisco: Berrett-Koehler, 1998.

undertaken unless some profits were being made somewhere. For example, George Soros, who participated in the financial speculation in Southeast Asia, is reported to have earned 1 billion pounds speculating on England's currency in 1995.[11] The only possible explanation is that if those who produce nothing are earning, through speculation, more money that entitles them to more real wealth, then those who actually do produce something must be becoming entitled to increasingly less wealth.

In summary, it appears that the illusion that money can grow without physical limits results from three things. First, as long as the production of real goods and services increases, more money is needed to pursue them, so growth in money is justified. But such growth cannot, of course, continue forever on a finite planet. Second, as the number or price of financial assets grows, such as through speculative bubbles, demand for money grows as well, and supply can increase to meet this demand. The fact is, however, that financial bubbles inevitably burst. Third, holders of financial capital see their capital grow because speculation can transfer resources from those who produce to those who merely speculate. Such transfer of wealth has limits, though the limits are obscured by continued economic growth. Thus, the appearance that money is exempt from the laws of thermodynamics is an illusion that can be maintained only while scale is increasing, or the financial sector is expanding relative to the real sector. It remains impossible for real money to grow without limit.

THINK ABOUT IT!

What do you think would happen if a national government tried the same approach to seigniorage as Ithaca HOURs? For example, the government could impose 100% reserve requirements to prevent banks from creating money, award every new entrant to the economy some lump sum of money (perhaps by providing 18-year-olds sufficient money to pay for a college education or start a business), and lend money into existence at 0% interest for socially desirable projects.

---

[11]W. Greider, *One World, Ready or Not: The Manic Logic of Global Capitalism*, New York: Simon & Schuster, 1997.

BIG IDEAS to remember

- Money as medium of exchange, unit of account, store of value
- Barter; simply commodity production; capitalist circulation
- Exchange value vs. use value
- Virtual wealth

- Seigniorage
- Fractional reserve banking
- Money creation
- Money as public good
- Federal Reserve System
- Money and laws of thermodynamics
- Local currencies

Distribution

We have emphasized that ecological economics is concerned with three issues: the allocation of resources, their distribution, and the scale of the economy. We have seen how the ecological sustainability of the Earth is related to the size or scale of the macroeconomy. We have also explored the economist's meaning of efficient allocation in our discussion of microeconomics and the basic market equation. We then looked at the macroeconomic allocation problem in Chapter 14. But the second issue, distribution and the fairness thereof, has remained largely in the background.

■ PARETO OPTIMALITY

In dealing with allocation, we saw that economics defines efficiency as the Pareto optimal allocation of resources by the market. This definition assumes a given distribution of wealth and income. More specifically, an efficient allocation is one that best satisfies individual wants *weighted by the individual's ability to pay*—that is, by her income and wealth. Change the distribution of income and wealth, and we get a different set of efficient prices (since different people want different things), which define a different Pareto optimum. Because different Pareto optima are based on different distributions of income and wealth, economists are reluctant to compare them; one optimum is as good as another. We saw that a major reason for scale expansion—economic growth—has been to avoid the issue of distributive equality. As long as everyone is getting more from aggregate growth, the distributive issue is less pressing, at least as a cure for poverty. Besides, the efficiency of the allocation of aggregate growth loses its well-defined meaning (Pareto optimality) once we accept the legitimacy of changing distribution in the interest of fairness. Consequently, economics has tended to address distribution out of logical necessity but quickly sets it aside in the interests of political convenience.

Does a Pareto Optimal Allocation Assume a Given Scale as well as a Given Distribution?

If we take the concept of scale literally, as in the scale model of a house, to involve only a proportional change in all linear (scalar) dimensions, then we might say that a scale change is simply an increase or decrease in which all proportions remain constant. All relative prices, measuring unchanged relative scarcities, would also remain constant, defining an unchanging Pareto optimal allocation in terms of proportions. This seems to be what standard economists often have in mind. But is it possible to have everything grow in proportion? No, for two reasons. First, if something is fixed, it obviously cannot grow proportionally to everything else. What is fixed from the ecological economist's perspective is the size of the total ecosystem. As the economic subsystem grows, albeit proportionally in terms of its internal dimensions, the ecosystem itself does not grow. The economy becomes larger as a proportion of the total system—what we have called an increase in its scale, meaning size relative to the ecosystem. Natural capital becomes more scarce relative to manmade capital.

Of course, if the economy were to expand to encompass Earth's entire ecosystem (the model of "economic imperialism" in Chapter 3), the scale issue would disappear. In this sense the neoclassical economist's claim that if only all externalities were perfectly internalized then prices would automatically solve the scale problem (in the process of allocating everything in creation) makes sense. But it's a rather utopian point, like Archimedes' boast that he could move the Earth, if only he had a fulcrum and a long enough lever.

The second difficulty, long noticed by biologists and some economists, is that if you scale up anything (increase all linear dimensions by a fixed factor), you will inevitably change the relative magnitudes of non-linear dimensions. Doubling length, width, and height will not double area; it will increase area by a factor of four and volume by a factor of eight. Biologists have long noted the importance of being the right size. If a grasshopper were scaled up to the size of an elephant, it could not jump over a house. It would not even be able to move, because its weight (proportional to volume) would have increased eightfold, while its strength (proportional to a cross-sectional area of muscle and bone) would have increased only fourfold.

Returning to our example of a house, doubling the scale will increase surfaces and materials fourfold and volumes to be heated or cooled eightfold. Relative scarcities and relative prices cannot remain the same. The

answer to our question, does the notion of Pareto optimal allocation assume a given scale as well as a given distribution, appears to be yes. Size
cannot increase proportionally because (1) there is a fixed factor, namely
the size of the ecosystem, and (2) it is mathematically impossible even for
all relevant internal dimensions of the subsystem to increase in the same
proportion. In sum, it seems quite true that an optimal allocation assumes
a given scale, just as it assumes a given distribution.

Economics prides itself on being a "positive science." Allocative efficiency is thought to be a positive, or empirically measurable, issue, even
though, as we just saw, it presupposes a given distribution. Whether or
not the scale of the economy is sustainable is also considered to be a positive issue involving biophysical constraints, although normative questions of conservation for the future and other species are not far below the
surface. Distributive equity, on the other hand, is a normative issue. This
is the main question addressed to distribution: Is it fair? Not, is it efficient?
or Is it ecologically sustainable? The question "Is it fair?" is directly and
unavoidably normative, and for that reason alone it is given minimal attention by the positivist tradition of economics.

But like other sciences, economics assumes certain cultural values.
First, the very criterion of objective efficiency, Pareto optimality, embodies an implicit normative judgment, namely that malevolence or invidious
satisfactions are not acceptable. If everyone but you becomes better off
and you remain no worse off, the Pareto criterion tells us that is an objective increase in social welfare. But if everyone else is better off except you,
and you are an envious person, then you will be less happy than before,
even though your absolute situation is no worse. Economists must make
either the (false) positive judgment that people are in fact not invidious
and jealous or the (true) normative judgment that envy at another's good
fortune is a moral failing rather than a welfare loss.

There is a second reason that economics is less positive than some
think. Redistribution can be efficient in the sense of increasing total social
utility, yet economists make the value judgment that this kind of efficiency
should not count. For example, redistributing a dollar from the low marginal utility use of the rich to the high marginal utility use of the poor will
increase total utility to society and is in that sense efficient. The Pareto criterion forbids such interpersonal comparisons and summations of utility.
Some argue that the major function of the Pareto criterion was precisely
to sterilize the egalitarian implications of the law of diminishing marginal
utility,[1] a law that economics cannot afford to give up, as we saw in our
discussion of demand curves (see Chapter 9).

---

[1]J. Robinson, *Economic Philosophy*, Middlesex, England: Penguin, 1962.

If we admit interpersonal comparisons of utility, then distribution has efficiency implications as well as fairness implications. The extreme individualism of economics insists that people are so qualitatively different in their hermetical isolation from one another that it makes no sense to say that a leg amputation hurts Smith more than a pin prick hurts Jones. If we are all isolated individuals, we can rule out such obviously realistic human characteristics as envy and benevolence. Man as atomistic individual is the *Homo economicus* of neoclassical economics. Ecological economics' concept of the nature of man is "person-in-community," not isolated atom. *Community* here means community both with other humans and with the rest of the biosphere.

■ THE DISTRIBUTION OF INCOME AND WEALTH

Ecological economics distinguishes between the distribution of income and the distribution of wealth and between the functional and the personal distribution of income.

Wealth is a stock of assets, measured at a point in time, that is, cash in the bank, plus the market value of bonds, corporate shares, land, real estate, and consumer durables as of a given date. Income is the flow of earnings from these assets, plus the earnings of your own labor power (or human capital), between two dates, that is, over a period of time, usually a year. Labor power is not usually counted as capital because one cannot sell it all at once to another person (short of slavery) but can only rent it for certain durations. Income and wealth are thus two different magnitudes, measured in different units and distributed differently over the population.[2] Wealth is usually more concentrated than income. And financial wealth is even more concentrated than wealth in general. In 1989, the top 1% owned 48% of financial wealth, while the bottom 40% had negative net worth. Virtually all of the growth in wealth between 1983 and 1989 in the U.S. went to the top 20%. The bottom 80% was excluded from this growth (Table 16.1), and the bottom 40% saw their wealth decline in real terms. Inequality of wealth declined somewhat from 1989 through 1998. However, from 2001 to 2004, median incomes fell by nearly 7%, while mean incomes increased by 10%. In that period, median net worth fell by 0.7% in spite of skyrocketing home prices, and median financial wealth fell by an astonishing 26.5% while mean financial wealth increased. All are indicators of

---

[2]Wealth is measured in dollars (for example) and income in dollars/time. These magnitudes are as different as miles (distance) and miles per hour (speed).

Table 16.1

U.S. PERCENTAGE SHARE OF WEALTH AND INCOME BY PERCENTILE GROUP

Year	Percentile Shares			Gini Coefficient
	Top 1%	Next 19%	Bottom 80%	
Net Worth (Wealth)				
1983	33.8	47.6	18.7	0.799
1989	37.4	45.3	16.2	0.832
1992	37.2	46.6	16.3	0.823
1995	38.5	45.8	16.1	0.828
1998	38.1	45.3	16.6	0.822
2001	33.4	51.3	16.6	0.826
2004	34.3	50.4	16.3	0.829
Income				
1983	12.8	39.0	48.1	0.480
1989	16.4	39.0	44.5	0.521
1992	15.7	40.7	43.7	0.528
1995	14.4	40.8	44.9	0.518
1998	16.6	39.6	43.8	0.531
2000	20.0	38.6	41.4	0.562
2003	17.0	40.9	42.1	0.540

Source: E. N. Wolff, Top Heavy, The Twentieth Century Fund Report, *New York: New Press, 1995, p. 67 (years 1983–1992)*; E. N. Wolff, Recent Trends in Wealth Ownership, 1983–1998, *Working Paper No. 300, Table 2, Jerome Levy Economics Institute, April 2000 and E. B. Wolff,* Recent Trends in Household Wealth in the United States: Rising Debt and the Middle-Class Squeeze, *SSRN eLibrary.*

dramatic increases in inequality.[3] By 2006, after-tax income inequality was the highest ever recorded.[4]

Economics has a theory that explains income, as discussed next, and one that explains the prices of assets (though not entirely, as the "price" of entrepreneurship is a residual) but no theory at all to explain the distribution of wealth among individuals. It is the historical result of whose ancestors got there first, of marriage, of inheritance, plus individual ability and effort, and just plain luck.

■ THE FUNCTIONAL AND PERSONAL DISTRIBUTION OF INCOME

Income distributed among people, regardless of its source, is called the personal distribution. Income is also distributed according to how much of total income goes to wages, interest, rent, and profit—the functional

---

[3]See E. Wolff, *Recent Trends in Household Wealth in the United States: Rising Debt and the Middle-Class Squeeze*, SSRN eLibrary, 2007.

[4]A. Sherman, *Income Gaps Hit Record Levels in 2006, New Data Show: Rich–Poor Gap Tripled Between 1979 and 2006*, Washington, DC: Center on Budget and Policy Prioirities, 2009.

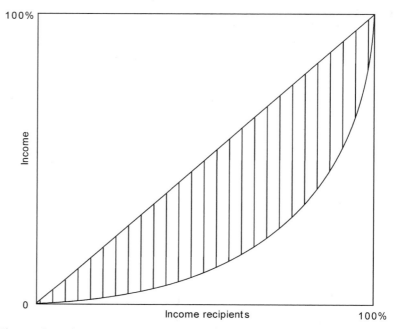

Figure 16.1 • The Lorenz curve. Because the Lorenz curve is in percentages, its shape does not depend on units of measure. It is therefore useful for making comparisons across countries and over time.

distribution. The idea behind the functional distribution is that income is not first created, then distributed. Rather, it is distributed as it is created among the factors combining to create it.

Remember from the circular flow diagram (see Figure 2.4) that supply and demand in the factors market determine the prices of factors—wages, interest, rent, with profit as a residual. Factor prices times the total amount of each factor used yields the functional distribution, usually expressed as percentage of total income going to landowners (rent), laborers (wages), capitalists (interest), and entrepreneurs (profit). Prices of each factor times the amount of the factor owned by each individual yields the personal distribution of income. The amount of each factor owned by each person, including labor power, is the personal distribution of wealth. Therefore, the personal distribution of wealth times the rental price of each type of wealth asset determines the personal distribution of income.

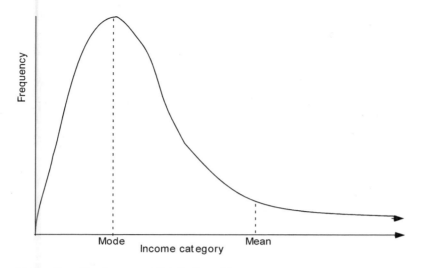

Figure 16.2 • The frequency distribution of income.

■ MEASURING DISTRIBUTION

Although economists have no good theory by which to explain the distribution of wealth and income, they do have useful ways of measuring and describing it statistically.[5] One useful representation is the **Lorenz curve**, shown in Figure 16.1. The *x*-axis shows the number of income recipients in terms of cumulative percentages, from lowest to highest income. The *y*-axis shows the percentage of total income. The lengths of the axes are equal, so that when closed in, they make a square.

The Lorenz curve plots the percentage of total income going to each percentage of income recipient. We know that 0% of income recipients will get 0% of the income, and that 100% of income recipients will get 100% of the income, so we already know the two extreme points on any Lorenz curve. If each percentage of the population received the same percentage of the income (i.e., the bottom 20% got 20% of total income, the bottom 70% got 70% of income), we would have perfect equality. The Lorenz curve would be the 45-degree line connecting (0, 0) with (100, 100). But suppose the bottom 80% of recipients get 44% of the income. That gives us another point, one that lies well below the 45-degree line. If we fill in many points between the extremes, we get a curve shaped like the one in Figure 16.1. The closer the curve to the 45-degree line, the more equal the distribution; the farther away, the less equal. The shaded

---

[5]For a clear and insightful exposition, see J. Pen, *Income Distribution*, New York: Praeger, 1971.

area defined by the curve and the 45-degree line measures inequality. In the limit, if one person got 100% of income and everyone else got 0%, the Lorenz curve would coincide with the axes and look like a backwards L.

The ratio of the shaded area (between the curve and the 45-degree line) to the total triangular area under the 45-degree line is called the **Gini coefficient**. For perfect equality the shaded area is zero, and consequently the Gini coefficient is 0; for perfect inequality the shaded area takes up the whole area under the 45-degree line, and consequently the Gini coefficient is 1. Values of the Gini coefficient for U.S. wealth and income distribution are given in Table 16.1.

The *Gini coefficient* is used to measure the inequality of the distribution of wealth or income across a population. A Gini coefficient of 1 implies perfect inequality (one person owns everything), and a coefficient of zero indicates a perfectly equal distribution.

A more familiar statistical description is the common frequency distribution, shown in Figure 16.2. The *x*-axis shows income category, and the *y*-axis shows number of members in each income category (frequency). Income distribution does not follow a normal distribution, as does height or many other personal characteristics. Rather, it is highly skewed, with the mode well below the mean and a very, very long tail to the right needed to reach the top income.

If we wanted to show the maximum income on Figure 16.2, we would need a fold-out extending the horizontal axis by the length of a football field. Graphical representations generally do not capture the extreme inequality at the upper income range. Income categories are frequently truncated at a maximum category of "$100,000 and over," where "over" means four orders of magnitude over.

Another interesting way to look at income distribution is to consider a football field, where the zero yard line represents the poorest person, the the 50 yard line the person with median income, and 100 yard line the richest person in the U.S. Incomes are depicted as the height of a stack of $100 bills. At the 50 yard line the median personal income (in 2005) of $25,149 is represented by a one-inch stack of bills. Around the 99 yard line, we see a stack of bills about a foot high—$300 million. Nearing the 100 yard line, the top hedge fund manager in 2008, a recession year, earned $2.5 billion, a stack of bills 1.7 miles high. Bill Gates once earned a stack nearly 30 miles high![6]

Moreover, these data are just for the United States. The distribution of wealth and income *between* countries is far greater than that found *within* countries.

What is the proper range of inequality in the distribution of income? Surely it is impossible to have one person owning everything and everyone else owning nothing. Maybe we could have everyone else getting a subsis-

---

[6]L. Story, Top Hedge Fund Managers Do Well in a Down Year, *New York Times*, March 24, 2009. See also http://www.lcurve.org.

tence wage, and the fortunate one person enjoying the entire social surplus above subsistence. But most people would not consider that fair, even though possible. At the other extreme, few people think a perfectly equal distribution—a Gini coefficient equal to zero—would be fair either. After all, some people work harder than others, and some jobs are more difficult than others. Fairness in a larger sense would require some income differences. There is a legitimate case to be made that differences in distribution provide a socially useful incentive for industriousness and innovation.

Is there a legitimate range of inequality, beyond which further inequality becomes either unfair or dysfunctional? What might such a range be? Plato thought that the richest citizen should be four times wealthier than the poorest. Ben Cohen and Jerry Greenfield, of Ben and Jerry's ice cream fame, at one time reportedly pledged that the highest-paid executive would receive no more than five times the salary of the lowest-paid employee. Maybe Plato, Ben, and Jerry were wrong, though, and maybe a factor of 10 would be better. Or 20 or 50. Currently the acceptable ratio is not defined, and in 1999 in the United States, the *typical* CEO earned 475 times the *typical* worker.[7] Ecological economics does not accept the current notion that real total output can grow forever. If the total is limited, then the maximum for one person is implicitly limited. The issue of a proper range of inequality in distribution is therefore critical for ecological economics, even though it has not yet received due attention. The standard economist's effort to keep distribution at bay forever by eternal growth is not a satisfactory solution.

Finally a word on the functional distribution of income. For industrial countries, the division varies around the following: wages = 70%, profits = 20%, interest = 8%, and land rent = 2%. For ecological economics, what is striking is that essentially none of the value of the total product is attributed to natural resources or services. Even land rent is mainly a locational premium, not a payment for resources in situ or natural services—one more piece of evidence that the flow of low entropy from nature is treated as a free good. If we think of two social classes struggling to divide the pie, we have laborers getting 70% and capitalists, business owners, and landowners together getting about 30%. This division represents a kind of balance of power in the social struggle. Neither side wants to include nature as a participant in production, which would require that nature's services be paid according to their scarcity and productivity.

Even if one wanted to pay for nature's contribution, who would collect on nature's behalf? There is no social class analogous to labor or capital

---

[7] J. Reingold and F. Jesperson, Executive Pay: It Continues to Explode—and Options Alone Are Creating Paper Billionaires, *Business Week*, April 17, 2000. Online: http://www.businessweek.com/careers/content/jan1990/b3677014.htm.

Box 16-2   DISTRIBUTION AND TAXATION

Income distribution has changed markedly over the years in the United States. Paul Krugman refers to the period up to the Great Depression as "the Gilded Age," when both wealth and income were highly concentrated in the hands of the few. Beginning just prior to World War II up through 1943, there was an impressive reduction in income inequality known as the great compression (which Krugman attributes largely to President Franklin Roosevelt's New Deal), followed by about 40 years of "Middle-Class America." Beginning the 1980s, about the time of Ronald Reagan's election, income inequality began to increase steadily in "the Great Divergence."[a] A graph on Paul Krugman's blog (see footnote) depicting the income share of the top 10% of U.S. society shows these trends very clearly.

A slightly different graph in Figure 16.3 shows the income share of the top 0.1% of U.S. society, along with the highest marginal tax bracket. It's pretty evident that the highest marginal tax bracket is inversely correlated with pre-tax income, suggesting that the more income the wealthy are allowed to take home, the more capital they accumulate, and the greater their income in future years. Middle-Class America was associated with top marginal tax brackets of 70% and higher. Currently, the top tax rate is less than half that amount. Hedge fund managers are taxed at the capital gains rate of 15% per year, but due to a tax loophole, they are able to delay paying taxes on their income for as long as 10 years.[b]

[a]P. Krugman, The Conscience of a Liberal blog. Online: http://krugman.blogs.nytimes .com/2007/09/18/introducing-this-blog/.

[b]J. Anderson, Managers Use Hedge Funds as Big I.R.A.'s, New York Times, April 17, 2007.

that has an interest in seeing to it that nature's services are properly accounted and paid for. Historically the landlord class may to some extent have played the role of defender of nature's services, but that class hardly exists anymore, and few lament its demise. The government is the biggest landholder in the U.S., and it has followed a policy of cheap resources in order to benefit and ease the tensions between labor and capital. The existing classes, labor and capital, see it in their mutual interest not to share with a third party. Since in reality there is no third party, all that would be necessary is to pay into a fund a scarcity rent for natural resources and services, and then redistribute the fund back to labor and capital, perhaps on the same 70–30 division. This would get the cost accounting and prices right and improve the efficiency of allocation, without necessarily affecting the distribution. Alternatively, since many of the goods and services provided by natural capital are nonmarket goods, the scarcity rent

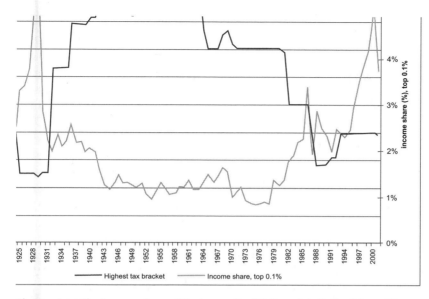

Figure 16.3 • The income share of the top 0.1% of U.S. society (left axis), and the highest marginal U.S. Tax break (right bracket) between 1913 and 2002.

could go toward supplying other nonmarket goods. The government could do this directly or could subsidize the private production of such goods. The rent could also be redistributed progressively by financing the abolition of regressive taxes.

■ CONSEQUENCES OF DISTRIBUTION FOR COMMUNITY AND HEALTH

The existing distribution of wealth is not only a precondition for efficient allocation; it is also a fundamental dimension of justice in society. As such, it affects us more directly than we might at first think. Evidence indicates that inequality of income distribution (independently of absolute poverty) has a substantial effect on rates of morbidity and mortality.[8] The relatively poor have higher incidences of death and sickness than the relatively rich, regardless of the absolute level of income of the relatively poor. The main reason investigators suggest is the extra stress associated with being relatively poor, being at the bottom of a dominance hierarchy. This extra stress is caused by less control over the circumstances of one's life, greater risks of job loss, a lower level of social standing and respect, and more

---

[8]See. G. Wilkinson, *Mind the Gap*, New Haven, CT: Yale University Press, 2001.

frequent experiences of disrespect and shame, with consequent anger and violence. Life at the bottom is more threatening, and the threat often comes from stressful relations with people higher up, including bosses, landlords, and government officials. Stress, of course, has well-known negative direct physiological effects on health.

In addition to these direct effects, inequality has indirect social effects on health. It is more difficult to form friendships across wider income gaps, as well as more difficult to form civic associations when wealth levels and economic interests are very disparate. Lack of friends and civic cohesion is also correlated with ill health. Treating people as atomistic, isolated individuals, unaffected by social relationships, literally makes them sick. As seen from our discussion of Max-Neef's human needs matrix (see Table 14.1), we are persons-in-community, related to each other internally—that is, our personal identity is constituted largely by our relation to others in the community. We are not independently defined entities held together only by external relations of the cash nexus. When these identity-constituting social connections become strained and corrupted by excessive inequality, we get sick more often and we die younger. We are also less happy.

■ INTERTEMPORAL DISTRIBUTION OF WEALTH

Every bit as important as the distribution of wealth and income within a generation is the distribution of resources between generations. However, while people have pondered the distribution of resources within a generation for millennia, the concern for distribution between generations is more recent. For the vast majority of human history, natural resources appeared limitless and technological advance was slow. People had approximately the same resource endowment as their great-grandparents had enjoyed, and they expected their great-grandchildren to inherit the same endowment as well. As the pace of technological change and fossil energy use accelerated with the Industrial Revolution, change became noticeable from one generation to the next, and people began to expect a better life for their children than they themselves had enjoyed. The "Protestant work ethic" asked people to work hard and invest for their children. At least up through the 1960s, the question most economists asked was: How much consumption should this generation sacrifice for the ever-growing well-being of the next?[9]

---

[9]E.g., J. Robinson, *Essays in the Theory of Economic Growth*, London: Macmillan, 1968; E. Phelps, Second Essay on the Golden Rule of Accumulation, *American Economic Review* 55(4): 793–814 (1965).

However, the onset of the atomic age made it apparent that technological advance had the capacity to bring harm as well as good. Growth in population and per-capita consumption raised the specter of resource depletion. Worsening pollution caused alarm, and ecologists began to worry that many systems were nearing irreversible, catastrophic thresholds. The relevant question was no longer: How much should we sacrifice to make the future even better off? Now it was: How much should we sacrifice to keep the future from being worse off than the present? Paradoxically, at least in the United States, a culture change was occurring at about the same time. The work ethic was no longer "work hard, live frugally, and invest in the future" but rather "work hard, borrow heavily, and consume as much as possible now." As a result, personal savings rates in the U.S. plunged to historic lows and rapidly approached zero early in the twenty-first century, while federal deficits reached historic highs.

Should people strive to make the future better off than the present? Do we have at least an obligation to make sure it is not worse off than the present? There are no easy answers to the "appropriate" distribution of wealth between generations. Even a brief survey of philosophies is beyond the scope of this text. We will quickly examine two alternative approaches: the ecological economics approach, based on ethical judgments concerning obligations to future generations (intergenerational justice), and the more mainstream approach in economics that argues for an "objective" decision-making rule (intergenerational allocation).

The Normative Approach of Ecological Economics

Ecological economists generally take the position that intergenerational resource distribution is an ethical issue. The generation into which someone is born is based entirely on chance. There is therefore no moral justification for claiming that one generation has any more right to natural resources, the building blocks of the economy, than any other. At the very least, future generations have an inalienable right to sufficient resources to provide a satisfactory quality of life. The current generation thus has a corresponding duty to preserve an adequate amount of resources. What is adequate depends on both technological and ecological change, both of which are characterized by pure uncertainty (ignorance). How we choose to deal with uncertainty is also an ethical decision.

What does this mean in practical terms?

Renewable and nonrenewable resources are fundamentally different and must be treated separately. An equal distribution of finite nonrenewable resources among a virtually infinite number of future generations would imply no resource use by any single generation. But there is no point in leaving resources in the ground forever, never to do anyone any good, so an upper limit to exhaustible resources for any one generation might be determined by the waste absorption capacity of the environment. As long as the use of the resource generates waste no faster than the ecosystem can absorb it, the use of exhaustible resources by one generation will not reduce renewable natural capital. Keeping fossil fuel use within such limits would automatically limit our ability to extract other mineral resources.

Even with a limited ability to extract nonrenewable but recyclable resources, each generation would have a further obligation to efficiently recycle such resources or at least minimize the generation and dispersion of "garbo-junk" as much as possible to make such resources as intergenerationally nonrival as possible. If existing technologies make our well-being dependent on nonrenewable resources—as is currently the case—then we are simply obliged to develop substitutes for these resources. One option would be to capture marginal user costs, the unearned income from nonrenewable resources, and invest them toward developing such substitutes.[10]

Renewable resources as fund-services provide essential life-support functions, and these functions clearly must be maintained. Renewable resources as stock-flows must also be harvested at sustainable levels. No one created renewable resources, and therefore no single generation has the right to reduce the amount of the resource a future generation can sustainably consume, suggesting resource stocks must be at least as large as that which provides the maximum sustainable yield. As we saw in Chapter 12, sustainable management of renewable resources in a manner that "optimizes" both stock-flow and fund-service benefits will in general maintain these resources far from any catastrophic ecological thresholds. It is worth bearing in mind that as nonrenewable resources are finite, the exhaustion of these resources is a finite loss to future generations. Renewable resources, as both stock-flows and fund-services, produce a finite flow over an immeasurable number of future generations, and their irreversible loss therefore imposes a perpetual cost to the future.

The "Positive" Approach of Neoclassical Economics

Conventional economists, in contrast, favor an objective decision rule to determine the intergenerational allocation of resources. The problem thus becomes simply a technical one of comparing future benefits and costs with those that occur in the present. From this point of view, the market can tell us the value of things in the future relative to things today, and therefore the market can solve the problem of intergenerational allocation.

---

[10]For practical guidelines on investing scarcity rents, see S. El Serafy, "The Proper Calculation of Income from Depletable Natural Resources." In Y. J. Ahmad, S. El Serafy, and E. Lutz, eds., *Environmental Accounting for Sustainable Development*, Washington, DC: World Bank, 1989.

Intertemporal Discounting. How does the market reveal future values? Where adequate financial markets exist, people can borrow money today at interest, which requires them to pay back more money in the future. The fact that people engage in this activity reveals that people prefer things now rather than in the future, and economics must respect people's preferences.

There are three basic reasons why people might prefer things now to things in the future. First, people may simply be impatient. Anyone who goes into interest-bearing debt to purchase something is willing to sacrifice a greater quantity in the future for a smaller quantity now. Some of this impatience may come from uncertainty—no one knows for sure if he or she will be alive tomorrow, so why not eat, drink, and be merry today? This rationale for discounting is known as the **pure time rate of preference (PTRP)**.

Second, for things that reproduce, it makes sense that a given quantity in the future would be worth less than a given quantity now. For example, a handful of seed corn now can become a bushel of marketable corn a few months from now, so if growing corn was risk-free and required no resources or effort, then a handful now would be at least as valuable as a bushel in a few months. Of course, growing corn is risky and requires land, labor, and resources. However, market goods (in this case, seed corn) can be sold for money. Investing the money earned from the sale of the seed corn in an insured bank is not very risky and for the individual investor basically requires no further resources or labor. As we explained in Chapter 10, this rationale for discounting is known as **opportunity cost**, the lost opportunity to invest. If money is a substitute for any other resource, then we should give more weight to any resource today over the same resource tomorrow.

Third, the economy has grown fairly steadily for hundreds of years. People therefore expect that they will be richer in the future than they are today. Just as an extra $1000 provides less utility to Bill Gates than to a pauper, the law of diminishing marginal utility means that money in the future will be worth less than the same amount of money today. This is sometimes referred to as the "richer future" argument.

In general, this process of valuing the future less than the present is known as **intertemporal discounting**, introduced in Chapter 10. Businesspeople explicitly discount the future when making investment decisions, and mainstream economists argue that people automatically apply this concept to all of their purchase decisions. As a result, they conclude, the market efficiently allocates goods between the present and future.

What's more, if intertemporal discounting leads to allocative efficiency in the market, then it should also be applied to nonmarket investments. For example, one of the biggest nonmarket decisions we face today is how to deal with global climate change. Virtually all economic analyses of climate change place a lower weight on future costs and benefits than on present ones. These analyses look at different policy scenarios and for each sum up the present costs and benefits with discounted future costs and benefits to arrive at a **net present value (NPV)**. NPV basically tells us what present and future costs and benefits are worth to us *today* (not to the future *tomorrow*), which implies that future generations have no particular right to any resources, and we have no obligation to preserve any. Under this type of benefit-cost analysis, the higher the NPV relative to required investments, the better the project.

Such analyses can carry a great deal of weight as society decides how to address some of the most pressing problems we now confront. The central importance of the discount rate in determining the outcome of such analyses means the topic deserves our attention.

Box 16-3	INTERTEMPORAL DISCOUNTING AND GLOBAL CLIMATE CHANGE

Policy makers seeking an objective decision-making tool for resolving the problems of global climate change have turned to economists. Economists typically respond to the problem by creating complex models of future costs and benefits of climate change and compare these to the costs of mitigation measures in a cost-benefit analysis designed to calculate net present value. Not surprisingly, analyses using a fairly high discount rate find that future damages from global warming do not justify efforts today to reduce greenhouse gases. The 6% discount rate used in one study would have us believe that we should not invest $300 million today to prevent $30 trillion (a rough estimate of today's global GNP) in damages in 200 years.[a] A similar study using a 2% discount rate, in contrast, finds that we should make substantial investments now to reduce the impacts of global warming in the future. Similarly, what we decide to do with an old-growth forest that supplies a small but steady flow of benefits forever if left intact or a large, one-time return if it is clear-cut will depend on the discount rate we choose.

A frequently asked question is: Does a higher or lower discount rate favor the environment? For a given fishery or mine, as we have seen, higher discount rates increase the intensity and rate of exploitation and therefore are bad for the environment. But a higher interest rate (discount rate) slows down aggregate growth in GNP and throughput, thus easing pressure on the environment. In terms of evaluating a given project, a high discount rate favors projects whose costs are mainly in the future and whose benefits are in the present, rather than those whose costs are in the present with benefits in the future. Most issues in economics are not simple, and that certainly holds for discounting.

In many such models, the choice of a discount rate may be the single most important factor, yet respected economists addressing the same problem use dramatically different rates and arrive at dramatically different results. Are such models actually objective decision-making tools?

---

[a]*30 trillion is a number that's hard to wrap your mind around. Putting it into perspective, 30 trillion seconds is slightly less than a million years (951,294 years, to be precise).*

Discounting Reconsidered

We have already explained why intertemporal discounting can make sense for the individual and for market goods. We must now examine whether it also makes sense for society and for nonmarket goods.

We saw why individuals might have a pure time rate of preference: People are impatient; they don't live forever; possessions can be lost, destroyed, or stolen, and opportunities disappear. A reasonable individual may discount the future for any one of these reasons—why should I pay money now to reduce damages from global warming that will occur only after I am dead?—but the same logic does not apply to society. Relative to the individuals of which they are composed, societies are immortal, and uncertainties are averaged out. For this reason, there is, in fact, fairly wide consensus within the economics profession that social discount rates should indeed be lower than individual discount rates. The **social discount rate** is a rate of conversion of future value to present value that reflects society's collective ethical judgment, as opposed to an individualistic judgment, such as the market rate of interest.

When it comes to the opportunity cost of capital, however, the consensus changes. Financial capital does function as a productive asset, and if we have it now instead of in the future, we have the opportunity to invest it in productive activities that will increase the quantity of market goods in the future. There are a number of important issues we must bear in mind, however.

First, the real value of money can grow only if the production of goods and services that money can acquire also grows, and we know that the production of goods and services cannot grow forever on a finite planet. While there may always be some areas that are growing, justifying a discount rate for the individual, the economy as a whole cannot grow indefinitely, in which case a social discount rate into the indefinite future may be inappropriate.

Second, we must recognize that many investments are "profitable" because we ignore many of the costs of production. We know that all human productive activities use up natural resources and return waste to the environment, and these costs of production are often ignored. Many of these costs, such as contributions to global warming, have greater impacts on future generations. Thus, ignoring costs to future generations allows us to earn higher returns on investments. We then use these higher returns to justify the fact that we ignore costs imposed on future generations. Even in the short run, then, it seems that market-determined interest rates are not suitable discount rates.

Related to the opportunity cost of capital is the argument that the future will be richer than the present because of investments we make now. Of course, if the economy does not continue to grow, the future will not be richer, and if we deplete our natural resource stock, there is every chance the future will be poorer. In fact, measures such as the ISEW suggest that society is already growing poorer, not richer, if we take into account external costs. Also, if we believe that natural capital must be treated separately from manmade capital (because they are complements rather than substitutes and natural capital has become the limiting factor), then the decline in natural capital, coupled with the law of diminishing marginal utility, suggests we should apply a negative discount rate to natural capital. At the very least, we might consider applying a positive discount rate only to goods and services that are actually highly fungible with money—that is, that can be converted into money and back again with little effort. Basically, this would mean that we should discount only market goods and services.

Third, there are only finite opportunities for productive investment in an economy; investments, like other things, show diminishing marginal returns. For example, someone borrows money to explore for oil, and someone else borrows money to build a car factory. The next person to borrow money to explore for oil will have fewer places to explore and therefore will expect lower returns. The next person to borrow to build a car factory will face a more saturated market and therefore expect to sell fewer cars. As more and more people borrow to invest and opportunities are used up, the returns on investments can be expected to decrease, ultimately falling to zero. More likely, if interest rates are determined in the market by the supply and demand of money for borrowing and returns on investments, a balance will be reached in which investors cannot afford to pay high enough interest rates for consumers to be interested in deferring consumption. Theoretically, then, in a perfect market situation, the opportunity cost of capital at the margin will just equal the PTRP of the existing generation. (Obviously, future generations cannot take part in financial markets any more than they can in any other market.) However, a high PTRP means consumption will be high and investment and growth low,[11] and a low PTRP implies the op-

---

[11]Returns on investments will be high, but the total amount invested will be low, and therefore growth will be low.

posite. Thus, if we allow the market interest rate to determine the discount rate, there would theoretically be an inverse correlation between discount rate and economic growth, the exact opposite of what would justify the "richer future" rationale for discounting.

Box 16-4 DISCOUNTING, PSYCHOLOGY, AND ECONOMICS

Economists argue that economics is the science of human preferences, so human preferences must be respected. If people value the present more than the future, we must respect that. The question is: Do people *exponentially* discount the future? While it is true you may prefer to have something now rather than the same thing in 5 years, how do you value something that happens 100 years in the future compared to 105 years? If you heard that global warming was going to result in the deaths of 50 million Bangladeshis in 125 years, would that make you feel only half as bad as finding out it would actually kill those 50 million Bangladeshis in 100 years? If you are like most people, you would feel equally bad in either case, yet influential economic models of the impacts of global warming really do assume we would care only half as much about those deaths if they occurred 25 years later.

Empirical studies show that people do discount the future but do not do so exponentially. We might give more weight to what happens now than to what happens in the near future, but we are nearly indifferent between the same outcome occurring at different times in the more distant future. One approach to modeling this type of behavior mathematically is known as **hyperbolic discounting**. While this precise formulation of intertemporal discounting may not be perfect, evidence suggests it is far more representative of human preferences than exponential discounting. While the approach was first introduced over 30 years ago and has gained increasing attention in the last few years, it is still fairly rare to see it in use.

An increasing number of studies in the area of Behavioral Economics are finding that the traditional economic assumptions of human behavior are often seriously flawed. If economics is serious about becoming the science of human preferences, it would do well to pay more attention to how humans really behave.[a]

---

[a]*For a good introduction to the field of Behavioral Economics, see R. Thaler and C. Sunstein. Nudge: improving decisions about health, wealth, and happiness. Yale University Press, New Haven. 2008.*

Finally, many economists argue that technology is the driving force for economic growth. Not only does technology ensure we won't run out of resources and the economy won't stop growing, but it offers yet another reason to discount the value of resources in the future. Technology is likely to develop substitutes for natural resources. When these substitutes become available, the resources they replace will lose value. Therefore, they will be worth less in the future than they are today. After all, hasn't oil largely replaced coal, and haven't fiber optics replaced copper in many uses? However, technology ultimately complements resources and can never completely replace them. Some 150 years ago, oil had little value. Today, it is an integral part of an overwhelming number of industrial processes and products. As we saw in Chapter 5, we are actually developing new uses for oil and other raw materials faster than we are developing replacements, again suggesting that the value of raw materials will increase in the future, not shrink.

What can we say about discount rates in the end? They do make sense for individuals in the short run. For some small-scale, short-term social projects, they may also make sense. However, justifications for discounting the future on a large scale and over long time horizons are questionable at best.[12] **Intertemporal allocation** is the apportionment of resources across different stages in the lifetimes of basically the same set of people (the same generation). Discounting can make sense for someone efficiently allocating resources intertemporally. But as we lengthen the time period we are more and more talking about different people (different generations) and less and less about the same people at different stages in their lives. **Intertemporal distribution** is the apportioning of resources across different generations, across different people. Distribution is fundamentally different from allocation, and, consequently, justice replaces efficiency as the relevant criterion for policy when time periods become intergenerational.

BIG IDEAS to remember

- Pareto optimality
- Role of scale and distribution in defining Pareto optimal allocation
- Income distribution vs. wealth distribution
- Functional vs. personal distribution
- Social limits to range of inequality
- Lorenz curve
- Gini coefficient
- Inequality and health
- Intertemporal distribution vs. intertemporal allocation
- Discounting and net present value
- Pure time rate of preference (PTRP)
- Individual vs. social discount rates

[12]See A. Voinov and J. Farley, Reconciling Sustainability, Systems Theory and Discounting, *Ecological Economics* 63:104–113 (2007) for a more detailed discussion.

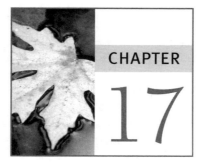

The IS-LM Model

We have now explored three of the major issues with which macroeconomics is concerned: gross national product (GNP), money, and distribution. We questioned the appropriateness of GNP as the desirable end for economic policy and emphasized the importance of a just distribution as a desirable end but said little about policies for attaining whichever ends we choose to pursue. In this chapter, we examine the policy tools at the macroeconomist's disposal that can help us attain an economy with sustainable scale, just distribution, and efficient allocation.

Of course, to know how policies work, we have to know how the macroeconomy works. One way of doing this might be to build on microeconomic principles to construct a model in which supply and demand of all goods and services balances simultaneously. This approach would extend the basic market equation presented in Chapter 8—MUxn*MPPax = MUyn*MPPay—into a general equilibrium model encompassing all goods ($x, y, z \ldots$), all commodities ($a, b, c \ldots$), and all consumers ($n, m, o \ldots$). Such a model can easily become overwhelming. A thousand simultaneous equations with a thousand unknowns is hard to come into mental contact with. It does show that everything depends on everything else, which is interesting and usefully humbling, but it is also crippling from a policy perspective to have to face the implication that in order to predict anything, you have first to know everything. But a smaller system of two or three or five especially important aggregate sectors interacting through two or three or five simultaneous equations that reflect key behavior can aid the understanding and give basic policy insights. This is the kind of model that most macroeconomists have sought. They still look at the whole economy, but they divide it into fewer but larger aggregate sectors than does the general equilibrium model of microeconomics.[1] A model of this type, first offered in 1937 by Sir John Hicks[2] and now called the IS-LM model, has proven to be a good "two-digit" compromise between completeness and simplicity. It has become the workhorse model in macroeconomics. Below we will explain this model and then discuss its applications to ecological economics.

The model divides the economy into two sectors: the real sector (dealing with national income, savings, investment, rates of productivity of capital, government spending, taxation, etc.) and the monetary sector (money supply, interest rates, demand for liquid cash balances). The real sector reflects the theories and insights of classical economics, and the monetary sector reflects the insights of John Maynard Keynes, which in 1937 were still quite new. The model seeks to explain how the interdependent behavior of consumers and savers, lenders and borrowers, and monetary authorities interact to determine the level of national income and the rate of interest.

BOX 17-1   THE QUANTITY OF MONEY THEORY OF INCOME

Another way of relating the real and monetary sectors in an aggregate way is through the "identity of exchange," $MV = PQ$, where Q is quantity of final commodities sold to households, P is average price of exchange, M is stock of money, and V is velocity of circulation of money (number of times an average dollar is spent per year on final goods and services). Since by definition $V = PQ/M$, the equation of exchange is an identity or truism. To the extent that V is a constant or slow to change, reflecting stable payment habits and settlement periods, the identity becomes the "quantity of money theory of income," stating that changes in PQ are proportional to changes in M. If the economy is at full employment, it will be very hard to increase Q in the short run, and the change in PQ will be mainly a change in P (i.e., inflation). Historically M and P have often moved in direct proportion, yielding a quantity of money theory of the price level.

---

[1] Our measures of the two most basic magnitudes of macroeconomics, GNP and money, are too dialectical and uncertain to be able to support exact calculations implicit in complicated models. As Oskar Morganstern remarked in his classic *On the Accuracy of Economic Observations*, "economics is a two-digit science."

[2] J. Hicks, Mr. Keynes and the "Classics," *Econometrica* 5(2) (April 1937).

■ IS: THE REAL SECTOR

Let's begin with the real or classical sector. The real sector is in equilibrium when the supply of goods by firms is just equal to the demand for
goods by households (the lower half of the circular economy in Figure
2.4). Of course, the demand for goods by households is determined by
their income—the money firms pay households for their factors of production (e.g., labor)—and the supply of goods is determined by the firms'
employment of those factors of production (the upper half of the circular
economy in Figure 2.4). In equilibrium, income (Y) equals output (GNP).
Remember from the circular flow diagram in Figure 2.5 that the equilibrium condition for the continued flow of national income at a given level
is that leakages equal injections. In the simplest case, the leakage is savings (S) by households, the new injection is investment (I) by firms.
Therefore, the equilibrium condition for the real sector is S = I.

But how do S and I get determined? Let r be the interest rate and Y be
national income (GNP). In equilibrium, income paid to the factors of production will just equal the output of goods provided by those factors of
production, and the income will be used to purchase the output. Savers
(i.e., households) will save more if their income Y is higher than if it is
lower. Also savers will save more with a higher interest rate r than with a
lower one. Investors (i.e., firms) will borrow and invest more if the interest rate is lower and if income is higher.[3] In other words, savings is some
function of the interest rate and national income, say S = S(r, Y). Likewise,
investment is some different function (representing the behavior of firms
instead of households) of the same two variables, say I = I(r, Y).

In equilibrium,

$$S = I$$

or

$$S(r, Y) = I(r, Y)$$

The above equation is satisfied for all combinations of r and Y such that S
= I, that is, such that savers and investors are both satisfied.

There are many such combinations of r and Y; we have only one equation with two unknowns. Plotting all the combinations of r and Y that re-

---

[3]Throughout this chapter, if you get confused about the relationships between savings, investments, interest rates, and incomes, think in extremes. For example, who will save more, someone who earns $10 million per year or someone who earns minimum wage? Will firms invest
more in the huge U.S. economy or in the small Haitian economy? Will people save more money
at interest rates of 500% per year or −50% per year? Which of these interest rates would induce
firms to invest more (remember that firms must either borrow money to invest it or forgo the interest they would make by lending their own money)?

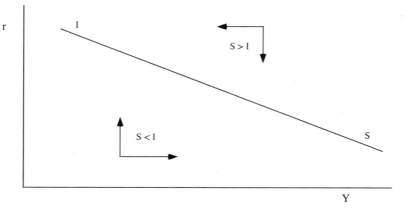

Figure 17.1 • The IS curve: At low (high) levels of income Y, there is a correspondingly low (high) level of savings. At high (low) rates of interest r, there is a low (high) demand for investment. Therefore, at low (high) levels of Y, savings and investment will be in equilibrium only when r is high (low). If interest rates are too high for a given level of income, savings (leakages) will be greater than investment (injections). Firms producing more goods than people consume reduce production, and the economy shrinks. Firms with excess capacity borrow less, so the price of borrowing (the interest rate) falls to clear the market. The converse is true when investment is greater than savings.

sult in S = I gives us Hicks' so-called IS curve, short for I = S (Figure 17.1). To reiterate, this is the combination of r and Y that leads to equilibrium in the real sector: leakages (savings) equal injections (investment), and the demand for goods is just equal to the supply.

Why is the IS curve drawn with a negative slope? Businesses will borrow money to invest only if they can make sufficient returns from the investment to pay off the loan plus interest and still have money left over for profit. A businessperson would not borrow money at 6% interest to invest in a project expected to return 5% annually on the investment but would borrow at a rate of 4%. As interest rates go down further, more and more investments become profitable, and therefore more investments are made. More investment leads to higher Y. Therefore, high interest rates lead to low rates of investment and low income, while low interest rates lead to high rates of investment and high income. Savings, in contrast, is probably determined more by income than by interest rates.[4] When income is low, all money has to be spent simply to meet basic consumption needs, and none is available to save. As income increases, basic consumption

---

[4]Savings rates should also increase as interest rates increase, as under these circumstances savings yield higher returns, and consumption has a higher opportunity cost. But empirical evidence does not support this. One reason may be that if savers are motivated by attainment of a target future amount, a higher interest rate would mean less saving is needed to reach the future target.

Box 17-2 **A GRAPHIC DERIVATION OF THE IS CURVE**

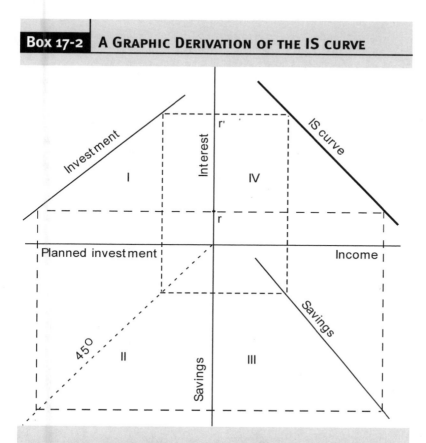

Figure 17.2 • A graphic depiction of the derivation of the IS curve.

Figure 17.2 illustrates one way to derive the IS curve. Quadrant I shows the basic relationship between interest rates and investment—high interest rates lead to low levels of investment, and low interest rates lead to high levels of investment. There is a negative correlation between interest rates and investment, as depicted on the graph. Quadrant III shows the relationship between savings and income. Poor people must spend all their income to meet their basic needs and cannot afford to save anything. As income increases, people begin to save, so there is a positive correlation between income and savings, as depicted on the graph. We know that in equilibrium (which is what the IS curve depicts), investment equals savings. Quadrant II contains a 45-degree line that allows us to translate a given rate of investment from quadrant I to an identical rate of savings in quadrant III. Quadrant IV shows the relationship between income and interest in a real sector equilibrium.

If we start with interest rate r, we can see from quadrant I that this will correspond to level of investment I. Dropping a line down from I in quadrant I to the 45-degree line then across to quadrant III lets us determine the equilibrium level of savings S in quadrant III. We can see from quadrant III that this level of savings corresponds to income Y. The point in quadrant IV where income Y meets interest rate r gives us one point on the IS curve. If we do the same for interest rate r', we have two points on the IS curve. We see that a low level of interest leads to equilibrium only when income is high, and a high level of interest leads to equilibrium when income is low. Perhaps the simplest way to remember this relationship is that at low interest rates, investment will be high, and high investment leads to high income.

It really helps to understand this curve if you derive it yourself.

needs require a smaller percentage of income, and more is left over to save, so in general higher incomes lead to greater savings.

Combining these two tendencies, we would expect that at high levels of income when lots of money is being saved, investors will borrow all that money to invest only if interest rates are low. At low levels of income, savings are low, and unless interest rates are high, businesses will demand more money than is being saved. For some readers, a diagrammatic explanation for the negative slope of the IS curve will be easier to follow (Figure 17.2).

Macroeconomics does not assume that the economy is always in equilibrium, but it does assume that it is at least moving in that direction. For example, we know that if r rises, then savers will try to save more, and investors will be less willing to borrow and invest, leading to a condition in which planned S > I. In other words, savers want to save more than investors want to invest at the new higher r. This will have two impacts. First, leakages will be greater than new injections, causing income to fall. Second, savers earn interest on their savings because investors are willing to pay that interest to borrow the money. Interest is the price of money. When the supply of savings is greater than the demand for savings, the interest rate must fall. The mechanism is the same as for any other good, as explained in Figure 9.2. At a lower Y savers save less, and at a lower r investors borrow more, and both r and Y continue to fall until I again equals S at a lower income (Y) and higher interest rate (r) than before. If the interest rate falls, then investment will become greater than savings, and adjustment will occur in the opposite manner. These dynamics are indicated by the arrows in Figure 17.1.

■ LM: THE MONETARY SECTOR

We turn now to the monetary sector and the LM curve, which shows the levels of income (Y) and interest (r) at which the demand for money balances (money held by people) equals the supply of money. We must first ask why individuals want to hold money balances when they could easily exchange them for real assets. From our earlier discussion the answer is clear: people hold cash balances to avoid the inconvenience of barter. Keynes referred to this as the **transaction demand for money**. He also spoke of a related **liquidity preference**, meaning that, other things being equal, people prefer liquid assets to "frozen" assets because they are so easily convertible into anything else, therefore fungible. Money is the most liquid of all assets. But of course other things are seldom equal, and the cost of holding wealth in the form of fungible money is to forgo the interest that could be had by lending the money or the utility from spending it on a real asset or commodity. Yet if too much of your wealth is tied up in nonliquid forms, you will have difficulty making necessary transactions in a timely manner and meeting unexpected contingencies. The higher the national income, the more need for transactions and consequently the more money everyone will need (a higher transactions demand for money), and the higher the interest rate will have to be to induce owners of those transaction balances to sacrifice liquidity by lending them.

The demand for money balances (DM) thus depends on r and Y, by means of a relation of liquidity preference (L). Thus,

$$DM = L(r, Y)$$

The equilibrium condition is that the demand for money equals the supply of money (SM):

$$DM = SM$$

What determines the supply of money? In earlier times it was the geology and technology of gold or silver mining (a part of the real sector),

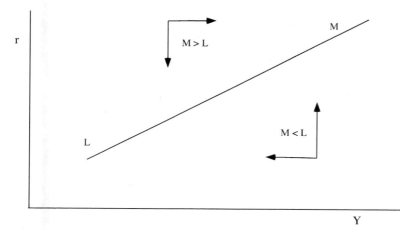

Figure 17.3 • The LM curve.

but today we have not real commodity money but fiat or token money, controlled by the government through the private banking sector, as discussed in Chapter 15. For simplicity, the model usually takes SM as given by the government, equal to M. Thus,

$$L(r, Y) = M$$

is the equilibrium condition for the monetary sector, and the LM curve consists of all combinations of r and Y such that the aggregate demand for cash balances is equal to the given money supply (Figure 17.3). Since we have one equation with two unknowns, we cannot get unique values of the unknowns, but we can determine all the combinations of r and Y that satisfy our one equation.

Box 17-3 **THE FEDERAL RESERVE BANK**

In the U.S., money is *not* controlled by the democratically elected government but rather by the Federal Reserve Bank (the Fed), a nonelected "branch" of government. Decisions concerning monetary policy are decided upon by a seven-member board of governors, with lesser influence by the directors of the 12 regional Federal Reserve Banks. Members are appointed by the president (with Senate approval) for 14-year staggered terms, and the chair and vice chair are appointed for 4-year terms. Despite the importance of monetary policy in the functioning of our economy, the system is specifically designed to insulate the Fed from pressure by democratically elected politicians. The Fed is not expected to respond to voters. This does not mean that the Fed does not have a constituency to which it feels responsible, as we shall discuss later.

Why is the LM (short for L = M) curve drawn with a positive slope? Let's ask ourselves what are the consequences on the interest rate (r) of an increase in income (Y). A larger Y means a larger volume of transactions and will cause a greater demand for transaction balances. This will lead to a higher r to compensate for the loss in liquidity from lending those balances. Thus, a higher Y will require a higher r for money holders to again be satisfied (for L to equal M), hence the positive slope of the LM curve. This relationship seems to be sufficiently clear that a more detailed graphic explanation is unnecessary.

When the monetary sector is out of equilibrium, what specific mechanisms drive it toward equilibrium? Say the monetary authority increases the money supply, so there is more money available than people actually desire to hold at the existing interest rate—that is, M > L. Excess money is used to buy bonds and other nonliquid interest-bearing assets (which we will refer to jointly as "bonds" for convenience). More money chasing the same number of bonds will drive up their price.

There are many types of bonds, but in the simplest case, when someone buys a bond, they are paying something now to receive a fixed amount when the bond matures. For example, if I pay $50 today for a $100 bond that matures in 10 years, my rate of return is about 7.2%. An increase in the money supply might drive the price of the bond up to $60, which provides a rate of return of only 5.24%. The higher the price for a bond, the lower the interest rate on that bond. Thus, an increase in the supply of money increases the demand for bonds and drives down the interest rate.

At lower interest rates, there is less opportunity cost to holding money and hence a higher demand for money. Lower interest rates also stimulate investment, leading to economic growth, which further stimulates the demand for money. The result is a new equilibrium at lower interest rates and higher income. A decrease in the money supply of course leads to the opposite result. These forces are illustrated by the arrows in Figure 17.3.

■ COMBINING IS AND LM

Putting the IS and LM curves together lets us determine a unique combination of r and Y (namely r*, Y*) that satisfies both the S = I condition of the real sector and the L = M condition of the monetary sector (Figure 17.4). The point of intersection is the only point common to both curves, the only point that gives equilibrium in both real and monetary sectors. Basically we now have two simultaneous equations determining two unknowns, r and Y.

The **IS-LM model** is used in a comparative statics[5] way to analyze the effect on r and Y of changes in the underlying determinants—namely, propensity to save, the efficiency (productivity) of capital investment, and liquidity preference. Of particular interest to policy makers is the impact of policy variables on r and Y—namely, government expenditure, taxation, and the money supply. Each of these changes results in a shift in one of the curves and consequently in a move along the other curve toward a new intersection point. What we are really interested in, then, is how the economy moves toward equilibrium after policies or outside (exogenous) changes push it away.

---

[5]Comparative statics is the analysis of what happens to endogenous variables in a model (in this case, r and Y) as a result of change in exogenous parameters (in this case, propensity to save, efficiency of capital investment, and liquidity preference). It compares the new equilibrium variables with the old ones, without explaining the precise dynamic path leading from the old to the new equilibrium.

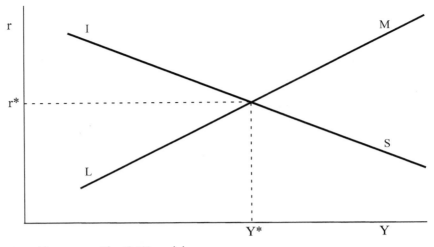

Figure 17.4 • The IS-LM model.

■ EXOGENOUS CHANGES IN IS AND LM

First let's look at some exogenous changes, those that are basically independent of fiscal and monetary policy and therefore outside the IS-LM model. Consider an increase in the marginal propensity to save. Such a change in savings rate might result from fears of an economic downturn that would lead to lower wages and greater unemployment. (We might hope that people might one day simply decide to consume less in order to protect the environment.) In either case, people decide to save more and spend less of their disposable income. This means that now S > I for all the combinations of r and Y on the IS curve. We need a new IS curve for which S = I again. If people save more at every r, this means S > I, or leakage greater than injection, so the flow of income will fall to the level at which S = I again. How does this happen? Of course, if the marginal propensity to save increases, then the marginal propensity to consume must decrease. As people consume less, businesses will be unable to sell their goods, leading to unplanned accumulations of inventory. This in turn will lead businesses to reduce planned investment and production—perhaps laying people off. Unemployment resulting from layoffs further decreases consumption, requiring another round of adjustment, lowering Y still more. At the same time, an increase in the supply of money due to higher savings and a decrease in the demand for money due to lower investment drives down the costs of money, that is, the interest rate (which may ameliorate to some extent the decline in investment). When the dynamics have played out and S = I again, every r will be paired with a smaller Y in the new IS than in the old one. The IS curve will have shifted to the left. The new intersection with LM will occur at a lower r* and lower Y* than before. Even though people are saving a larger fraction of their income, they will end up having a smaller income out of which to save, with the result that S will increase by less than the marginal propensity to save. The final result, when S = I again, may well be that S will be lower than the level at which we started. Thus, the effort of everyone to save more in the aggregate could result in everyone actually saving less— the so-called paradox of thrift. In such a case, a higher savings rate induced by fear of recession could itself cause a recession—a self-fulfilling prophecy.

Now suppose there's an increase in the efficiency of investment (an increase in the marginal productivity of capital), thanks to a new invention. For example, many people claim that this is exactly what has happened in today's "new economy," in which information technology is said to have increased productivity. This would increase I, so that I > S now along the old IS curve. With I > S, injections are greater than leakages out of the circular flow, so the flow of income will grow until S = I again. The new IS curve will have a higher Y for each r. The curve shifts rightward. The new equilibrium occurs with a higher Y* and a higher r*. An improvement in the marginal efficiency of capital raises both income and the interest rate.

Finally, turning to the LM curve, suppose there was an increase in liquidity preference, so that L > M. Such a change could result from increasing uncertainty over future economic conditions and a desire by people to be prepared for the unforeseen with cash on hand. Alternatively, the deregulation of banking in the United States during the mid-1970s allowed certain checking accounts to pay interest. This reduced the opportunity cost

Box 17-4   JUNK BONDS AND TIMBER COMPANIES

Seemingly abstract things like interest rates on bonds and Wall Street transactions can affect real economic production and the provision of environmental services. For example, during the 1980s, hostile takeovers and the introduction of junk bonds on Wall Street led to deforestation on the West Coast. How did this happen? Mergers, when two companies join together, and acquisitions, when one company purchases another, are a normal part of corporate activity in the U.S. Sometimes, however, one company does not wish to be taken over by another. For example, mergers and acquisitions (M&A) focused primarily on short-term profits can weaken or destroy the company being acquired, leading to massive layoffs.

Reasonably enough, managers not eager to be laid off will be opposed to a merger, and under such circumstances, takeover attempts are "hostile." One company acquires another through the purchase of a controlling share of stocks. As soon as someone starts buying enough stock to control a company, the stock price rises. A company threatened by a hostile takeover can attempt to defend itself by repurchasing its own stock, driving up the price of stock even further. To get enough money for a hostile takeover, the company attempting the takeover can offer high-yield, high-risk bonds known as "junk bonds" in Wall Street jargon.[a]

The best target for takeover is a company that has lots of assets in a nonliquid form that can be liquidated after takeover to pay off the junk bonds but cannot be sold quickly to defend against the takeover. Timber companies have valuable assets in the form of forests that can be liquidated after takeover but cannot be sold quickly to buy back stock and prevent a hostile takeover. This made them popular takeover targets during the 1980s.

A classic example is Charles Hurwitz's acquisition of Pacific Lumber in the mid-1980s. Pacific Lumber fought the takeover, but using a combination of junk bonds and short-term loans, Hurwitz won out, acquiring the company's 196,000 acres of forest, including the largest unprotected stands of virgin redwood in the world. Hurwitz was saddled with an enormous debt and crushing interest payments. To repay the debt, Hurwitz liquidated much of the forest stock, including many old-growth redwood groves. Some illegal cuts were conducted on weekends and holidays to avoid state regulators. Wall Street innovations during the 1980s accelerated the decimation of the nation's last remaining virgin forests and of the environmental services those forests once provided.[b]

[a]*Different companies (and cities and countries) have different credit ratings based on their financial soundness. Bonds from financially sound companies are themselves very sound, bonds from less sound companies are not, and the risk of default is higher.*

[b]*N. Daly, Ravaging the Redwood: Charles Hurwitz, Michael Milken and the Costs of Greed,* Multinational Monitor 16(9) (1994). Online: http://multinationalmonitor.org /hyper/issues/1994/09/mmo994_07.html.

of holding money and therefore probably increased the liquidity preference as well. In either case, for any income and associated level of needed transaction balances, there is a greater willingness to hold those balances, to hold more than strictly needed. It takes a higher r to induce holders of money to lend. Consequently each level of Y will be associated with a higher r on the new LM than on the old one. At a higher r, investment is likely to decrease, leading to a decrease in Y. The new LM will shift upward. The new equilibrium will occur at a higher r* and lower Y*. An increase in liquidity preference raises the interest rate and lowers national income.

The above analysis of changes in propensity to save, efficiency of investment, liquidity preference, and money supply is summarized in Figure 17.5.

■ IS-LM AND MONETARY AND FISCAL POLICY

Changes in the propensity to save, the efficiency of investment, and liquidity preferences are not brought about directly by policy interventions; they are affected by psychology and technology and as a result are difficult, if not impossible, to predict. However, policy makers do have two sets of economic levers by which they can influence these variables: monetary policy and fiscal policy.

What does the IS-LM model tell us about different monetary and fiscal policy levers? The analysis of monetary and fiscal policy in macroeconomics can be worked out by tracing the effects on IS or LM of changes to the money supply and of government taxing and spending.

Monetary policy basically affects the money supply. When the mone-

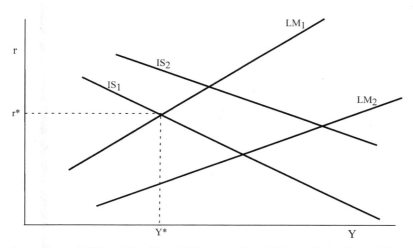

Figure 17.5 • Shifting of the IS and LM curves. The shift from IS$_1$ to IS$_2$ could be the result of either a decrease in the marginal propensity to save or an increase in the marginal efficiency of investment. The shift from LM$_1$ to LM$_2$ could be caused by either a reduction in liquidity preference or an increase in the money supply. Can you work out the changes in r and Y resulting from an increase or decrease in each of the four parameters, others remaining constant? What about changes in two or more parameters at the same time?

tary authority (the Federal Reserve, in the U.S.) increases money supply, the LM curve shifts downward and M > L, which drives interest rates down, as explained earlier. Lower interest rates stimulate the economy (consumers save less and businesses invest more), and income grows. If the money supply is increased by too much, there can be too much money chasing too few goods, and inflation threatens. Reducing the money supply drives interest rates up, shrinks the economy, and can help control inflation.

Fiscal policy is basically government expenditure and taxation. When the government spends money, industry has to produce more goods and services to meet the increased demand. This drives up income and also increases the demand for investments, driving up interest rates. The IS curve shifts to the right. Decreasing government spending has the opposite effect.

There are three ways the government can finance expenditures. First, it can impose taxes. When the government increases taxes, people have less to spend, decreasing demand and leading to less investment. The economy shrinks, and interest rates go down. However, if the government spends the entire tax increase, the stimulus of increased expenditure outweighs the contraction caused by taxes, since some of the tax money now being spent would have been saved, in which case the economy would grow while interest rates go up. Second, the government can use debt financing and borrow money. The government borrows money by selling bonds. An increase in the supply of bonds drives down the price and drives up the interest rate, as just explained on page 329. Third, the government can use its right to seigniorage to simply print and spend money, which increases the money supply. As we noted previously, the increased money supply will further stimulate the economy but will have a countervailing impact on interest rates.

Seigniorage-financed fiscal policy seems the logical choice for stimulating the economy, but it carries the threat of inflation. Governments could dramatically increase their ability to use seigniorage if they increased reserve ratios to 100%, as suggested by Frederick Soddy so long ago. The government would then be able to print and spend money when the price index started to fall, and to tax and destroy money if inflation threatened to become a problem. The government would also be able to target monetary policy much more effectively, using it to address issues of scale, distribution, and allocation.

The impact of fiscal and monetary policy depends on how much excess capacity (unemployed labor and underused capital) exists in the economy. Consider a bowling alley in a small, isolated town where the government is undertaking a large project to stimulate economic growth. When unemployment is high, wages may be fairly low, and few people have disposable income to spend on bowling. As a result, the bowling

Box 17-5

MONETARY AND FISCAL POLICY FOR A STEADY-STATE ECONOMY

A steady-state economy requires a nongrowing throughput of low-entropy matter-energy from nature, through the economy, and back to nature as waste. In addition to cap and trade policies to directly limit throughput, what type of monetary and fiscal policies can indirectly help us achieve this?

Assuming initially a fixed relationship between GNP and throughput, a steady-state economy requires a constant real money supply. Soddy's 100% reserve requirements (also supported by respected economists Irving Fisher and Frank Knight) would prevent banks from creating money. Only savings deposits could be loaned, and for a time period less than or equal to the time period of the savings account. Money would return to the role of a public utility (rather than the result of private commercial activity of lending and borrowing) and the right to seigniorage to the government. However, with a steady-state money supply, there is no additional seigniorage to be had. Governments can still use money creation and destruction as policy tools, but the two must balance. Interest-free government loans to achieve policy goals would create money, but loan repayment would destroy it, with no increase in long-term money supply. The repayment period should depend on the extent to which the loans support public goals. Under a steady-state policy, when government spends money into existence to alleviate economic downswings, it must simultaneously legislate future tax increases to destroy that money. Money creation and destruction should be counter-cyclical, leading to economic stability, rather than pro-cyclical, as occurs in the current system. The risk of the government creating money faster than the economy grows, thus causing inflation, is reduced when the goal is to reduce the size of the economy to a sustainable level.

What about fiscal policy? The ecological crises we currently face, ranging from global warming to biodiversity loss, are even more serious than World War II and justify marginal tax levels at a similar level—well over 90%. Most economists consider high taxes a disincentive to economic activity, which is exactly what is needed as we move toward a steady state. High taxes also increase economic stability and facilitate government investment in essential public goods, such as those generated by ecological restoration.

Once we have achieved sustainable throughput, technological advance may still allow growth in the real value of market goods and services, in which case government creation of new money could again be feasible. The biggest challenge lies in shifting from the old system to a new one without too much disruption. Incremental changes should be treated as scientific experiments testing the underlying theories and values and allowing us to improve upon them in a process of adaptive management. Fractional reserve requirements can be increased gradually, beginning during recessions, when banks voluntarily loan less than is allowed. In a recessionary environment, it also makes sense for governments to spend money into existence—in essence printing non-interest-bearing currency instead of the interest-bearing government bonds used in debt-financed expenditures.

alley is virtually empty. If the government funds a large project in town, some people are directly employed by the project, and they spend much of their money in town, inducing other local businesses to hire to meet the increased demand. People use their extra income to go bowling, and the bowling alley's income grows.

Now imagine that the government implements the same project in a town with very low unemployment. Bowling is popular, and the alley is full every night. The government needs employees for the project, but increasing demand when supply is low drives up wages, the price of employees. Disposable income increases, but every new bowler at the alley is simply **crowding out** another bowler, who would have to leave the alley. The alley might like to expand, but the government is borrowing money to finance its project, driving up interest rates, making it too expensive for the alley owner to expand. The alley can raise its prices with the increased demand, but it must also pay higher prices for its labor force and therefore can only break even. When an economy is at full employment, the bowling alley owners might be much better off with an expansionary monetary policy that lowered interest rates so they could expand. In contrast, if the government lowered interest rates when the alley had excess capacity, expansion would do the owners no good at all.

The failure of lowering interest rates to stimulate economies with low demand is known as a **liquidity trap**. In general, the economy is somewhere between the extremes of depression and operation at full capacity (i.e., most bowling alleys are full sometimes, but very few are always full). While increased government expenditure leads to some degree of crowding out and increased interest rates, it also increases income.

Table 17.1 summarizes the impacts of fiscal and monetary policy on interest and income. In each case, the impact is the opposite for the opposite policy.

Inflation and Disinflation

If we looked only at the IS-LM model, and if our goal was continued economic growth, the superior policy option would be clear: keep increasing the money supply to lower interest rates and stimulate investment, and use fiscal policy when necessary to stimulate demand. However, when we

▪ Table 17.1

EXPECTED IMPACTS OF BASIC MONETARY AND FISCAL POLICIES ON INTEREST RATE AND INCOME

Policy	Interest Rate	Income
Monetary expansion can be accomplished by: • Reduced reserve requirements • Selling bonds on the open market • Lowering the discount (interest) rate	(−) When economy is weak (high unemployment, low investment), monetary policy may have little or no impact on interest rates.	(+) When economy is weak, no impact on income. Known as the liquidity trap.
Tax increase	(−) Taxes (especially progressive income taxes) help stabilize the economy.	(−) Taxes collect more money when income grows and less when it shrinks.
Increased government expenditure can be spent on market or nonmarket goods.	(+)	(+)
• Financed by deficit spending	(+) Impact on interest rate may be small when economy is weak, large when economy is operating at full capacity.	(+) Income will increase when economy is weak but may not increase when it is already operating at full capacity; latter condition is known as crowding out.
• Financed by taxes	(+) Increase in interest rate is less than occurs with deficit spending.	(+) Growth rate is less than occurs with deficit spending.
• Financed by seignorage The opposite of seigniorage is when the government takes tax money out of circulation, destroying it without replacing it.	0	+ Likely to cause inflation under crowding-out conditions, with no real growth in income.

first presented the LM model, we saw that the real money supply is equal to the nominal money supply divided by prices—that is, the real money supply equals M/P. There is, therefore, another path toward equilibrium between supply and demand for money in response to an increase in money supply: price inflation. A larger nominal money supply divided by higher prices can lead to no change in real money supply. The closer the system is to full output (i.e., no excess capacity), the less output is likely to increase in response to lower interest rates, and the more likely monetary expansion is to result in inflation. **Inflation** is an increasing general level of prices (not a state of high prices).

Why are governments and monetary authorities so worried about inflation? Are their concerns justified? How does inflation affect the real economy? The first point to make is that people appear not to like infla-tion, which alone is some justification for trying to avoid it. Many econo-mists argue that inflation is regressive, but empirical support for this argument is difficult to find.[6] Empirical evidence does show, however, that real wages can fall substantially during prolonged episodes of high inflation.[7] In addition, during episodes of high inflation, it is likely that the wealthy and educated are better able to take advantage of investments and contracts that protect their money than the poor. Thus, with contin-uous high inflation, the poor may well lose ground to the rich.

Hyperinflation, often defined as inflation greater than 50% per month, can also destabilize the economy. In hyperinflation, money fails not only as a store of value but also as a medium of exchange. Impacts of moderate inflation depend to a large extent on whether it is expected or unexpected.

If everyone expects a certain rate of inflation, and their expectation comes to pass, then inflation is incorporated into contracts and causes very few problems. The only groups one would expect to lose from an ex-pected inflation are holders of money (which pays no interest) and peo-ple on fixed incomes. However, with expected inflation, most people will hold less money, and incomes are likely to be inflation adjusted. **Disin-flation** is a decrease in the rate of inflation. **Deflation** is a decline in the overall price level. Unexpected inflation, disinflation, and deflation have entirely different outcomes than expected inflation.

The most useful way to assess the impacts of unexpected inflation and disinflation is to compare debtors and creditors. When there is unex-pected inflation, any loans with nominal interest rates (i.e., interest rates that are not pegged to inflation) will be worth less and less every year. Debtors benefit and creditors suffer. For example, people in the 1960s got 30-year house mortgages at around 6%. When inflation in the 1970s climbed to over 12%, some homeowners ended up paying back less than they originally borrowed. In general, unexpected inflation systematically redistributes wealth from creditors (generally the rich) to debtors (gener-ally the poor). Most governments are net debtors, and therefore benefit, as do the future generations that are expected to pay off the government debts. However, a country cannot have unexpected inflation forever—eventually it becomes expected, or else it becomes hyperinflation, with its accompanying problems.

---

[6]A. Bulir and A. M. Gulde, Inflation and Income Distribution: Further Evidence on Empirical Links, IMF Working Papers, no. 95/86. Washington, DC: International Monetary Fund, 1995.

[7]B. Braumann, High Inflation and Real Wages, Western Hemisphere Department Series: Working Paper WP/01/50, May 1, 2001.

What happens when the government tries to cause disinflation or deflation? Obviously, just as unexpected inflation benefits debtors, unexpected disinflation must benefit creditors. In 1980, a 30-year mortgage at 14% didn't look so bad when inflation was 13% annually, and people expected their incomes to rise by at least that rate. By 1986, however, inflation (and wage increases) had fallen to less than 2%, and creditors were collecting a 12% annual real return on their 1980 loans. Thus, existing debtors suffer and existing creditors benefit from disinflation.

Other impacts of disinflation depend on whether it is brought about by fiscal or monetary policy. Theoretically inflation can be reduced by decreasing aggregate demand or increasing aggregate supply, but policy usually acts on demand. Fiscal policy can decrease aggregate demand only through greater taxation or reduced expenditure, both of which should lower the real interest rate, to the benefit of new debtors. Other distributional impacts depend on the specific policy used. For example, demand could be reduced by reducing subsidies for big business or by reducing transfer payments to the poor.

THINK ABOUT IT!

Under President Reagan there was a big emphasis on supply-side economics, increasing income by providing incentives for production (i.e., supply). Policy measures for achieving this include investment subsidies, reduced capital gains taxes, and reduced taxes for the rich. Can you explain why these policies would theoretically increase supply and reduce inflation?

The monetary authority, on the other hand, can act to reduce demand only by reducing the money supply, which increases real interest rates, to the detriment of debtors. Interest-sensitive sectors of the economy, such as farming and construction, also lose out. If losers are forced into liquidation or bankruptcy, they may be forced to sell their assets at bargain prices, and it is the well-to-do who maintain the liquidity necessary to purchase those assets. Thus, recessions may generate corporate mergers and increased concentration of the means of production.

The claim made for disinflationary policies is that in the short term the economy suffers, but in the long term stable money allows for steady growth and higher real wages. The problem is that short-term suffering can be severe, especially when monetary policy is used to decrease demand. While the jury is still out on the distributional impacts of moderate inflation, the distributional impacts of unemployment caused by disinflationary policies, as we will see below, are clear.

Unemployment

In the world of microeconomics, involuntary unemployment should not exist. Prices are set by supply and demand, and when the demand for labor is low, the price falls. At a lower wage, fewer people are willing to work, and supply falls accordingly, returning the system to equilibrium. Clearly, however, unemployment is a persistent problem in modern economies. We particularly want to examine two issues: the link between unemployment and inflation and the implications of unemployment for distribution.

Some unemployment is inevitable. People are constantly entering and leaving the labor market, changing jobs, and moving from place to place. Businesses go bankrupt or suffer downturns and lay people off. It always takes time to find a new job. This is known as "frictional" or "natural" unemployment. According to theory, if policy makers tried to reduce unemployment below this level, the result would be greater demand for a fixed number of workers. Workers would have more bargaining power and would demand higher wages, thereby causing inflation.[8] Thus, a widespread euphemism for "natural" unemployment is NAIRU, the non-accelerating inflation rate of unemployment.[9] There is considerable disagreement over what NAIRU actually is. James K. Galbraith argues that economists are really quite practical—their estimates of NAIRU simply reflect actual unemployment.[10]

THINK ABOUT IT!

When the prices of most commodities fall, output falls as well. While fewer people may be willing to work for lower wages, is that the same as an actual decrease in the supply of labor? Should labor, which is to say people, be treated solely as a commodity?

But the link between low unemployment and inflation is not clear empirically. Why not? We offer two explanations. First, in the era of globalization, large corporations are free to move their capital and production to other countries. Even when unemployment is low, corporations can counter demands for higher wages by a local workforce with the threat of moving to a lower-wage country. This explains how the low unemployment of the 1990s in the U.S. was accompanied by stagnant wages and a

---

[8]This theory was originally introduced by Milton Friedman in his 1967 American Economics Association presidential address.

[9]In the 1960s, economists found an inverse empirical relationship between unemployment and inflation, which was dubbed the Phillips curve. But during the 1970s, a number of economies experienced increasing unemployment and increasing inflation simultaneously.

[10]J. K. Galbraith, Well, Excuuuuse Me! *The International Economy*, December 1995.

diminished share of national income going to wage earners.[11] Second, we must point out that income from production is divided between wages, profit, and rent. Increased bargaining power by wage earners need not lead to "wage-push" inflation; it could instead simply increase the share of income going to wage earners and decrease the share going to rent or profit. Does increased bargaining power by owners lead to "profit-push" inflation?

In summary, then, low unemployment increases the bargaining power of wage earners, which translates into higher wages (though this effect is diminished by globalization). Higher wages can cause inflation, which then erodes the higher wages, or it can change distribution patterns between wages and profit. High unemployment, in contrast, increases the bargaining power of corporations and leads to redistribution toward the owners of capital. Whatever the validity of the theory behind NAIRU, it is quite clear that monetary authorities pay close attention to unemployment as an indicator of inflationary pressures. For example, when unemployment falls too low, the Fed tends to raise interest rates to reduce investment, employment, and demand. Distributional impacts of inflation are uncertain, but unemployment caused by disinflationary policies has clearly negative impacts on some of the poorest sectors in society.

Finally, it is worth reiterating that increasing unemployment can set up a vicious cycle. As people lose jobs, they lose money to purchase goods and services. With less demand, businesses respond by reducing supply, perhaps laying off more workers to do so, and further reducing demand. Many fiscal policies such as welfare payments, unemployment insurance, and other transfer payments are designed to diminish this impact, adding stability to the economy. Economic stability is a public good and an important policy objective.

The Impact of Policies on Scale, Distribution, and Allocation

Now that we understand the basic elements of fiscal and monetary policy, we can turn to their particular applications. How we apply these policies, of course, depends on what we wish to achieve. Mainstream macroeconomists primarily pursue continuous economic growth, with a lesser emphasis on distribution. Allocation is left to microeconomic forces. Ecological economists are concerned primarily with the impact of macroeconomic policies on scale (i.e., growth) but with a different goal than mainstream economists: to make sure that the costs of additional growth

---

[11]R. J. Gordon, The Time-Varying NAIRU and Its Implications for Economic Policy, NBER Working Paper No. W5735, May 1997.

in material throughput are not greater than the benefits. Ecological economists assume that eventually the costs will exceed the benefits if they haven't already. They also place much more importance on distribution than mainstream economists. In short, ecological economics strives to create an economy in which there is no growth in physical throughput, while not only avoiding the suffering caused by recession or depression but also eliminating existing poverty. The allocation of resources between market and nonmarket goods and services can play an extremely important role in this regard.

Macro-allocation

As we discussed earlier, free markets work very well at allocating resources among market goods but very poorly at allocating nonmarket goods, typically failing to provide them in satisfactory quantities. Many policy makers already recognize this point, as can be clearly seen in government budgets, the bulk of which are spent on public goods such as defense, health care, education, road systems, bridges, streetlights, national parks, and so on.[12] In fact, few institutions besides government allocate resources toward nonmarket goods, and only the government is able to use policy to reduce demand and hence expenditure for market goods and shift it toward nonmarket goods.

For simplicity, we refer to the allocation between market and nonmarket goods as **macro-allocation**, and allocation between market goods as **micro-allocation**. Probably the most important macro-allocation question is how much ecosystem structure should be converted to economic goods and services and how much should be conserved to provide nonmarketed ecosystem goods and services.

Macro-allocation is the allocation of resources between market and nonmarket goods and services.

In the private sector, monetary policy directly affects only the *market* economy, by stimulating or discouraging investment in the production and consumption of *market goods* for profit. Why is this so? Monetary policy acts primarily through its impact on interest rates and hence on borrowing and lending. The private sector invests little in nonmarket goods, since such goods generate no profit that can be used to pay back loans. Therefore, lower interest rates will not affect the production of nonmarket goods by the private sector. Not only will monetary expansion do nothing to provide public goods and open access resources, it can actually increase the degradation of these resources if the production of the market goods is accompanied by negative externalities affecting the environment. Re-

[12]National defense is generally considered a public good, though arms races, nuclear weapons, and excessive defense expenditures may do more to undermine national security than to ensure it. To the extent that disease is communicable and individuals are made uncomfortable by the suffering of others, health care is also a public good.

turn to our example of the bowling alley. If lower interest rates induce it to expand, it will not expand into a void and may expand into some ecosystem—a wetland, for example—that currently provides valuable nonmarket services to the local community. As we discussed earlier, such negative externalities are an inevitable outcome of market production.

Therefore, if our policy objective is sustainable scale, monetary expansion is very problematic. Even if the economic scale is well within the constraints imposed by the ecosystem, monetary expansion acts on market goods, which do not always offer the highest marginal contribution to human well-being. The microeconomic law of the equimarginal principle of maximization thus applies not only to the scale of the economic system relative to the ecosystem that sustains it but also to the division of market and nonmarket goods produced by an economy. In ecological economics, macro-allocation takes precedence over micro-allocation.

Theoretically, government money in a democratic society will be directed toward the goods and services that provide the greatest marginal utility for society as a whole. As we have discussed, an important role of government expenditure is to provide nonmarket goods.

It is important to distinguish between two classes of nonmarket goods, which have different effects on scale. Manmade nonmarket goods affect scale to the same degree as market goods. If the government project in the bowling alley town is a big government building, it may also encroach upon some valuable ecosystem and destroy the services it provides. In contrast, the government could restore wetlands that sustain biodiversity, promote seafood production, and protect against catastrophic storm surges such as those that devastated New Orleans. Protecting and restoring the ecosystems that provide nonmarket environmental services can effectively decrease scale, or at least help ensure that we do not surpass optimal scale. As the world becomes more full, the marginal benefits from protecting and restoring ecosystem funds, and hence the nonmarket services they provide, will increase relative to those from market goods and manmade public goods. As this happens, and if politicians come to understand the benefits and public good nature of ecosystem services better, more and more federal money should be allocated toward providing such services.

It is important to recognize, however, that government expenditure on ecosystem funds can still increase scale. How is this so? Once the initial expenditure enters the economy, **multiplier effect** occurs. Money spent to restore ecosystem funds will in turn be spent by its recipients on market goods—workers restoring the wetland may spend their money on bowling, pressuring the bowling alley to expand; construction workers expanding the bowling alley may spend their additional income on TVs, causing that industry to expand; and so on. The larger the multiplier, the larger the impact on the market sector of the economy, and the less control the government has over composition. Tax increases decrease the multiplier by leaving workers with less income to spend on market goods. A smaller multiplier increases the ability of the government to affect macro-allocation, and reduced income reduces scale. Taxes can also be used to discourage undesired behaviors, such as pollution, and subsidies can be used to encourage desired ones, such as environmental preservation. The full impact of taxation on scale and macro-allocation depends on how the taxes are spent, but taxes can certainly play an extremely important role in achieving an optimal scale—a point we will examine at greater length in Chapter 22.

Another important point must be made here. Under traditional analysis of the IS-LM curve, fiscal policy when the economy is operating at full capacity results in crowding out (remember the full bowling alley) and should be avoided. However, in terms of macro-allocation and scale, full output conditions can increase the effectiveness of fiscal policy. With full employment, if the government spends money to restore wetlands, interest rates and labor costs go up, and it is more difficult for the bowling alley to expand. (Fortunately, scenic wetlands offer a recreational alternative to bowling that does not displace ecosystem services.) Government expenditure on restoring ecosystems under such conditions will therefore have an unambiguous impact on reducing scale.

What are the distributional impacts of fiscal and monetary policy? Fiscal policy in the form of taxation and government transfers can be easily and effectively distributed as desired. Government transfers such as welfare, unemployment insurance, Medicare, Medicaid, and Social Security all play an important role in distribution. Corporate welfare programs (which outweigh transfer payments to the poor[13]) affect distribution in the opposite direction. Public goods are equally available to all, and their provision improves distribution. In terms of income, progressive taxation can help reduce gross inequalities in income distribution, a necessary condition morally and practically if we are to achieve a sustainable scale. Monetary policy can also play an important but narrower role in distribution. High interest rates caused by tight monetary policy can lead to unemployment, and they favor creditors over debtors, as discussed earlier in the section on inflation and disinflation. Low interest rates have the opposite effect.

In summary, in terms of ecological economic goals, monetary policy is a blunt instrument directed only toward the production and consumption of market goods, with limited flexibility in terms of distribution and

---

[13]C. M. Sennott, The $150 Billion "Welfare" Recipients: U.S. Corporations, *Boston Globe*, July 7, 1996.

macro-allocation. Expansive monetary policy increases scale. Fiscal policy has far greater flexibility in terms of scale, distribution, and macro-allocation.

■ IS-LM IN THE REAL WORLD

While the IS-LM model is very useful, it has important limitations.[14] The model is deceptively simple and does an inadequate job of conveying the real-life complexity of monetary and fiscal policies. While the model shows the general impacts of such policies, it fails to incorporate the issues of uncertainty, time lags, and structural changes, as well as the difficulty of choosing the appropriate policy variables to manipulate.

Economists typically have a poor understanding of what is happening in the economy at any given moment. Is unemployment too high? Is the economy growing too fast, threatening inflation? Are we headed for recession? Viewing the same data concerning the economy, economists often disagree on how to interpret them and how to react. For example, in 2008 the Fed argued that the recession the U.S. had entered was caused by a liquidity crisis, and dropped interest rates to nearly zero. When this failed to increase lending and investments, the government stepped in with massive fiscal expenditures. Some believed that the fiscal stimulus package was too large and would cause serious inflation, others that it was too small and that deflation was a more serious concern. In mid-2009, economists were debating whether the economy was heading out of recession or heading deeper in. Part of the problem is that the economic system is evolving rapidly in response to technological, environmental, cultural, and structural changes.

Compounding the difficulty of an inadequate understanding of the economy are the time lags involved in policy. There are two types of lags: lags in decision making (the inside lag) and lags between the time the decision is made and the policy takes effect (the outside lag). In fiscal policy, decisions such as tax cuts and expenditure increases typically are debated at length. Both legislative and executive branches must agree, and appropriate legislation must be passed. The decision lag therefore can be substantial. Once the decision to increase or decrease expenditure has been made and carried out, the outside lag may be relatively short, as such policies have an immediate effect on aggregate demand (though the full effect

---

[14]See J. R. Hicks, IS-LM: An Explanation, *Journal of Post-Keynesian Economics* III(2):139–154 (Winter 1980–1981). Hicks, the originator of the model, expresses reservations about how well it fits the real world once expectations and dynamics are recognized. He considers the model useful for understanding the past but less so for understanding the future. We agree but feel that many of Hicks' caveats apply to all equilibrium models and that, rare though it is among scholars, Hicks was too hard on himself.

of the multiplier will take some time). Tax cuts or increases, on the other hand, have much slower results and are often not even felt until the next tax year.

The Fed, in contrast, generally has a much shorter decision-making lag. The Federal Open Market Committee (FOMC), responsible for Fed policy, meets about eight times a year.[15] Policy is generally decided at the meeting, and open market transactions can take place almost immediately. However, the most relevant impact of these policies is on interest rates and their effect on investment and consumption decisions. Investment decisions are rarely spur of the moment; they generally have a long gestation period. Thus, the Fed has a short decision-making lag and a long lag before the policy takes effect.

These lags are very important to consider when deciding on a policy. It is quite possible that by the time a decision is made and the resulting policy takes effect, the problem the policy was designed to address will have disappeared, and a policy with the opposite effect may even be needed.

Another problem is disagreement over what type of policy should be pursued and what the impact will be, especially for monetary policy. The Fed usually tries to manipulate one of two targets: the money supply or the interest rate. Not only is there considerable debate over which course the Fed should pursue, there are serious obstacles to achieving either goal. For example, as Fed chairman Alan Greenspan admitted in congressional testimony, "We have a problem trying to define exactly what money is. . . . The current definition of money is not sufficient to give us a good means for controlling the Money Supply."[16]

Psychology can also make it difficult to manipulate interest rates. As we discussed earlier, interest rates are ultimately determined by the bond markets. Bonds, of course, mature in the future, and the amount someone is willing to pay for a bond depends on their *expectation* of future inflation. The Fed might implement an expansionary monetary policy to bring down interest rates, but if bond marketers believe this expansion will instead induce inflation or force monetary contraction in the near future, it could paradoxically drive interest rates *up*.

A final problem with policy in countries with independent monetary authorities is the difficulty in coordinating between monetary and fiscal policy. This problem can become acute when the monetary authority and the government have different policy objectives. The elected government is concerned mainly with growth and employment, two issues that affect voters and hence their elected representatives. In contrast, the Fed is con-

---

[15]The Federal Reserve Act mandates that the FOMC meet at least four times a year, and since 1981 it has met eight times a year.

[16]Congressional testimony, February 17, 2000.

cerned mainly with "sound" money (i.e., low inflation) and often seeks to prevent inflation, even when the policies used cause unemployment and misery.

Box 17-6   WHY IS THE FED SO ANTI-INFLATION?

From our discussion of inflation, it would seem that inflation is less harmful than unemployment induced by anti-inflationary policies. Why is the Fed so anti-inflation? In answering this question, it is worth bearing in mind who the natural constituency of the Fed is. Most members of the FOMC are bankers or Wall Street professionals, and the Fed seems to listen closely to the concerns of these groups. These two groups form the bulk of the wealthy creditors, who benefit from low inflation and disinflationary episodes and who are unable to increase their share of national income as readily during inflationary periods.[a]

[a]W. Greider, Secrets of the Temple: How the Federal Reserve Runs the Country, New York: Simon & Schuster, 1987.

Despite its shortcomings, the IS-LM model is a vast improvement over prior models. It is a two-sector general equilibrium model, the sectors being the real sector and the monetary sector. Before Hicks' model, economists often tried to explain the interest rate as a purely monetary phenomenon (liquidity preference and money supply) or a purely real phenomenon (savings and investment). There was a money rate of interest and an investment rate of interest and confusion about which set of factors "really" determined the interest rate. Hicks showed that the real and monetary sectors simultaneously interact to determine both the interest rate and national income. But Hicks said nothing about the ecosystem and biological rates of growth. In 1937 the world was still considered "empty." Thus, the IS-LM model treats all economic growth as identical; it does not distinguish between government expenditures on market goods, manmade public goods, or investments in ecological restoration, nor does it address distribution.

■ ADAPTING IS-LM TO ECOLOGICAL ECONOMICS

How might the IS-LM model be adapted to ecological economics? Remembering our basic vision of the macroeconomy as a subsystem of the finite and nongrowing ecosystem, the most obvious suggestion would be to impose an external constraint on the model representing the biophysical limits of the ecosystem. For example, we could assume a fixed throughput intensity per dollar of Y (i.e., GNP), so that a given Y in money terms implied a given physical throughput. Then we could estimate the maximum ecologically sustainable throughput, convert that into the equivalent Y, and impose that as an exogenous constraint on the model. Based on Figure 17.4, it would be represented by a vertical line at the Y corresponding to maximum sustainable throughput. It would not be a function of the interest rate at all.[17] Let's call the vertical line EC for "ecological capacity." It reflects a *biophysical* equilibrium, not an *economic* equilibrium. It is ignored by the actors whose behavior is captured in the IS and LM curves.[18]

The most obvious approach is not always the best, but it is usually a good place to start. Also, this approach closely parallels the macroeconomist's representation of full employment of labor as a perpendicular at the level of Y corresponding to full employment at an assumed labor intensity of GNP. Our EC line represents "full employment" of the environment at an assumed throughput intensity of GNP. Later we will discuss further the assumption of fixed throughput intensity.

Let's consider the three possible positions of the biophysical equilibrium relative to the economic equilibrium, shown in Figure 17.6. The first case represents the "empty world" scenario. The biophysical limit is not binding. The distance Y*C may be thought of as excess carrying capacity. Most macroeconomists who use the IS-LM model would have this case in mind, if indeed they thought at all about EC. If the distance Y*C is large, then for practical purposes of short-run policy there is no point in conceiving or drawing the EC line.

The second case is the "full world" (or overfull) scenario. The economic equilibrium has overshot the biophysical equilibrium. The distance CY*, the overshoot, is caused by unsustainable drawdown of natural capital. Thus, CY* would represent capital consumption counted as income. As natural capital is consumed, the EC line eventually has to shift even farther to the left, increasing the overshoot. Most ecological economists believe this to be a rather accurate description of the present state of affairs. Most conventional economists do not worry about long-term capital drawdown and shifting the EC curve farther to the left be-

---

[17]A. Heyes, A Proposal for the Greening of Textbook Macro: "IS-LM-EE," *Ecological Economics* 32(1) (2000); P. Lawn, *Toward Sustainable Development: An Ecological Economics Approach*, Boca Raton, FL: Lewis Publishers, 2001. Heyes and Lawn have proposed an EE curve corresponding to the ecological limits discussed here that would be a function of the interest rate. Several technologies produce income, some of which require or degrade more natural capital than others. Less natural capital–intensive technologies require investments and are thus more likely at lower rates of interests. One problem is that the investments themselves would require natural capital.

[18]However, it need not be ignored by government, which affects the IS curve. The government is perfectly capable of investing in environmental services produced by natural capital and other nonmarket goods. But it is completely ignored by monetary policy (the LM curve), which acts on the economy through its effect on interest rates and hence market goods.

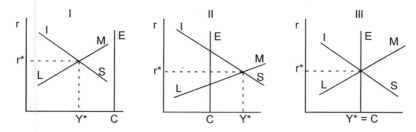

Figure 17.6 • The biophysical equilibrium relative to the economic equilibrium.

cause they believe that knowledge is shifting EC to the right and thereby restoring the empty world situation.

The third case represents a big coincidence under our assumptions. For the economic equilibrium to coincide with the biophysical equilibrium would require either extraordinary good luck or purposeful coordination and planning. There is nothing in the model to make it happen, just as, currently, there seems to be nothing in our institutions or behavior that would make it happen. In Chapters 21–24, we will discuss policy changes that could theoretically lead to this outcome.

Recall that we previously discussed the concept of full employment, which we might represent by an FE limit for labor similar to the EC limit for natural capital. Ideally, a FE labor line should coincide with the ISLM equilibrium point—make IS = LM at full employment. If FE is beyond the intersection of IS and LM, then policy makers might pursue FE through growth in Y. But what if FE is beyond EC? The problem is no longer to pursue FE by growth in Y but instead through structural change, such as shifting factor intensity away from fossil fuels and manmade capital (both of which rapidly draw down natural capital) and toward labor. We have already explained that when IS = LM beyond EC, we are likely to draw down natural capital, and it is implicit in the acronym NAIRU (the non-accelerating inflation rate of unemployment) that going beyond FE results in inflation. Why the difference? Why doesn't moving beyond EC also simply cause inflation? The answer is that natural resources are either free or cheap to begin with; they are not appropriately priced by the market mechanism, and excessive use therefore does not affect the price signal.

It remains true, however, that the assumption of constant throughput intensity of Y is troublesome. We know that throughput intensity of Y changes with new technology and with shifts in the mix of goods that make up Y, even if probably not with factor substitution of capital funds for throughput flows. Differing assumptions about throughput intensity of Y can at least be represented by a shift in the EC perpendicular. If productivity increases resulting from technological advances outpace resource exhaustion (e.g., new technologies require less raw material and fossil energy), we would expect EC to shift to the right, while if resource exhaustion outpaces technology (e.g., there is more slag per unit of useful ore or more environmental damage per barrel of oil extracted), we would expect a shift to the left. (As you have no doubt realized by now, curve shifting is not an uncommon device in economic analysis.) However, in terms of practical policy recommendations, perhaps the best approach would be simply to impose the ecological constraint as a limit on throughput. For any given technology, a fixed limit on throughput will also limit Y, but over time, new technologies and a different mix of goods and services can allow Y to increase without increasing throughput and threatening the life-support functions of the ecosystem.

BIG IDEAS to remember

- Macroeconomic model vs. general equilibrium
- Real vs. monetary sectors
- IS = LM
- MV = PQ
- Transaction demand for money
- Liquidity preference
- Relation of bond prices to interest rate

- IS-LM analysis of monetary and fiscal policy
- Comparative statics
- Crowding out
- Inflation and disinflation
- Unemployment
- Macro-allocation
- IS-LM adapted to show economic and biophysical equilibriums

Conclusions to Part IV

Chapters 14–17 offered a whirlwind tour of the macroeconomy. The real economy consists of the physical transformation of low-entropy matter-energy from nature into forms that enhance human welfare but in the process inevitably reduces the flow of welfare-enhancing services from ecosystem funds. GNP is an inadequate measure of the real economy, because it lumps together both goods and bads. The monetary sector of the economy functions as a lubricant for the allocation process. Who is entitled to seigniorage is a policy variable in the monetary system affecting distribution, as well as allocation between private and public goods. Current seigniorage policies in most countries favor the wealthy and the private sector. Monetary systems can also affect scale, and interest-bearing, debt-based money creation is incompatible with a steady-state economy, which will ultimately require a nonincreasing money supply. The law of diminishing marginal utility tells us that distribution, both within and between generations, can be an important tool for increasing human welfare. Macroeconomic policy levers include government expenditure, taxes, money supply, and interest rates. These policy levers are currently used to promote continuous economic growth but could instead be used to attain the goals of sustainable scale and just distribution—goals essentially ignored by the market microeconomy. We next turn our attention to international economics, to see how it affects the policy levers we just discussed.

PART V

International Trade

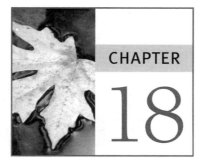

International Trade

I t is a lot easier to grow grapes and make wine in Italy than in Norway. It is easier to hunt reindeer in Norway than Italy. To the extent that Norwegians want Chianti and Italians want reindeer steaks, there is an obvious basis for international trade. Economists never needed to prove that. Nor, given climate and geography, is it hard to understand why Italians traditionally drink Chianti and eat fish, while Norwegians traditionally drink aquavit and eat reindeer and whales. But we all like to experience other peoples' traditions, tastes, and capacities. International trade allows that, to the mutual benefit of all. In this view, national production for national consumption is the cake; international trade is the icing. Of course, no one objects to trade in this sense. But this is not an accurate picture of either the reality of trade today or the trends and goals of globalization. However, before we tackle the difficulties of globalization, let's look more closely at the classical argument for free trade, an argument that goes well beyond the common-sense picture just sketched.

■ THE CLASSICAL THEORY: COMPARATIVE ADVANTAGE

The case for free trade, like the case for exchange in general, hinges on the assumption that trade is voluntary. Both parties to a voluntary exchange must be better off, in their own estimation, after the trade than before—otherwise they would not have made the trade. Under what conditions is this happy result likely to be the case? The most obvious condition is that of **absolute advantage** in cost differences, reflected in the example above. If country A can produce something at a lower absolute cost than country B, while B can likewise produce something else at a lower absolute cost than A, then there is a reason for voluntary exchange, assuming A wants some of B's product and B wants some of A's product in the first place.

Initially economists thought that if one country produced all goods more cheaply than another, there was no basis for mutually advantageous trade between them; the more efficient country would only harm itself by trading with the less efficient country. David Ricardo demonstrated that this was not so—that both countries could benefit from trade even if one country had an absolute advantage in all tradable goods.[1] The key to understanding this is to focus on comparative advantage rather than absolute advantage.

Ricardo demonstrated this using a two-country, two-goods numerical example. We will retrace his logic using just such an example shortly. But first it is worthwhile to whet our appetite by noting how proud economists are of this result. Trade theorist R. Findlay referred to comparative advantage as the "deepest and most beautiful result in all of economics."[2] Surveys have shown that about 95% of economists support the policy of "free trade." Princeton economist Paul Krugman said, "If there were an Economist's Creed it would surely contain the affirmations 'I believe in the Principle of Comparative Advantage,' and 'I believe in free trade.'"[3] With that fanfare, let's bring the comparative advantage argument onstage.

Consider a world in which we have two countries, A and B, each producing two goods, coal (C) and wheat (W). Since we want to demonstrate the gains from trade without appealing to absolute advantage, confining our reasoning only to comparative advantage, let's impose on ourselves a veil of ignorance about absolute cost differences. Specifically, let's say that country A measures costs in terms of a-units, and country B measures costs in terms of b-units, and that we know nothing about the relation of a to b. Perhaps $a > b$, or $a < b$, or $a = b$. Therefore, we cannot compare costs across countries, as required to know absolute advantage. But we can compare the cost of C and W within each country. The comparative or relative cost of W to C in country A can be calculated and compared to the comparative or relative cost of W to C in country B. That information will be sufficient to demonstrate the possibility of mutual gains from trade between A and B. The argument will depend only on comparative advantage (internal cost ratios), and not on cross-country comparisons of cost (absolute advantage) because the latter have been made impossible by assuming incomparable units for measuring costs in the two countries. Table 18.1 summarizes the unit costs of C and W in A and B.

A country has an *absolute advantage* if it can produce the good in question at a lower absolute cost than its trading partner. It has a comparative advantage if it can produce the good in question more cheaply relative to other goods it produces than can its trading partners, regardless of absolute costs.

---

[1] D. Ricardo, *On the Principles of Political Economy and Taxation*, 3rd ed., London: John Murray, Albemarle-Street, 1821 (originally published in 1817). Online: http://www.systemics.com/docs/ricardo/principles.html.

[2] R. Findlay, "Comparative Advantage." In J. Eatwell et al., eds., *The New Palgrave: The World of Economics*, New York: Norton, 1991, p. 99.

[3] P. Krugman, Is Free Trade Passé? *Economic Perspectives* 1(2):131 (1987).

■ Table 18.1

UNIT COSTS OF COMMODITIES

	Country A	Country B	Total Output
Coal	1a	1b	2C
Wheat	1a	4b	2W
Total Resources	2a	5b	

Each country needs both coal and wheat, so before trade, each country allocates its resources between the two commodities and produces as much of each as permitted by its unit costs, as shown in the table. A has total resources of 2a. With unit costs of both coal and wheat equal to 1a, it can produce one unit of coal and one unit of wheat with its endowment of 2a resources. Country B has a total resource endowment of 5b. B allocates 4b resources to produce one unit of wheat and with the 1b remaining can produce one unit of coal. Total world output before trade is 2W + 2C.

Ricardo has a look at this situation and realizes that we can do better by allowing specialization and trade according to comparative advantage. He notices that W is four times as expensive as C in country B and only one time as expensive as C in country A. In country B coal is cheaper relative to wheat than it is in country A. Therefore B has a comparative advantage in coal. Similarly, wheat is cheaper relative to coal in country A than it is in B. Consequently A has a comparative advantage in wheat.

Now let each country specialize in the production of the commodity in which it has a comparative advantage and trade for the other commodity. A will specialize in wheat, allocating all of its 2a resources to wheat, and with a unit cost of wheat of 1a will end up producing 2W. B will allocate all 5b units of its resources to production of coal and with a unit cost of coal of 1b will produce 5C. Total world product after specialization and trade is 2W + 5C. The world has gained an extra 3C with no extra resources. The world is better off. And we have shown this without ever having to compare costs of wheat in country A with costs of wheat in country B or costs of coal in country A with costs of coal in country B. Only comparative advantage (a difference in internal cost ratios) matters, not absolute advantage. Suppose we suddenly learned that in absolute terms a = 5b, so that country B has an absolute advantage in both C and W. Does that change our conclusion? Not a bit.

Although the world is better off by 3C, one might still wonder which country gets the 3C or how it is divided between them. We can see that the world as a whole is better off, but how do we know that each country is better off? Might not one country suffer while the other gains a lot? Ricardo had an answer. How the total gains from specialization and trade, 3C, will be divided between A and B depends on the terms of trade, the relative price at which W exchanges for C between the two countries. That will depend on supply and demand and bargaining power, so we cannot say exactly how the gains will be divided. But we can be sure, says Ricardo, that neither country will be worse off after trade. How do we know this? Because, since trade is voluntary, neither country will accept terms of trade less favorable than the terms of its own internal cost ratio, the terms on which it can "trade with itself." A will not trade 1W for less than 1C, because it could do better by reallocating its own resources back from W to C again. Likewise, B will not give more than 4C for 1W. So the terms of trade will fall somewhere between the limits of 1C/1W and 4C/1W. Anywhere between those ratios both countries gain a part of the extra 3C.

■ KINKS IN THE THEORY

Ricardo's demonstration is indeed interesting and impressive. To more than double coal production, with no sacrifice of wheat production and no additional resources, is a neat trick. Within the world of its assumptions, the logic of the comparative advantage argument is unassailable. But it is time to have a closer look at the assumptions. What are the assumptions, and might they veil some costs that need to be subtracted from the gain of 3C?

First, it is not really true that the extra 3C is produced with no additional resources. There is the obvious increase in the rate of depletion of coal mines and of pollution resulting from burning the coal. What the trade economists mean by "no extra resources" is simply no additional labor or capital. Recall the neoclassical production function discussed earlier, in which output is a function only of labor and capital inputs. But, contrary to the neoclassical assumptions, the resource cost of extra output cannot simply be ignored. Therefore some deduction from the value of 3C is needed to account for extra depletion and pollution. But let us assume that is done and there are still net gains from trade and specialization.

Second, we have neglected or abstracted from transport costs, implicitly assuming them to be zero. Wheat has to be shipped from A to B and coal from B to A. If the energy costs of that transportation were 3C or greater, then the world would gain nothing. It is worth noting that transportation is energy intensive, and currently energy is directly subsidized in many countries; in addition, many of its external costs are not internalized in its price. Consequently, international trade is indirectly subsidized by energy prices that are below the true cost of energy. But let us suppose that there is still a net gain from trade after fully counted transport costs are deducted.

Third, we also assumed that in each country the cost of specialization was negligible. But there are two important costs to recognize. First, in A all coal miners must become wheat farmers, and in B all wheat farmers must become coal miners. Making such a shift is costly to all whose livelihood is changed. Also in the future the range of choice of occupation has been reduced from two to one—surely a welfare loss. Most people derive at least as much life satisfaction from how they earn their income as from how they spend it. Economists practically identify welfare with increased range of choice among commodities but are strangely silent about the welfare effects of the range of choice among jobs. Second, after specialization countries lose their freedom not to trade. They have become vitally dependent on each other. One's access to essential items depends on the cooperation of people on the other side of the world, who, however, admirable they may be, have different customs, different values, different interests, and different types of government. Remember that the fundamental condition for trade to be mutually beneficial is that it be voluntary. The voluntariness of "free trade" is compromised by the interdependence resulting from specialization. Interdependent countries are no longer free not to trade, and it is precisely the freedom not to trade that was the original guarantee of mutual benefits of trade in the first place.

True enough, as Ricardo pointed out, if the terms of trade become too disadvantageous, the country getting the worse end of the deal can opt out and despecialize—put some of its wheat farmers back in the coal mines, restore land degraded by mining so that it can again grow wheat, reinvest in mining equipment, reactivate mining legislation, re-employ exporters and importers, and so on. But this is both costly and socially disruptive in reality. The model assumes specialization is reversible, while in fact it is closer to being irreversible—that is, reversible but at a high cost. Countries specializing in nonessential products—bananas, sugar, cocoa, and so on—are especially vulnerable to hardship from having lost their freedom not to trade. It is easier to drive a hard bargain if you are selling essentials and buying nonessentials than vice versa. Clearly there should be some further deduction from the 3C resulting from the above costs of specialization.

THINK ABOUT IT!

Many countries specialize in only one or few commodities for export. For example, Ecuador specializes in bananas, Ghana in cocoa, and Cuba in sugar. What happens to the price of these commodities when lots of other countries start producing them? What happens to the price when growing conditions produce a global bumper crop? Look up price trends for these commodities. Do you think specialization in agricultural commodities is a good development strategy?

■ CAPITAL MOBILITY AND COMPARATIVE ADVANTAGE

But suppose we still have a net gain. Can we then conclude that the argument for free trade on the basis of comparative advantage holds true? In fact, there is an often-overlooked provision of the theory, one of great relevance today in the debate over globalization, that suggests that it does not hold true. Note that our numerical demonstration of comparative advantage implicitly assumes factor (labor and capital) immobility between country A and B. Only coal and wheat crossed borders. Labor and capital stayed at home and were reallocated domestically between coal and wheat according to the principle of comparative advantage. Since it is usually the capitalist who makes the investment allocation decisions, let us just focus on capital and its mobility or lack thereof. Clearly our model implicitly assumes that capital is mobile between sectors within each country but immobile between countries. Capitalists cannot even compare costs or profitability between A and B because, thanks to our veil of ignorance, they cannot compare a-units with b-units. Therefore they could not possibly know whether investment in the other country would be profitable or not.

One other way to show the implicit assumption of immobile capital in modern texts is to note that the examples, like ours, are usually in terms of barter—wheat for coal, with no money involved. Barter trade is always necessarily balanced. No monetary or short-term capital flows are needed to balance the differences in exports and imports of bartered goods. The **current account** is the difference between the monetary value of imported and exported goods and services. If imports are greater than exports, the current account is in deficit. If exports are greater than imports, the current account is in surplus. **Capital accounts** are the difference between monetary flows, used to purchase various assets, into and out of a country. Such assets include stocks, bonds, real estate, and other assets that remain in the country. When more money flows into the country than out, the capital accounts are in surplus. The current account of the balance of payments is always balanced in barter, so there is no need for a compensating imbalance on capital account. Therefore, barter examples assume balanced trade, and balanced trade means no capital mobility.

Immobile capital does not mean that producer goods cannot be exchanged on current account. Machines and tractors, "materialist capital," can be traded just like shoes and sugar. What is immobile is capital in the "fundist" sense, money, or liens on the future product of the deficit country. Immobile capital in the fundist sense is the same thing as balanced trade on the current account, or trade that requires no compensating transfer of fundist capital on capital account. In other words, immobility of capital does not prevent Brazil from importing machines and tractors and paying for them with exports of shoes and sugar. It just prohibits Brazil from importing machines and tractors *faster* than it can pay for them by exporting shoes and sugar, and thereby paying for the extra machines and tractors by issuing liens against its future production of shoes and sugar.

The capital immobility assumption is often hidden by examples in terms of individuals who specialize and trade, rather than nations. A favorite example is the lawyer and her secretary. The lawyer happens to be a champion speed typist. She has an absolute advantage over the secretary both in typing and of course in practicing law, since the secretary is not a lawyer. But the lawyer nevertheless finds it advantageous to hire the secretary to do the typing, and both benefit from this exchange. Although the lawyer is a better typist than the secretary, she is not much better. But she is a much better lawyer. So the secretary has a comparative advantage in typing (although an absolute disadvantage), and the lawyer has a comparative (and absolute) advantage in law. So they specialize according to comparative advantage. The lawyer is not so silly as to spend her time typing when she could be billing clients at $300 per hour. Where is the assumption of capital immobility in this example? It is hidden by the obvious fact that it is impossible for the productive energy and capacity of the secretary to be transferred to the absolutely more efficient person of the lawyer. The lawyer is not a vampire who can suck the lifeblood and energy out of the hapless secretary in order to use it more efficiently. In other words, productive capacity, "capital," is immobile between the lawyer and her secretary, so the logic of comparative advantage works.

Between countries, however, it is not impossible for productive capacity, capital, to be transferred from one country to another in response to absolute advantage. Capital mobility has to be explicitly ruled out for comparative advantage argument to work between countries. Although each country as a whole benefits under comparative advantage trade, not every citizen or group of citizens will benefit. Everyone could benefit, but only if winners were to compensate losers—that is, only if coal miners in B compensate wheat farmers, and wheat farmers in A compensate coal miners for the costs of changing their livelihood. As we move to absolute advantage (capital mobility internationally), some entire nations may lose, but the world as a whole will gain and could compensate the losing country, just as within countries the government could compensate the losing industry. Such compensation usually does not take place within nations and almost never takes place between nations. Within nations there are at least institutions of community that could carry out internal transfers. At a global level there are no such institutions. The move from comparative advantage to absolute advantage seems to be part of the general retreat from the Pareto to the Hicks-Kaldor (often referred to as "potential Pareto") welfare criterion. The latter represents a retreat from the condition that no person be made worse off to the weaker condition that winners could, if they so choose, compensate losers and still be better off. Similarly, the former is a retreat from the condition that no nation be made worse off to the weaker condition that winning nations could compensate losing nations, if they choose to, and still be better off. In both cases the weaker criterion takes compensation as only potential, not actual, and in the case of trade the compensation condition is seldom even mentioned.

■ ABSOLUTE ADVANTAGE

Ricardo, to his credit, was very explicit about his assumption of immobile capital between countries. If capital were mobile internationally, it is obvious that capitalists would seek the greatest absolute profit and consequently the lowest absolute cost of production. In our example, if capital were mobile and the capitalists knew that $1a = 5b$, so that B had an absolute advantage in both coal and wheat, they would invest in B and forget about A. Comparative advantage would be irrelevant. If capital is mobile, absolute advantage is the relevant criterion. The only reason the capitalist would ever be interested in comparative advantage is if capital were immobile internationally. If capital cannot follow absolute advantage abroad, the next best thing is to follow comparative advantage specialization at home and trade for the foreign product. Comparative advantage is a clever second-best adaptation to the constraint of international capital immobility. But without that constraint, it has no reason to be, and absolute advantage is all that counts.

Why is the capital immobility assumption so important? Because it is utterly counterfactual in today's world. Capital is mobile all over the world in trillion-dollar amounts at the speed of light.

Why is this overlooked? The nice thing about the comparative advantage argument is that both countries benefit from free trade, and gains are mutual—at least in theory, if not always in fact. The problem with absolute advantage is that both countries do not necessarily gain. If one country has an absolute disadvantage in both commodities, it will lose jobs and income as capital moves abroad. But under absolute advantage, world production will still increase. In theory, it would increase by more than under comparative advantage. This is because in moving from comparative to absolute advantage, we relax a prior constraint on world product maximization—namely, the condition of capital immobility. Mobile capital can seek out more productive opportunities than it could when it was confined to its country of origin. But while there is certainly a case for absolute advantage, it lacks the politically very convenient feature of guaranteed mutual benefit that was part of the traditional comparative advantage argument for free trade. All countries still could be better off if there were some regulatory institution whereby winners could compensate losers. But that would no longer be "free" trade. The focus of policy has shifted away from the welfare of nations to the welfare of the globe as a whole. And that brings us to the issue of globalization, or more specifically of globalization versus internationalization as alternative models of world community.

> **THINK ABOUT IT!**
> *If the marginal costs of economic growth already exceed the marginal benefits, what are the implications of globalization for the welfare of the planet as a whole?*

■ GLOBALIZATION VS. INTERNATIONALIZATION

Internationalization refers to the increasing importance of relations between nations: international trade, international treaties, alliances, protocols, and so on. The basic unit of community and policy remains the nation, even as relations between nations, and between individuals in different nations, become increasingly necessary and important.

Globalization refers to global economic integration of many formerly national economies into one global economy, by free trade, especially by free capital mobility, and also, as a distant but increasingly important third, by easy or uncontrolled migration. Globalization is the effective erasure of national boundaries for economic purposes. National boundaries become totally porous with respect to goods and capital and increasingly porous with respect to people, viewed in this context as cheap labor or in some cases cheap human capital.

Ricardo and the classical economists who argued for free trade based on comparative advantage were basically nationalists and retained their fundamental commitment to the nation even as they advocated internationalization.

In sum, globalization is the economic integration of the globe. But exactly what is integration? The word derives from *integer*, meaning one, complete, or whole. Integration means far more than interdependence—it is the act of combining separate but related units into a single whole. Integration is to interdependence as marriage is to friendship. Since there can be only one whole, only one unity with reference to which parts are integrated, it follows that global economic integration logically implies national economic disintegration—parts are torn out of their national context (disintegrated) in order to be reintegrated into the new whole, the globalized economy. As the saying goes, to make an omelet, you have to break some eggs. The disintegration of the national egg is necessary to integrate the global omelet.

■ THE BRETTON WOODS INSTITUTIONS

At the end of World War II there was a conference of nations held in Bretton Woods, New Hampshire, for the purpose of reestablishing international trade and commerce, which had been disrupted by the war. The international diplomats and economists, led by John Maynard Keynes of England, and U.S. Secretary of the Treasury Henry Morgenthau and his aide Harry Dexter White were successful in negotiating the charter that set up the **Bretton Woods Institutions,** the **International Monetary Fund (IMF)** and the **International Bank for Reconstruction and Development (World Bank).** These two institutions are made up of member nations. They were founded on the federal model of internationalization, as just discussed, not the integral model of globalization. Their founding, almost 60 years ago, was a wonderful achievement of international cooperation and diplomacy. It symbolized the end of an era of economic depression, followed by war and destruction, and the beginning of a hopeful era of peace and production in which swords would be beaten into plowshares.

The atmosphere of eager optimism was expressed by Morgenthau,[4] who envisaged "a dynamic world economy in which the peoples of every nation will be able to realize their potentialities in peace . . . and enjoy, increasingly, the fruits of material progress on an earth infinitely blessed with natural riches." Morgenthau went on to say that "prosperity has no fixed limits. It is not a finite substance to be diminished by division." The same note was sounded by Keynes: "In general, it will be the duty of the Bank, by wise and prudent lending, to promote a policy of expansion of the world's economy. . . . By expansion we should mean the increase of resources and production in real terms, in physical quantity, accompanied by a corresponding increase in purchasing power." The "empty world" vision of the economy was dominant. Notions of ecological limits to growth were not on the horizon, much less on the agenda. The founders of the Bretton Woods Institutions felt that they had far more pressing issues to deal with in 1945, and they were surely right. But the world has changed a lot in 60 years. Population has roughly tripled, and resource throughput has increased more than ninefold,[5] moving us a long way from the "empty" toward the "full" world.

---

[4]Morgenthau and Keynes quotes are from B. Rich, *Mortgaging the Earth*, Boston: Beacon Press, 1994, pp. 54–55. For an interesting history of the Bretton Woods Conference see also the biography of Keynes by R. Skidelsky, *John Maynard Keynes*, Vol. III, "Fighting for Freedom, 1937–1946," New York: Viking Press, 2000.

[5]Calculated from data in J. B. Delong, *Macroeconomics*, Burr Ridge, IL: McGraw-Hill, 2002, chap. 5. The data are labeled GNP, but in the text Delong refers to it as "material output," what

The division of labor between these institutions was that the IMF would focus on short-term balance-of-payments financing (the current account of the balance of payments), while the World Bank would concentrate on long-term lending (the capital account).[6] The word *reconstruction* in the World Bank's proper name referred to reconstruction of war-torn countries. That function, however, was largely taken over by the Marshall Plan, leaving the World Bank to focus almost entirely on lending for the development of underdeveloped countries.

One of Keynes' ideas, rejected by the conference in favor of the present IMF, was for an International Clearing Union for settling trade balances. In Keynes' plan, all trading countries would have an account with the Clearing Union, denominated in an international monetary unit called the "bancor," which would be convertible into each national currency at a fixed rate, like gold. Clearing of accounts would be multilateral; that is, if country A had a surplus with B and a deficit with C, and the two partially cancelled out, then A would have to settle the difference only with the union.

The innovative feature of Keynes' plan was that interest would have to be paid to the union on credit balances as well as debit balances, at least when the balances exceeded a certain amount. In other words, there would be a penalty for running a balance of trade surplus, as well as for running a deficit. All nations would have an incentive to avoid both a surplus and a deficit, to balance their trade accounts, leading to less international debt and reduced capital flows. We think this proposal merits reconsideration today. It is superior to massive IMF bailouts of debtor countries that continue to run trade deficits. It also puts pressure on countries, like China, that are addicted to running huge trade surpluses.

■ THE WORLD TRADE ORGANIZATION

The **World Trade Organization (WTO)** is of more recent birth than the IMF and World Bank, having its origins in the General Agreement on Trade and Tariffs (GATT) rather than in the Bretton Woods conference.[7] The purpose of GATT was to reduce tariffs and other barriers to international trade. The WTO is often lumped together with the World Bank and IMF, however, because the three institutions have common policy goals of free trade, free capital mobility, and export-led growth—in other words, globalization. To the extent that the World Bank and the IMF push a pol-

---

we call throughput. There is not a 1:1 correlation between GNP and throughput, but Delong's figures show a nearly tenfold increase in GNP from 1950 to 2000. A ninefold increase in throughput from 1945 to the present is probably in the ballpark.

[6]Current accounts and capital accounts are described in detail in Chapter 20.

[7]An international trade organization was proposed at Bretton Woods but was ultimately rejected by the U.S. and replaced by GATT.

icy of globalization, they run into conflict with the internationalist model of world community underlying their charter, a model very different from that of globalization, as we have already emphasized. The WTO's commitment to globalization is evident in the statement of its former director-general, Renato Ruggiero: "We are no longer writing the rules of interaction among separate national economies. We are writing the constitution of a single global economy."[8] This is a clear affirmation of globalization and rejection of internationalization as defined earlier. Of course, different directors-general may alter policies.

But meanwhile, if the IMF-WB are no longer serving the national interests of their member countries, according to their charters, whose interests are they serving?

What are the advantages of an internationalized economic system over a globalized one? One is that nation-states that can control their own boundaries are better able to set their own monetary and fiscal policy, as well as such things as environmental standards and a minimum wage. Markets hate boundaries, but policy requires boundaries.

It is worth remembering that international free trade is not really trade between nations but rather trade between private firms or individuals residing in different nations. Their transactions are carried out for the private benefit of the contracting parties, not for the larger benefit of their national communities. The policy of free trade represents the assumption that if these transactions benefit the private contracting parties, then they will also benefit the larger collectives (nations) to which each party belongs. "What's good for General Motors is good for the USA," to recall a famous statement.

Let's apply the assumption to another collective, the corporation. Do corporations allow their individual employees to freely contract with employees of another corporation in pursuit of their own self-interest? This happens, but if the employees are caught, they are usually sent to jail. All transactions initiated by employees are supposed to be in the interests of the corporation, not the employee, and must be judged so by officers of the corporation. Corporations regulate the trades negotiated by their employees. Why should nations not regulate trade negotiated by their citizens? Advocates of an internationalized economy argue that nations must have the ability to regulate their corporate citizens, just as a corporation regulates the behavior of its employees.

Advocates of a globally integrated economy argue that nations are obsolete and that they have been responsible for two world wars in the twen-

---

[8]From a speech to the United Nations Conference on Trade and Development's (UNCTAD) Trade and Development Board in October 1996.

tieth century. True enough, it is important to remember the real evils of nationalism. We agree with the Bretton Woods delegates, however, that the answer to nationalism is internationalism, not globalism. "A world with no boundaries" is a good song lyric, but taken literally, it makes policy in the interest of local community impossible. We see the world community as a community of local and national communities—as a community of communities. Consequently, the problems we see with the Bretton Woods Institutions arise not from their historical charter but from their institutional tendency to forget their charter and substitute globalism for internationalism.

■ SUMMARY POINTS

What are the critical points to take home from this chapter? First, globalization is entirely different from internationalization. Under globalization, all national economies are integrated into one global economy and must obey the laws laid down by a global economic institution, currently the WTO. Under internationalization, relations between countries grow increasingly important, but the nation remains the basic unit of community and policy.

Second, the concept of comparative advantage is largely irrelevant in a world of mobile financial capital. We must instead look at absolute advantage. While absolute advantage is likely to lead to greater overall gains in terms of market production, it will also produce both winners and losers on an international level. If we include the ecological costs of greater economic growth, there are likely to be far more losers than winners. Without the elegant theoretical conclusion of comparative advantage that free trade is a win-win affair for all countries, it is necessary that we look at the empirical evidence regarding each country's absolute advantage along with ecological impacts to determine winners and losers under a regime of global integration. In the next chapter, we focus on globalization and its likely consequences.

BIG IDEAS to remember

- Comparative vs. absolute advantage
- Capital mobility and comparative vs. absolute advantage
- Globalization vs. internationalization
- Bretton Woods Institutions (IMF and World Bank)
- World Trade Organization (WTO)

CHAPTER

19

Globalization

Why are so many smart people such ardent advocates of globalization? Mainly, it is because **globalization** results in more efficient allocation of resources toward market goods and services, resulting in faster rates of global economic growth. But there are other consequences, also intended: (1) an increase in international competition, in which countries must compete against each other for a share of global markets; (2) more intense national specialization according to the dictates of competitive (not comparative) advantage; (3) worldwide enforcement of trade-related intellectual property rights; and (4) control over local and national affairs by an international institution. Empirical evidence suggests another important consequence, an unintended one: an increased concentration of wealth within and between countries. It is this last consequence, perhaps, that has sparked the strongest opposition to globalization. These consequences of course are in addition to the impacts of economic growth on scale.

In this chapter we look at each of these consequences in a bit more detail within the context of the policy goals of ecological economics: efficient allocation, just distribution, and sustainable scale.

■ EFFICIENT ALLOCATION

Advocates of globalization claim that free trade is efficient, producing the much-touted gains from trade. But efficiency depends on a number of critical assumptions and conditions, including these:

1. There must be a large number of nearly identical firms.

2. Information must be freely shared.

3. There must be strong incentives to internalize costs.

Are the conditions for efficient production being met? We address this question for each condition listed above. We also ask whether globalization maintains the free choice for satisfying employment and the role that choice of employment plays in enhancing human welfare.

Perfect Competition vs. Transnational Corporations

The assumption of perfect competition, which requires a very large number of nearly identical firms, is a cornerstone of neoclassical economic theory. Perfect competition weeds out the inefficient and ensures the efficient allocation of resources within both national and international markets.

Globalization forces firms to compete against other firms worldwide. However, in general, only very large businesses have the resources to enter foreign markets. WTO rules actively forbid countries from promoting national small businesses if this can be construed as discrimination against large foreign corporations. Large corporations with economies of scale, or willing to accept low profits in an effort to gain market share, can easily underprice local businesses and either bankrupt or acquire them, thereby reducing the total number of firms. In fact, global mergers and acquisitions have been most intense in the areas of financial services and telecommunications, precisely the economic sectors in which WTO agreements were first completed.[1]

Global competitiveness may therefore be incompatible with market competition in a given nation. As a rule, many economists agree that if 40% of a given market is controlled by four firms, the market is no longer competitive. Such concentration is not at all unusual in the agricultural sector: in the U.S. Midwest, four firms control well over 40% of the trade in most major agricultural commodities,[2] and the top four agrochemical corporations reportedly control over 55% of the global market.[3] Nonetheless, in 1999 the U.S. government approved the merger of the two largest international grain trading corporations, Cargill and Continental Grain, in spite of explicit concerns that the merger might result in monopsony power [4] (over 80% of international trade is controlled by only ten firms).[5]

---

[1]L. Wallach and M. Sforza, *Whose Trade Organization?: Corporate Globalization and the Erosion of Democracy*, Washington, DC: Public Citizen, 1999.

[2]W. Heffernan, Report to the National Farmers Union: Consolidation in the Food and Agriculture System, Columbia: University of Missouri, 1999. Note that it can be very difficult to assess market concentration.

[3]Action Group on Erosion, Technology and Concentration, Concentration in Corporate Power: The Unmentioned Agenda, ETC communique #71, 2001. Online: http://www.rafi.org/documents/com_globlization.pdf.

[4]Monopoly is a single seller, and monopsony is a single buyer. In this case, the U.S. Department of Justice was more concerned about prices to farmers than prices to consumers

[5]G. van Empel and M. Timmermans, Risk Management in the International Grain Industry,

Ironically, the U.S. assistant attorney general in charge of reviewing the merger reportedly suggested that even more consolidation would be needed to maintain the competitiveness of American agriculture in global markets.[6]

If mergers are indeed needed for firms to remain "competitive" in a global economy, will this lead to a more efficient allocation of resources? This is a highly controversial question. Some economists argue that only coercive monopolies, defined as businesses that can set their "prices and production policies independent of the market, with immunity from competition, from the law of supply and demand,"[7] are a problem and that such monopolies arise only as a result of government intervention. Concerning monopolies not enforced by the government, "It takes extraordinary skill to hold more than fifty percent of a large industry's market in a free economy. It requires unusual productive ability, unfailing business judgment, unrelenting effort at the continuous improvement of one's product and technique. The rare company which is able to retain its share of the market year after year and decade after decade does so by means of productive efficiency—and deserves praise, not condemnation."[8] From this point of view, trusts that emerge in free markets are more efficient and always subject to competition. However, in U.S. agricultural markets, for example, we've seen the share of food dollars going to farmers dwindle from nearly 40% in the second decade of the twentieth century to under 10% over the last 100 years.[9] As Chicago School economist and Nobel laureate Ronald Coase pointed out, firms are islands of central planning in a sea of market relationships.[10] The islands of central planning become larger and larger relative to the remaining sea of market relationships as a result of mergers. More and more resources are allocated by within-firm central planning and less by between-firm market relationships. Of the 100 largest economic organizations, more than half are corporations. One-third of the commerce that crosses national boundaries does not cross a corporate boundary; it is an intrafirm, nonmarket, transfer. Is there

---

Commodities Now, December 2000. Online http://www.commodities-now.com/cnonline/dec2000/article3/a3-pl.shtml.

[6]Reported in A. Cockburn and J. St. Clair, How Three Firms Came to Rule the World, in *Counterpunch*, November 20, 1999. Online: www.counterpunch.org. A competitive market is defined as one with enough firms that all firms are price takers and none are price makers. Mergers in highly concentrated markets by definition make those markets less competitive, not more.

[7]A. Greenspan, Antitrust. In A. Rand, ed., *Capitalism: The Unknown Ideal*, 1961.

[8]Ibid., p. 70.

[9]B. Halweil, Where Have All the Farmers Gone? *World Watch*, September/October 2000.

[10]R. Coase, The Nature of the Firm, *Economica* 4(16):386–405 (1937).

any reason that central planning should work better for a large corporation than it does for a nation?

Patents and Monopolies

At the global scale, intellectual property rights are tied to trade. Why? It is hard to trade property if the legal right to the property is in dispute. As a result, the Agreement on Trade Related Aspects of Intellectual Property (TRIPs) requires all WTO signatories to protect intellectual property rights (IPRs) for 20 years,[11] with violators subject to trade sanctions and fines.

In Chapter 10, we discussed two major inefficiencies associated with patents. First, information is a nonrival good, and making it excludable leads to inefficiencies. Second, patents are nothing more than temporary, government-enforced monopolies, and such monopolies are inherently inefficient. We also discussed the counterargument for patents, that unless we provide the economic incentive of monopoly ownership for a significant period of time (20 years, they suggest), little new knowledge and innovation will be forthcoming.

THINK ABOUT IT!

Do you remember why it is inefficient to make nonrival goods excludable? Do you remember why monopolies are inefficient?

Although patents have existed in England since the seventeenth century, in the United States and France since the 1790s, and in most of Europe since the 1880s, international patents are a fairly recent phenomenon. They have been practical only since the International Patent Institute was established at the Hague in 1947. Until the 1980s, patents played an unimportant role in international commerce. Empirically it seems to be the burst of inventions that has stimulated demand for greater patent protection, more than vice versa.

In 1790 Samuel Slater, the "Father of American Industry," essentially stole the design for the first American textile factory from Richard Arkwright, the English industrialist.[12] Currently corporations and individuals from developed countries own 97% of all patents, and the WTO provides mechanisms for enforcing these patents globally. Is this likely to encourage or discourage a new Samuel Slater, a father of industry in a country that truly needs it? At the very least, it is difficult to argue that technology

---

[11]Wallach and Sforza, op. cit.

[12]The Story of Samuel Slater, Slater Mill Historic Site, Online: http://www.slatermill.org/html/history.html.

has foundered so much since the 1970s that we need a substantial expansion of patent protection under the WTO to stimulate its advance.

> **THINK ABOUT IT!**
>
> *Make a list of the most important contributions, discoveries, or inventions in your major field of study. How many of them were motivated by patentability of intellectual property? There might be interesting differences between fields of study.*

As economist Joseph Schumpeter emphasized, being the first with an innovation already gives one a temporary monopoly by virtue of novelty. In his view, these recurring temporary monopolies were the source of profit in a competitive economy whose theoretical tendency is to compete profits down to zero. This is the very condition of economic efficiency—why thwart it?

This is not to say that we should abolish all intellectual property rights. Such an action might create more problems than it would solve. But we should certainly begin restricting the domain and length of patent monopolies rather than increasing them so rapidly and recklessly. And we should become much more willing to share knowledge along with the cost of producing it. Shared knowledge increases the productivity of all labor, capital, and resources—things that are inherently scarce, rival, and excludable. Knowledge is not inherently scarce and is the quintessential public good—nonrival and nonexcludable—even though patents make it artificially excludable.

One important and practical policy implication of these considerations is that international development aid should consist far more of freely shared knowledge and far less of foreign investment and interest-bearing loans. Let's recall the following words from John Maynard Keynes, one of the founders of the Bretton Woods Institutions:

> *I sympathize therefore, with those who would minimize, rather than those who would maximize, economic entanglement between nations. Ideas, knowledge, art, hospitality, travel—these are the things which should of their nature be international. But let goods be homespun whenever it is reasonably and conveniently possible; and, above all, let finance be primarily national.* [13]

Externalizing Costs

As we also discussed in Chapter 10, a necessity for efficient markets is that producers pay the costs of production, and they produce to the point where marginal costs are just equal to marginal benefits. This condition is

---

[13] J. M. Keynes, "National Self-Sufficiency." In D. Muggeridge, ed., *The Collected Writings of John Maynard Keynes*, vol. 21, London: Macmillan and Cambridge University Press, 1933.

BOX 19-1 | DISCOVERIES NOT MOTIVATED BY PATENT MONOPOLY PROFITS

No doubt many important inventions have been stimulated by patent rights. However, the heliocentric view of the universe, gravity, the periodic table of elements, electromagnetic theory, and the laws of optics, mechanics, thermodynamics, and heredity were all discovered without the benefit of intellectual property rights and the profit motive. Mathematics has been called the language of the universe. Where would our technology be without it? While no culture has ever allowed a patent on mathematical theorems, mathematicians keep producing new ones.[a] Nor has anyone ever had intellectual property rights to the English language, or to fire, the wheel, or money. Yet all these things somehow came into being. The invention of the shipboard chronometer, necessary for navigational calculation of longitude, was stimulated by a one-time prize, not a 20-year patent monopoly. Even economists work long and hard to produce economic theories that are not patentable. Alfred Marshall got no royalties from users of supply and demand and elasticity. J. R. Hicks expected no royalties, and got none, for developing the IS-LM model and the proper concept of income.

In fact, it is difficult to name a single modern invention that does not depend on ideas freely shared from their first conception. While patent rights have stimulated important inventions, that is less than half the story. In the words of Lawrence Lessig,

> Free resources have always been central to innovation, creativity and democracy. The roads are free in the sense I mean; they give value to the businesses around them. Central Park is free in the sense I mean; it gives value to the city that it centers. A jazz musician draws freely upon the chord sequence of a popular song to create a new improvisation, which, if popular, will itself be used by others. Scientists plotting an orbit of a spacecraft draw freely upon the equations developed by Kepler and Newton and modified by Einstein. Inventor Mitch Kapor drew freely upon the idea of a spreadsheet—VisiCalc—to build the first killer application for the IBM PC—Lotus 1-2-3. In all of these cases, the availability of a resource that remains outside of the exclusive control of someone else—whether a government or a private individual—has been central to progress in science and the arts. It will also be central to progress in the future.[b]

---

[a]D. S. Evans, Who Owns Ideas? The War over Global Intellectual Property, Foreign Affairs 81(6):160–166 (November/December 2002).

[b]L. Lessig, The Future of Ideas: The Fate of the Commons in a Connected World, New York: Random House, 2001. Online: http://music.barrow.org/2002/Q3/free/page3.htm.

not met when externalities exist, and there are several ways in which economic globalization increases the quantity and severity of externalities.

The goal of the WTO is to increase economic growth and the transport of goods between countries, and both growth and fossil fuel–based transportation are accompanied by significant externalities. At the national level, laws exist to reduce externalities, but the WTO has the power to challenge these laws and the ability to enforce its decisions. While technically countries are allowed to pass environmental legislation, the WTO often declares such laws barriers to trade, and even the threat of a WTO ruling can deter lawmakers. For example:

1. Challenged by Venezuela, the United States was forced to allow the import of gasoline that does not comply with U.S. Clean Air Act regulations.

2. The WTO ruled against the U.S. Endangered Species Act, which prohibits the import of shrimp from countries that do not mandate turtle excluder devices.

3. Under GATT, Mexico won a decision against the U.S. Marine Mammal Protection Act's dolphin-safe tuna provision. Under threats by Mexico of WTO enforcement action, President Clinton and Vice President Gore took the lead in getting Congress to weaken the offending law.

4. Australia's law strictly limiting the import of raw salmon, designed to prevent domestic stocks from contamination with foreign bacteria, was declared a barrier to trade. Scientific studies showed that the risk of infection existed, but the WTO ruled that the probability of infection also had to be shown to justify import restrictions.[14]

Chapter 11 of the North American Free Trade Agreement (NAFTA) specifically allows corporations to sue national governments of NAFTA nations in secret tribunals when government decisions or regulations affect their investments. In one of dozens of suits filed under Chapter 11, the Ethyl Corporation sued Canada when Canada banned the gasoline additive methylcyclopentadienyl manganese tricarbonyl (MMT) because of health risks. Canada settled out of court, not only paying Ethyl Corporation's court cost and $13.7 million for lost profits but also revoking the ban.[15]

---

[14]Unfortunately, many of the ecosystem goods and services threatened by economic growth and free trade are characterized by uncertainty and ignorance, in which case by definition probabilities of possible outcomes cannot be determined.

[15]M. Poirier, The NAFTA Chapter 11 Expropriation Debate Through the Eyes of a Property Theorist, *Environmental Law* Fall 2003.

In each of these cases, the overturned regulations had been put in place by relatively democratic governments to reduce negative externalities affecting nonmarket goods and services. A number of other environmental laws are currently threatened by the WTO and NAFTA.[16]

Standards-Lowering Competition

At the same time that international trade agreements make it difficult for countries to legislate against externalities, the need to compete for market share reduces national incentives to legislate against externalities in what is known as standards-lowering competition (a race to the bottom). The country that does the poorest job of internalizing all social and environmental costs of production into its prices gets a competitive advantage in international trade. More of world production shifts to countries that do the poorest job of counting costs—a sure recipe for reducing the efficiency of global production. As uncounted, externalized costs increase, the positive correlation between GDP growth and welfare disappears or even becomes negative. Recall the prescient words of John Ruskin: "That which seems to be wealth" becomes in verity the "gilded index of far reaching ruin."[17] The first rule of efficiency is "count all the costs," not "specialize according to comparative advantage."

One way to confront the race-to-the-bottom tendency is to argue for harmonization of cost-accounting standards across countries. This is certainly logical and in line with global integration. If all countries internalized external social and environmental costs to the same degree, there would be no incentive for mobile capital to move to the country that did not internalize these costs because such countries would not exist. It would be hard to negotiate such a global harmonization agreement. There are, in fact, good reasons why different countries have different cost-accounting practices. In any case, it might be argued that countries should measure costs according to their own values, not "international standards." The traditional comparative advantage argument is compatible with each country's measuring costs as it pleases. As we saw in Table 18.1, a-units and b-units, which might reflect totally different theories of value, need never be compared in the comparative advantage system. But with capital mobility and absolute advantage comes the necessity to compare a-units to b-units and the problem of standards-lowering competition to attract mobile capital.

---

[16]All examples are from Wallach and Sforza, op. cit.

[17]J. Ruskin, "Unto This Last." Online: http://www.nalanda.nitc.ac.in/resources/english/etext-project/economics/Ruskin.pdf.

BOX 19-2	WEALTH, POWER, AND EFFICIENCY

Another potential problem with economic globalization is an increase in rent-seeking behavior in the form of lobbying by large corporations and wealthy individuals to influence policy.[a] If large corporations are "islands of central planning," showing less growth than smaller corporations probably as a result of the inefficiencies of central planning, why do they continue to thrive? One possibility is that the concentrated wealth of large corporations readily translates into political power, and large corporations can use this power to promote policies that allow them to thrive in spite of any inefficiencies inherent to centrally planned economies.

Large corporations routinely help politicians set not only domestic rules of the game but also international rules. The trade advisor to President Nixon was a vice president of Cargill, the world's largest grain exporter. President Reagan relied on a Cargill employee to draft the U.S. agricultural proposal for GATT.[b] President Clinton appointed Monsanto CEO Robert Shapiro as a trade representative to the WTO. President George W. Bush relied on Enron CEO Kenneth Lay when designing energy policies. The WTO meetings in Seattle in 1999 were sponsored primarily by large corporations. Can we be sure that this advice and assistance come with no strings attached?

[a]Recall that rent is profit over and above the normal profits of operation.

[b]K. Lehman and A. Krebs, "Control of the World's Food Supply." In J. Mander and E. Goldsmith, eds., The Case Against the Global Economy, San Francisco: Sierra Club Books, 1996.

Specialization and Diminished Well-Being

Free trade and free capital mobility increase pressures for specialization according to competitive (absolute) advantage. Therefore, as noted earlier, the range of choice of ways to earn a livelihood becomes greatly narrowed. In Uruguay, for example, everyone would have to be either a shepherd or a cowboy in conformity with the dictates of competitive advantage in the global market. Everything else should be imported in exchange for beef, mutton, wool, and leather. Any Uruguayan who wants to play in a symphony orchestra or be an airline pilot should emigrate.

Most people derive as much satisfaction from how they earn their income as from how they spend it. Narrowing that range of choice is a welfare loss uncounted by trade theorists. Globalization assumes either that emigration and immigration are costless or that narrowing the range of occupational choice within a nation is costless. Both assumptions are false.

While the range of choice in earning one's income is ignored by trade theorists, the range of choice in spending one's income receives exaggerated emphasis. For example, the United States imports Danish butter cookies, and Denmark imports U.S. butter cookies. The cookies cross each other somewhere over the North Atlantic. Although the gains from trading such similar commodities cannot be great, trade theorists insist that the welfare of cookie connoisseurs is increased by expanding the range of consumer choice to the limit.

Perhaps, but could not those gains be had more cheaply by simply trading recipes? One might think so, but recipes (trade-related intellectual property rights) are the one thing that free traders really want to protect.

■ SUSTAINABLE SCALE

While globalization advocates laud efficiency, their goal is not simply more efficient production of what we now produce but rather ever greater production. If the goal of international trade is to promote growth in GDP with little or no attention paid to scale, in the long run, a "successful" trade regime will lead us beyond the sustainable scale for the global economy. This is true no matter how efficient the allocation of resources between countries. It should already be clear that greater externalities and standards-lowering competition pose threats to sustainable scale.

Two other issues bear mentioning. First, the integration into one global system gives us only one chance to see if the system works—we cannot learn from our mistakes. Second, the negative environmental impacts of our consumption may occur in another country, which makes them that much more likely to be ignored.

Learning from Our Mistakes

In the past, numerous civilizations have crumbled as they have surpassed ecological barriers. Examples are the civilization on Easter Island, the Mayan empire, and the early civilizations of the Fertile Crescent. Fortunately, these were isolated incidents in which only the local carrying capacity was overwhelmed, and today they can serve as examples of mistakes we cannot afford to make. However, as trade expands, local limits to scale become less relevant and global limits more so. While trade may decrease the chances of surpassing sustainable scale in any one area, it also means that if we do surpass it, we are more likely to do so for the planet as a whole. Consequently, it becomes more difficult to learn from our mistakes as we go. Thus, globalization requires us to get it right the first time.

Out of Sight, Out of Mind

Even if globalization did not lead countries with high environmental standards to lower them, international trade can make it easier to ignore the costs of economic growth. In recent decades, as the most developed nations saw their environments deteriorating, they passed laws to control some types of pollution and resource depletion. To some extent this led to greater efficiency, decreased consumption of polluting products, and improved technologies for controlling pollution, but in many cases it seems to have led to the relocation of polluting and resource extracting industries to countries without such laws.[18] The environments in the developed countries improved at the expense of the poorer countries. With the spatial connection between economic growth and environmental damage severed, many people seem to believe that the causal connection has been severed as well. Indeed, many economists now claim that environments in the developed countries improved precisely because of economic growth.

In reality, the net impact of relocation of environmentally damaging industries on scale can be highly negative. For example, when Australia's wet tropical rainforests were declared a World Heritage site (largely due to pressure from environmentalists) and the region's reasonably well-managed logging operations were shut down, total timber consumption in Australia did not decrease. Instead, Australia has substituted its own tropical timber supply with timber from tropical countries with worse logging practices. The net outcome is probably a greater loss of ecosystem services worldwide.

Similar, and perhaps more threatening, is the relocation of waste from toxic industries. The Basel Action Network (BAN) has documented the devastating effects of exporting electronic waste to China and other developing nations[19] and in 2007 a shipment of toxic waste from Europe to the Ivory Coast was disposed of in the city of Abidjan, causing numerous deaths.[20] There is even talk of exporting nuclear waste for disposal in Russia. Such exports allow the overdeveloped countries to reduce further degradation of their environments.[21] However, they may well result in less careful waste disposal than would have occurred in the country where the waste originated. Furthermore, simply transporting toxic wastes increases the danger of negative environmental impacts.

We already know that markets fail to signal many environmental costs.

---

[18]D. Rothman, Environmental Kuznets Curves—Real Progress or Passing the Buck? A Case for Consumption-Based Approaches, *Ecological Economics* 25:177–194 (1998).

[19]See http://www.ban.org.

[20]L. Polgreen and M. Simons, Global Sludge Ends in Tragedy for Ivory Coast, *New York Times*. October 2, 2006.

[21]Overdeveloped countries are defined as those whose level of per-capita resource consumption is such that if generalized to all countries could not be sustained indefinitely. See H. Daly, *Beyond Growth: The Economics of Sustainable Development*, Boston: Beacon Press, 1996.

In a democracy, when people are exposed to environmental externalities, they can signal their preferences through political institutions. If people in the developed democracies export their wastes or environmentally damaging industries to less democratic countries, we lose this signal of environmental scarcity. The first rule of cost internalization is to internalize costs to the firm that generates them. If we fail to do this, then we must at least internalize costs to the country under whose laws the firm was operating when it generated the externalities. The second rule could be enforced by prohibiting the export of toxic waste, as called for by the Basel convention on hazardous wastes, which the United States had not signed as of 2009.

Positive Aspects of Trade with Respect to Scale

We have thus far presented an incomplete picture of the impacts of globalization on scale. In the absence of international trade, appropriate scale (in terms of economic activity and human populations) would be determined at the national level and by the most limiting factor. For example, one country might have abundant agricultural land but inadequate mineral resources. In other countries, scale might be limited by land area, mineral resources, or fuel supplies, in yet others by waste absorption capacity, rainfall, or agricultural productivity. International trade can help alleviate the most limiting constraints on scale within each country. If international trade suddenly ceased, some countries would find themselves well beyond sustainable as well as desirable scale.[22]

Efforts to sustain a high standard of living for too large a population with limited resources would no doubt force some countries to liquidate natural capital—for example, by extending the agricultural frontier to lands that cannot sustain it or burning their forests to meet energy needs. Other countries might be forced to mine low-quality, highly polluting fossil fuels.

International trade can help sustain larger populations with higher levels of material consumption than isolated national economies alone could sustain. Unfortunately, this happy outcome is likely only if sustainable scale and equitable distribution are explicit and are the principal goals of international trade. It is more likely to occur under internationalization than globalization.

---

[22]Surpassing sustainable scale means that the sustaining ecological system must eventually collapse, and surpassing desirable scale means that the costs of additional growth outweigh the benefits.

■ JUST DISTRIBUTION

Finally, we turn to the impact of globalization on distribution. Proponents of globalization claim that it will bring about "a world free of poverty" (the professed goal of the World Bank), while opponents often argue that it will further concentrate wealth and power in the hands of the few. To determine the most likely outcome of globalization under existing institutions, we must assess both theoretical and empirical evidence. We conclude with a brief look at the most important of commodities: food.[23]

Absolute Disadvantage

In the presence of capital mobility, money will logically flow to wherever there is an absolute advantage of production and away from countries where this is none. The world's poorest countries may be poor precisely because they are inefficient at producing nearly everything. If this is true, then resources are likely to flow away from the poor countries, and the countries most likely to suffer from globalization are in fact the very poorest.

Does this conclusion have any empirical support? According to the IMF, most developing countries have failed to raise their per-capita incomes toward those of industrial countries. United Nations Development Programme statistics show that in the three decades prior to 1996, the share of income received by the world's poorest 20% fell from 2.3% to 1.4%, while the share going to the world's richest 20% increased from 70% to 85%. Still, these statistics refer only to relative income, not absolute income. The world's poorest 20% have seen some gains in income over the past 40 years.[24] Globalization on a significant scale, however, is a fairly recent phenomenon. What has happened to the poorest of the poor more recently?

The WTO came into being in 1995. Using the Global Development Network Growth Database,[25] we calculated that of the 20 poorest countries for which data were available, 8 had actually suffered a loss in real per-capita income between 1995 and 2003. The top performer in the

---

[23]In spite of the recent trend toward privatization of water, we do not consider water a commodity. Water is not produced for sale; it is produced by nature. Privatization is basically an enclosure of the commons, and a very inefficient one since it invariably creates a monopoly. (How many different water companies can you buy water from? How many water lines lead into your house?)

[24]E. Kapstein, "Distributive Justice as an International Public Good: A Historical Perspective." In I. Kaul, I. Grunberg, and M. Stern, eds., *Global Public Goods: International Cooperation in the 21st Century*, New York: Oxford University Press, 1999.

[25]Online: http://www.nyu.edu/fas/institute/dri/globaldevelopmentnetworkgrowthdatabase .html.

group was Mozambique, just recovering from a devastating civil war, which recorded per-capita income gains of 58%, or $466 (1996 dollars). In contrast, the wealthiest 20 countries in 1995 had experienced an average increase in per-capita income of 17.6% by 2003. The worst-performing country in this group, Switzerland, saw an increase per capita of $1478 (1996 dollars). The worst-performing group of countries in terms of growth were the 40 Heavily Indebted Poor Countries (HIPCs), suggesting that World Bank and IMF lending programs have been less than successful. In absolute terms, one year's income growth in the United States could pay off the debt of the HIPCs and simultaneously double their per-capita income.[26]

If we lived on an infinite planet in which one person's consumption had no impact on anyone else, and if human nature did not lead us to measure our wealth in comparison to others, it would make no difference to the poor countries what happened in the wealthy ones. The fact is, however, that we do live on a finite planet, and we do compare ourselves to others. The increase in income in the wealthy countries is fueled by nonrenewable resource consumption (including nonsustainable depletion of potentially renewable resources), which means that these resources are not available for future improvements in the well-being of the poorest. And resource use generates a corresponding amount of waste and accompanying damage to public good ecosystem services that would otherwise benefit these poorest countries.

This observation draws our attention to a fact otherwise obscured by the data. Most of these poorest countries were involved in international trade in the one area where they might have an absolute advantage: the extraction and export of natural capital. The revenue they received from both export and domestic sales of these resources counted as part of their income. Without this revenue, income as measured by GDP would have fallen even more. Yet as you will recall, we earlier defined income as the amount you can consume in one period without affecting your ability to consume in subsequent periods. Thus, revenue from nonrenewable natural resource extraction cannot be counted entirely as income, and the situation of these poorest countries is even worse than it appears.

Of course, the evidence presented here says nothing conclusive about globalization. One could argue, as economists often do, that the problem was insufficient liberalization. Perhaps things would have been even worse in the poorest countries without globalization. What happened in the countries that most avidly pursued economic liberalization?

Many countries have shown periods of economic growth as their

---

[26]Data from World Bank World Development Indicators Database. Online: http://www.world bank.org/data/countrydata/countrydata.html. We assumed a typical 2.7% growth rate for the U.S.

economies have liberalized. While China is hardly a textbook example of free trade, it has certainly liberalized its markets and engaged heavily in global trade, achieving in the process record levels of economic growth. At the same time, it has seen record increases in inequality[27] and environmental degradation, which is likely to have disproportionate impacts on the poor. A Green Gross Domestic Product project was implemented in 2004 but suspended in 2007 when results proved politically unacceptable; adjusted growth fell to nearly zero in some provinces.[28] Argentina, on the other, hand was a poster child of neoliberal reform beginning in the 1970s. Though real per-capita income increased by over 40% from 1990 to 1998, poverty rates surged as well. The experiment ended in complete failure in 2001 when the economy collapsed, and poverty rates soared to 58%. Mexico and Turkey, also considered showcases for neoliberal policies, also suffered economic crises and growing poverty.[29] One cause of these failures is intense instability resulting from the volatility of international capital flows, as we will discuss in the next chapter.

Probably the most damning evidence of the distributional impacts of globalization comes from the science of global climate change. It is absolute economic growth, not relative growth, that contributes to climate change. The vast majority of absolute growth unquestionably flows to the wealthiest countries. While the economies of India and China have boomed along with their carbon emissions, the growth has been export led, and carbon emissions should be attributed to the consuming countries. While the already wealthy countries reap most of the benefits from growing carbon emissions, the   poorest countries will bear a disproportionate share of the costs.[30]

Standards-Lowering Competition and Labor

We discussed above how countries pressured by global competition may ignore external costs to the environment in a race to the bottom. To remain competitive, countries may similarly need to accept or even promote lower labor costs.

---

[27]D. T. Yang, "Urban-Biased Policies and Rising Income Inequality in China," *The American Economic Review* 89(1999):306–310.

[28]J. Kahn and J. Yardley, "As China Roars, Pollution Reaches Deadly Extremes," *The New York Times*, August 26, 2007.

[29]P. Cooney, "Argentina's Quarter Century Experiment with Neoliberalism: From Dictatorship to Depression," *Revista de Economia Contemporanea* 11(2007):7–37; I. Grabel, "Neoliberal Finance and Crisis in the Developing World—Argentina, Mexico, Turkey, and Other Countries—Statistical Data Included," *Monthly Review*, April 2002.

[30]IPCC, "Climate Change 2007—Impacts, Adaptation and Vulnerability: Contribution of Working Group II to the Fourth Assessment Report of the IPCC," Cambridge, Cambridge University Press, 2007.

In the United States and Europe, an implicit social contract has been established to ameliorate industrial strife between labor and capital. Specifically, a just distribution of income between labor and capital has been taken to be one that is more equal within these countries than it is for the world as a whole. Global integration of markets necessarily abrogates that social contract. There is pressure on American and European wages to fall because labor is much more abundant globally than nationally. By the same logic, returns to capital in these countries should increase because capital is more scarce globally than nationally. This could lead U.S. income distribution, already on par with India's, to become even more unequal. Theoretically, one might argue that wages would be bid up in the rest of the world. But the relative numbers make this a bit like saying that, theoretically, when I jump off a ladder, gravity not only pulls me to the ground, it also moves the ground toward me.

In general, if a country pursues economic growth by developing the export sector, it must be able to sell what it produces on the highly competitive global market, and costs must be kept low. In a world of mobile capital, absolute advantage in production is what counts. Most less-developed countries (LDCs) do not have advanced technologies that can lower production costs, a well-developed infrastructure that can lower transportation costs, or institutions that lower transaction costs or make investments particularly safe. Instead, they have two sources of absolute advantage: abundant labor and (in some cases) abundant natural resources.

To compete in export-oriented industrial production, LDCs can acquire and maintain an absolute advantage only by keeping wages and benefits down or allowing environmental degradation. The first option does little to help alleviate poverty and often requires the suppression of labor rights, while the second option typically has disproportionate impacts on the poor.

What happens if instead a country seeks to industrialize to serve the needs of the domestic market? Obviously the prerequisite for this to occur is that a domestic market actually exists. A market can exist only if there is purchasing power, and purchasing power requires wages. This is why Henry Ford (no friend of labor) chose to pay his workers $5.00/day (at the time an exceptionally high wage): he wanted them to be able to afford the cars they were producing. Thus, for LDCs, a focus on liberal international trade will tend to push wages down, while a successful focus on production for the domestic economy will require higher wages.

When confronted with this argument, people might point to the Asian Tigers (Taiwan, Korea, Singapore, Hong Kong, and more recently Thailand and Malaysia). These countries, like Japan, pursued export-oriented industrialization and saw dramatic increases in their standards of living.

Yet on closer examination, the historical record of the Asian Tigers actually bolsters the argument for developing the domestic market. First, these nations are characterized by highly protectionist policies and a high degree of government intervention, not by open markets. More to the point, their initial successes were greatly facilitated by strong domestic markets. In Korea, Taiwan, and Japan, agrarian reform preceded industrialization. In these predominantly agrarian societies, the transfer of land to small farmers allowed the farmers to accumulate and spend the surplus they generated. As Dr. Sun Yat-Sen stated in the Son Mm Chu-I (the Three Principles of the People), "industrialization should follow, not precede, the building up of the internal capacity to consume." Successful land reform in Taiwan doubled the purchasing power of the farmer at a time when Taiwan was an agricultural economy.[31] In their early industrialization efforts, the Asian Tigers focused on import-substituting industrialization (ISI). Captive markets allowed them to develop the skills necessary for global competition.[32]

On the other hand, production for the domestic economy can have a negative impact on real wages as well. If a country currently importing industrial goods decides instead to replace them, the new industries will have a hard time competing against established producers. Thus, ISI generally requires tariffs and quotas on imports. Such tariffs will stimulate demand for domestic goods (benefiting the producer) but drive up the price of imports in order to make the import substitute more competitive and lower real wages.

In terms of distribution relative to export-oriented economies, however, this impact may not be as dire as it seems. First, export-oriented countries typically undervalue their currencies to make exports cheap and imports expensive, so there may be little difference between ISI and export promotion with respect to consumer prices. In addition, the easiest goods to produce are often the cheapest, those purchased by the working masses (soap, aspirin, matches, etc.). With simple production technologies, the LDCs are not at a serious disadvantage when producing these goods, and tariffs can be kept quite low. Luxury goods, on the other hand, tend to be more technologically sophisticated and may require higher tariffs. However, it is the wealthier classes that purchase these goods, and they can better afford to pay the tariffs.

As the industry develops, two things happen. First, efficiency should improve, and tariffs can be reduced—especially if the producers know that

---

[31]Quoted in F. Harrison, *Five Lessons for Land Reformers: The Case of Taiwan*. Reprinted from *Land & Liberty*, May–June (1980). Online: http://www.cooperativeindividualism.org/harrison_taiwan_land_reform.html.

[32]E. Vogel, *The Four Little Dragons: The Spread of Industrialization in East Asia*, Cambridge, MA: Harvard University Press, 1991.

tariffs will be reduced and therefore have an imperative to improve efficiency. Second, any market based solely on a wealthy minority will quickly become saturated. Further industrialization will depend on larger markets, which can be created through higher wages for the labor force. Once industrial capacity is well developed and production techniques have been refined, a country may be able to compete on the export market without forcing wages down to the global minimum. This is essentially the strategy that was pursued by the newly industrialized Asian economies.[33]

It is also worth remembering that in the 1960s and 1970s, the countries engaged in ISI often showed the highest growth rates, and they were touted as examples for others to follow. Brazil's economic growth under this strategy was considered miraculous. Economic wisdom is surprisingly fickle.

Probably the most important lesson to take from this discussion is that one size never fits all. Different cultures are likely to need different approaches to development, and what works under one set of global economic conditions might fail miserably under another. Economists would do well to keep this in mind.

Food Security and Free Trade in Agriculture

Potentially the most serious source of unfairness in the liberalization of trade is the threat it poses to food security. "Free" trade in food threatens security in two important ways. First, the market system provides goods and services to those who have the money to purchase them. If in the future the WTO or other international agreements succeed in liberalizing trade in agriculture, poor citizens of LDCs will be competing with the rich citizens of ODCs (overdeveloped countries) for food. In his groundbreaking study of famines, Amartya Sen has shown that famines are generally the result of a lack of entitlements to food rather than a lack of food itself.[34] In the market economy, this simply means a lack of money to purchase food, even when actual supplies are abundant. The situation can occur because of unemployment or a decrease in the value of the goods some group produces relative to food. In the presence of international trade, the domestic sector must bid against the rest of the world for food purchases. If the economy suffers a recession or there is a currency devaluation, local ability to purchase food decreases relative to global ability, and food may be exported even as the local population starves. Clearly, international trade can be critical in addressing famines that are caused by food availability decline, but only if countries have the resources to pur-

[33]Ibid.

[34]A. Sen, *Poverty and Famines: An Essay on Entitlement and Deprivation*, 6th ed., New York: Oxford University Press, 1992.

chase that food. If agricultural markets are completely liberalized, it is easy to imagine Western nations importing food for their cattle from nations suffering famine.

The fact that farmers are typically the most disadvantaged group in many LDCs brings a second source of unfairness if agriculture is liberalized. LDC costs of agricultural production tend to be higher than those of large agro-industrial farms in ODCs (in part because so many negative externalities of industrial agriculture are not internalized). This means that trade liberalization will often decrease prices for food in LDCs. While lower food prices may help the urban poor and wage earners, it can also cause lower incomes and declines in welfare for the poorest group, namely farmers. Low food prices reduce incentives for domestic agriculture. Theoretically, under trade liberalization, these poor farmers should be able to grow cash crops for export. Unfortunately, cash crops often need higher levels of inputs and are far riskier than traditional food crops that have been bred for millennia to minimize the risk of failure. While average returns over several years may be higher with cash crops even with more frequent failures, people do not eat "on average"—they eat every day.

Box 19-3   PUBLIC LAW 480 AND FOOD SECURITY

If one country becomes dependent on others for food supply, they run serious risks to their autonomy. For example, the U.S. Public Law 480 provides food at subsidized cost to LDCs. Although it is nominally a gesture of benevolence, American politician Hubert Humphrey once said in reference to this law: "I have heard . . . that people may become dependent on us for food. I know that was not supposed to be good news. To me that was good news, because before people can do anything they have got to eat. And if you are looking for a way to get people to lean on you and to be dependent on you, in terms of their cooperation with you, it seems to me that food dependence would be terrific."[a]

Secretary of Agriculture Earl Butz also referred to food as a weapon and "one of the principal tools in our negotiating kit."[b] Understandably, countries that come to depend on food imports may not share Humphrey's enthusiasm for that dependence.

[a]Sen. Hubert H. Humphrey, in naming P. L. 480 the "Food for Peace" program, Wall Street Journal, May 7, 1982.

[b]USDA Secretary Earl Butz, 1974 World Food Conference in Rome.

■ SUMMARY POINTS

We are better served by a process of internationalization in which countries are free to act on their own information to address local problems of scale and distribution (areas where the market manifestly fails) in a culturally sensitive manner. We must also carefully analyze the actual and potential impacts of globalization on scale, distribution, and efficiency.

Evidence suggests that globalization may be undermining the conditions needed for efficient market allocation by creating fewer, bigger firms, more negative externalities, and more monopolies on nonrival information. More negative externalities and increased economic growth, coupled with a limit on the national ability to regulate externalities, is a threat to sustainable scale. Empirical evidence also suggests that globalization under the principle of absolute advantage may simply reinforce existing patterns of winning and losing, leading to even greater concentration of wealth, both within and between countries.

BIG IDEAS to remember

- Perfect competition vs. mergers, patents, and transnational corporations
- Trade vs. environment in the WTO
- Standards-lowering competition
- Wealth and power
- Specialization and welfare
- Trade and scale
- Distribution and globalization
- Food security and international trade

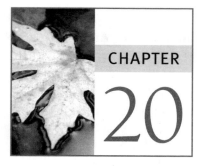

CHAPTER

20

Financial Globalization

National policies require national boundaries. Yet as globalization eradicates boundaries, even its advocates agree that "the world will move from closed, nationally controlled systems toward one open, global system under no one's control."[1] If no one is in control, no one makes national policies. We have already discussed some of the ways in which globalization undermines the ability of nations to determine their own policies concerning the environment. We will now turn our attention to the impact of globalization of financial capital.

Most financial capital flows are in the forms of electrons flitting from one computer to the next, but such flows can have serious impacts not only on the output of real goods and services but also on the need for and effectiveness of macroeconomic policies. According to the IMF, "globalization may be expected to increasingly constrain governments' choices of tax structures and tax rates."[2] The IMF also claims that it is the nation-state that must address issues of distribution and social welfare. Yet the effort to globalize and **liberalize** financial markets weakens the national policy levers necessary to achieve these goals while simultaneously increasing both the concentration of wealth and the risk of financial crises. This can have profoundly negative impacts on social welfare. To understand financial liberalization and how it can contribute to economic instability, we must briefly review balance of payments and exchange rates.

Liberalize means to loosen restrictions on trade and reduce regulation of markets.

---

[1]L. Bryan and D. Farrell, *Market Unbound: Unleashing Global Capitalism*, New York: Wiley, 1996.

[2]International Monetary Fund, *World Economic Outlook*, Washington, DC: IMF, 1997.

■ BALANCE OF PAYMENTS

There are two basic types of economic transactions between residents of different countries: the exchange of real goods and services and the exchange of assets. The net outcome of these transactions is measured by the **balance of payments (BOP)**. The BOP has two components, current accounts and capital accounts, corresponding to the two types of transactions.

The balance of payments (BOP) is the sum of the current account and capital account.

The current account measures the exchange of real goods and services as well as transfer payments—things consumed in the current period.

The capital account measures stocks, bonds, and capital abroad—stocks of capital yielding flows of revenue.

The **current account** measures the exchange of real goods and services as well as transfer payments. Goods and services are generally consumed in the current period, which is why it is called the current account. Real goods, of course, are market goods. Services include interest payments on loans, royalties on intellectual property, profits earned on investments abroad, and similar transactions. Transfer payments include money workers abroad send home to families, grants to foreign countries, and similar transactions.

Capital accounts include stocks, bonds, and property abroad. These items are not consumed. They are a stock of capital yielding a flow of revenue.

Money flowing into a country increases the balance on current account or capital account, and money flowing out of a country decreases it. If a country imports more goods and services than it exports, it runs a deficit in its current account, and if it exports more than it imports, it runs a surplus. Similarly, if foreigners purchase more assets in the home country than the home country purchases abroad, the home country runs a capital account surplus, and if the opposite is true, it runs a deficit. Note that a surplus in the current account can balance out a deficit in the capital account, and vice versa, to keep the balance of payments neutral.

One of the prime examples of these dynamics in today's world is trade between the U.S. and China. The United States purchases far more goods and services from China than it sells to China and thus runs an enormous current account deficit (over \$268 billion in 2008).[3] China then uses much of its corresponding current account surplus to purchase government bonds and other assets from the U.S., running a capital account deficit. From July 2008 to July 2009, China purchased \$250 billion in U.S. treasury securities.[4] In effect, China loans its current account surplus back to the U.S. It's somewhat ironic that the world's richest nation borrows so heavily from China to finance ever-greater consumption.

In general, countries try to keep their balance of payments neutral, running neither a surplus nor a deficit. In an accounting sense, formal bal-

---

[3]U.S. Census Bureau Foreign Trade Statistics, 2009. Online: http://www.census.gov/foreign-trade/balance/c5700.html.

[4]U.S. Department of Treasury, 2009. Online: http://www.treas.gov/tic/mfh.txt.

ance (capital plus current account) is guaranteed by a residual balancing item, formerly gold under the gold standard, but is now guaranteed by changes in a nation's foreign exchange reserves, short-term IMF credits (special drawing rights or SDRs), and IOUs from deficit countries called short-term capital flows.

■ EXCHANGE RATE REGIMES

An **exchange rate** is the amount of one currency you would have to pay to receive one unit of another. For example, in September 2009, one U.S. dollar could purchase about 6.8 Chinese renminbi. Any international exchange of goods, services, or assets requires an exchange of currencies, as those who sell generally want to be paid in their own currency. How the exchange of currency takes place depends on the exchange rate regime used by a particular country. The regime influences not only current and capital account balances but also economic stability and the effectiveness of domestic economic policy. There are two basic types of regime: fixed and flexible. When discussing exchange rates, *floating* is synonymous with *flexible*.

In a **fixed exchange rate regime**, the value of one country's currency is pegged (sometimes loosely) to that of another country. For example, from 1994 to 2004, the Chinese renminbi was pegged to the U.S. dollar at about an 8.3:1 ratio; that is, the People's Bank of China would sell as many dollars as anyone wanted for 8.3 renminbi each. Obviously, the central bank needed to have foreign currency on hand to make this exchange. Ideally, this currency would come from demand for Chinese products. When an American wanted to buy consumer goods from China, the central bank would also sell the required renminbi for $0.12 each. In reality, China and many other developing countries have some exchange rate controls, so people can't necessarily buy or sell as much of a given currency as they might like, but the general concepts described here still apply.

A problem can occur under a fixed rate regime when nationals of a fixed rate country consistently want to buy more (or fewer) goods, services, and assets from other countries than other countries want to purchase from them, that is, when the fixed rate country is running a BOP surplus (or deficit). If the fixed rate country runs a BOP surplus, as China does, it will accumulate excess reserves of foreign currency for which there is insufficient demand. In July 2009, China had over $2 trillion in foreign exchange reserves.[5] This will put pressure on the country to

---

[5]Bloomberg News, July 15, 2009, China's Foreign-Exchange Reserves Surge, Exceeding $2 Trillion. Online: http://www.bloomberg.com/apps/news?pid=20601087&sid=alZgI4B1lt3s.

revalue its currency—for example, offer fewer renminbi per dollar. If a fixed rate country runs a BOP deficit for too long, it will run out of foreign reserves, as happened to Argentina in 2001 after maintaining a fixed exchange rate for 10 years. In this case, there will be pressure to borrow money from abroad or devalue its currency—offer fewer dollars per Argentine peso. This is what happened to Argentina when it was forced to devalue its currency in January 2002.

In a **flexible exchange rate regime**, exchange rates are determined by global supply and demand for currencies, and central banks play no direct role. The only foreign currency available for purchase for foreign goods, services, and assets is that which foreigners spend on the purchase of domestic goods, services, and assets. If there is more demand for foreign currency than available, its price will be bid up, that is, the national currency depreciates, and when the opposite is true, the national currency appreciates.[6] Supply of and demand for foreign currency determine the exchange rate. This means that the BOP must always be zero—a current account deficit must be financed by a capital account surplus, and vice versa.

Most developed countries have flexible exchange rate regimes, but in reality, they are not perfectly flexible. In many cases, countries will grow concerned that their currencies are overvalued or undervalued and will buy or sell currency to correct the perceived imbalance. For example, in September 2000, the central banks of the G7 countries[7] coordinated policies to prop up the euro, which had fallen 30% against other currencies in the year since its release.[8] Some countries try to keep their exchange rate within a certain range, which is known as a managed float, somewhat of a hybrid between a fixed and flexible exchange rate.

Potentially, countries can manipulate the supply and demand for their own currencies. For example, high tariffs on imports will reduce the demand for imports, and hence the demand for foreign currencies. Alternatively, a country may simply control capital flows, directly determining the supply of its own currency and the demand for foreign currencies. In this case, national policies can determine the BOP under a fixed rate regime or

---

[6]Under a fixed exchange rate regime, when the domestic currency loses value it is known as devaluation, and when it gains value it is known as revaluation. In either case, this happens directly as a result of central bank policy. Under a flexible exchange rate regime, when market forces cause a currency to lose value, it is known as depreciation, and a gain in value is known as appreciation.

[7]G7 stands for the Group of 7, the seven most powerful industrialized nations in the world: the U.S., Canada, Germany, Japan, Italy, England, and France. After the 2007 financial crisis, the G7 was expanded to the G20, the 20 major economies in the world.

[8]BBC News, September 23, 2000, G7 Ready for Further Euro Action. Online: http://news.bbc.co.uk/hi/english/business/newsid_936000/936917.stm.

affect the exchange rate under a flexible rate regime. However, liberalization of trade and finance tends to limit or eliminate these options.

■ THEORIES ABOUT ECONOMIC STABILITY

Though economists disagree about the ultimate causes of the Great Depression, few disagree that a significant trigger was the stock market crash of 1929. Many stocks in the lead-up to the crisis had been purchased with bank loans which could no longer be repaid when the stocks plunged in value. Banks in turn were temporarily unable to repay depositors, leading panicked citizens to withdraw their money, causing many banks to collapse, freezing up credit flows. Industries could not borrow to invest and create jobs, and unemployed consumers were unwilling to spend, leading to a downward spiral in economic activity. The Great Depression led many economists to recognize that markets can fail to efficiently allocate goods and services and that government intervention was at times required. Many economists similarly came to believe that unregulated markets also fail in the financial system. Countries around the world used Keynesian monetary and fiscal policies to emerge from the Great Depression and imposed substantial regulations and government oversight on financial sectors to avoid another.

In the 1970s influential economists led by Milton Friedman began arguing that the Great Depression was actually caused by government monetary policy failing to provide adequate liquidity (i.e., money available for investment and consumption), not market failures, and that the financial sector, like the rest of the economy, was best left to self-correcting free market forces.[9] This school of thought can be broadly summarized as the Efficient Market Hypothesis. However, another school of influential economists continues to maintain that financial markets are inherently unstable in the absence of government regulation.[10] This school of thought can be broadly summarized as the Financial Instability Hypothesis.[11]

The Efficient Market Hypothesis (EMH) claims that the prices of stocks, bonds, other financial assets, and assets in general, are based on market fundamentals; they reflect all available information about their true value

---

[9]M. Friedman and A. J. Schwartz, *A Monetary History of the United States, 1867–1960*, Princeton, NJ: Princeton University Press, 1963; E. G. Fama, Efficient Capital Markets: A Review of Theory and Empirical Work, *Journal of Finance* 25(2):383–417 (1970); R. Lucas, *Studies in Business-Cycle Theory*, Cambridge: MIT Press, 1981.

[10]H. Minsky, *Stabilizing an Unstable Economy*, New Haven, CT: Yale University Press, 1986; J. Stiglitz, *Globalization and Its Discontents*, New York: Norton, 2002; G. Soros, *The Crisis of Global Capitalism: Open Society Endangered*, London: Little Brown and Co., 1998.

[11]In academic circles, both the Efficient Market Hypothesis and the Financial Instability Hypothesis refer to rather narrowly defined schools of thought. We are defining the terms fairly broadly here to encompass a range of views.

and are therefore equal to the net present value of all future returns. If un-informed investors ignorantly bid up or bid down the price of an asset, in-formed investors will profit by quickly bringing the price back to where it should be. In such a world, changes in stock prices randomly fluctuate around their true equilibrium value (the future is assumed to by risky, but not uncertain—see Box 6.1), speculative bubbles are highly unlikely, and market crashes are actually the result of changes in underlying values.[12] Government regulation not only is unnecessary but actually interferes with the efficiency of financial markets. Business cycles in fact are caused by unexpected government fiscal and monetary policy, in addition to technological changes or supply shocks. Such true believers in the market claim that deregulating the financial sector (see Table 20.1) increases eco-nomic growth, accountability, transparency, and economic stability.[13]

The Financial Instability Hypothesis (FIH), in contrast, asserts that the prices of investment assets are driven largely by speculation and are sub-ject to pure uncertainty in addition to insurable risk. Prices succumb to positive feedback loops (see Box 8.1) in which rising prices increase spec-ulative demand, increasing prices even further on the way up, with falling prices decreasing demand on the way back down.[14] Prices are mistaken for underlying values, and when prices are rising, overvalued assets are used as loan collateral, spurring ever more lending. Loans dedicated to consumption lead to economic growth, stimulating yet more loans for in-vestment, while loans for speculative investment lead to ever-higher asset prices.

In one formalization of the theory, there are three types of investors. **Hedge investors** are able to cover both capital and interest on their loans from returns on their investments, **speculative investors** can cover only interest and must continually roll over existing loans, while **Ponzi in-vestors** count on rising asset prices and ever more buyers to cover both interest and principal. When prices cease rising, as they ultimately must, Ponzi investors cannot pay back principal. Banks in turn can no longer turn over the loans of speculative investors, who must either sell their as-sets or also default. Rapid asset sales drive down prices, causing even more investors to sell and driving down prices further in another positive

Hedge investors are able to cover both capital and interest on their loans from returns to their investments.

Speculative investors can cover only interest and must continu-ally roll over existing loans.

Ponzi investors count on rising asset prices and ever more buyers to cover both interest and principal.

---

[12]See footnote 9, or for an extreme example of this argument, P. Garber, Famous First Bub-bles, *Journal of Economic Perspectives* 4:35–53 (1990).

[13]See G. Kaminsky and S. Schmukler, Short-Run Pain, Long-Run Gain: The Effects of Finan-cial Liberalization, NBER working papers no. 987, 2002; O. Obstfeld, The Global Capital Market: Benefactor or Menace? *Journal of Economic Perspectives* 12(4):9–30 (1998); G. Bekaert, C. Harvey, and C. Lundblad, Does Financial Liberalization Spur Growth? NBER working paper no. 8245, 2001.

[14]Financier George Soros refers to this process as reflexivity. *Irrational exuberance* was Alan Greenspan's term for the speculative upswing in prices, and *herd behavior* describes the tendency to buy when others are buying and sell when others are selling.

feedback loop. If returns on assets fall enough, hedge investors can no longer meet their loan payments, and they default as well.[15]

Financial instability of course has serious impacts on the real economy. Credit freezes up, businesses stop investing, and job markets worsen. In yet another positive feedback loop, the unemployed slow consumption, leading firms to produce less and lay off more workers, who in turn must default on any outstanding debts. The economy spirals downwards in a vicious circle. Financial markets naturally move towards disequilibrium, but government regulation can help temper this tendency.

Midway between the EMH and FIH is the emerging school of behavioral finance, which recognizes that people make systematic mistakes in their decision making. Asset prices may reflect these systematic mistakes, which allows them to diverge from underlying true values based on market fundamentals.[16]

■ GLOBAL FINANCIAL LIBERALIZATION

Beginning in the early 1970s, market economies around the world began removing government oversight of their financial sectors (see Table 20.1). Though subsequent financial crises led temporarily to increased regula-

■ Table 20.1

THE LIBERALIZED FINANCIAL SECTOR	
Capital accounts	• Banks and corporations are allowed to freely borrow money abroad. • Exchange rates are market determined and the same for current account and capital account transactions. • No restrictions on capital outflows (e.g., repatriation of profits).
Domestic financial sector	• No controls on interest rates or credit (e.g., subsidies or allocations for certain sectors). • Deposits allowed in foreign currencies. • Bank reserve requirements are minimal (e.g., less than 10%) or nonexistent. • Minimal importance of government-owned banks.
Markets in stocks and other securities	• Foreigners allowed to freely invest and repatriate profits. • Few restrictions on the types of financial instruments that can be bought and sold. • Minimal controls on leverage (borrowing money to finance investments with the hopes that interest payments on debt will be lower than returns on the investment).

Source: Adapted from G. L. Kaminsky and S. L. Schmukler, Short-Run Pain, Long-Run Gain: Financial Liberalization and Stock Market Cycles, Review of Finance *12:253–292 (2008).*

---

[15]H. Minsky, *Stabilizing an Unstable Economy* (New Haven, CT: Yale University Press, 1986).

[16]N. Barberis and R. Thaler, Richard. "A Survey of Behavioral Finance," in G. M. Constantinides, M. Harris, and R. M. Stulz, eds., *Handbook of the Economics of Finance* (New York: Elsevier, 2003), vol. 1, 1053–1128.

tion, the process accelerated in the 1980s, and most market economies had largely deregulated their financial sectors by the mid-1990s. What was the result of this liberalization?

In the recent era of liberalization, we often witnessed unpredicted changes in international capital flows in and out of countries, contributing to and resulting from sudden changes in asset values. Such changes have dramatic impacts on real economic variables in the countries affected. It appears that financial globalization allows national financial crises to spread from country to country, creating global economic crises. There are many examples of this.

1. The Latin American debt crisis, which started in 1982 when Mexico found itself unable to make payments on its debt, quickly spread to 39 other countries.

2. The Tequila crisis, which began in Mexico in December 1994, spread to Brazil and Argentina. The crisis led to emergency loans of $52 billion from the U.S. Treasury and the IMF.

3. The Asian financial "flu" started in Thailand in 1997 and quickly spread throughout Southeast Asia, then on to Africa, Russia, Poland, and Argentina.

Box 20-1	PETRO-DOLLARS AND THE LATIN AMERICAN DEBT CRISIS

One of the causes of the Latin American debt crisis was almost certainly the recycling of petro-dollars, the high profits from the 1973 and 1979 oil price increases invested in banks by the OPEC countries. These petro-dollars were loaned at very low interest rates to LDCs. Unfortunately, the interest rates were floating; that is, they moved up and down with changes in the global interest rate.

In 1981, the U.S. Federal Reserve Bank implemented a tight monetary policy in the U.S. to curb inflation induced by increased oil prices. Simultaneously, the Reagan government engaged in record levels of deficit spending. Both actions drove up interest rates on the dollar, from about 3% in the early 1970s to more than 16% a decade later. At the same time, these high interest rates in the U.S. increased demand for dollars, driving up the value of the dollar relative to other currencies by 11% in 1981 and 17% in 1982, further increasing the dollar-denominated debt burden.[a] Not surprisingly, the debtor countries had considerable difficulty repaying their loans under these terms.

---

[a]Federal Deposit Insurance Corporation, History of the Eighties—Lessons for the Future, 2001. Online: http://www.fdic.gov/bank/historical/history/index.html.

4. The sub-prime mortgage crisis started in the U.S. in 2007, then rapidly spread around the world. National governments have so far spent trillions of dollars to bail out financial sectors around the world.

All of these crises had certain elements in common. They caught the global markets by surprise, lowered economic output in the affected countries without any immediate change in physical productive capacity, spread from country to country, and had triggers beyond the control of the affected countries. Such crises have occurred at regular intervals for literally hundreds of years. While this means that crises are not solely a consequence of the current push for globalization, it is likely that the increase in the speed and quantity of global financial transactions can dramatically increase their frequency, impacts, and contagiousness. Even proponents of global financial liberalization recognize that economic instability increases with financial liberalization, particularly in the emerging market economies.[17] Unfortunately, no one understands the dynamics sufficiently to predict the next occurrence.

> **THINK ABOUT IT!**
> *Speculative bubbles have been a regular part of the economic landscape in market economies since the seventeenth century. Yet many economists continue to support the Efficient Market Hypothesis, which suggests that such bubbles should not occur in perfect free markets. Their response to each crisis is therefore to increase market liberalization. In conventional science, scientists recognize that one can never prove a theory to be true but that it is possible to prove one false. Do you think the Efficient Market Hypothesis could be proven false?*

■ THE ORIGINS OF FINANCIAL CRISES

Why do such financial crises occur, and why does financial liberalization seem to increase their frequency and severity? There are a number of theories, not mutually exclusive, which we'll review briefly. Speculative bubbles play an important part in many crises. Some asset, such as stocks, real estate, a national currency, or even tulip bulbs, is increasing in value, perhaps due to increasing consumer demand. Rising prices attract speculators, and speculative demand drives the price of the asset up further,

---

[17]See G. Kaminsky and S. Schmukler, Short-Run Pain, Long-Run Gain: Financial Liberalization and Stock Market Cycles, *Review of Finance* 12:253–292 (2008); O. Obstfeld, The Global Capital Market: Benefactor or Menace? *Journal of Economic Perspectives* 12(4):9–30 (1998); F. Mishkin, Financial Policies and the Prevention of Financial Crises in Emerging Market Countries, NBER working paper no. 8087, 2001.

bringing in yet more speculators. When prices are rising faster than interest rates, speculators **leverage** their investments with borrowed money. Higher asset prices in turn provide ever more collateral for loans. Even if speculators know an asset is overvalued, they may continue to invest on the assumption that other speculators will continue to drive up the price even further. Ponzi investors jump on board. Eventually the pool of investment money, much of it borrowed, is inadequate to continue driving up demand, and prices begin to fall. We have already explained the dynamics of the resulting crash. Under the Efficient Market Hypothesis, the resulting crash is considered a "market correction." Under the Financial Instability Hypothesis, the resulting crash is considered inevitable in the absence of regulations that limit leverage, and government intervention is typically necessary to bring the economy out of the resulting recession.

Leverage, or the purchasing of investments with borrowed money, allows speculators to dramatically increase their profit margins but also increases the chance of a system-wide crash. If an investor buys a million dollars in financial assets that pay a dividend of 12% per year, borrowing $900,000 at 7% to do so, she earns $12,000 from her own $100,000, plus 5% of $900,000, or $45,000, for a total gain of $57,000 on her initial investment of $100,000, giving a rate of return of 57%. However, if the investment fails to pay at least a 7% dividend, she may be unable to pay the interest on her loan and thus default on it.

In 2004, the U.S. Securities Exchange Commission, a government agency responsible for regulating the securities industry, decided to allow investment banks to take on far more debt, which is another way of saying that they were allowed to leverage their assets more aggressively. Investment banks quickly began to borrow as much as 32 times their total assets, often using this money to purchase securitized mortgages, providing banks with yet more money to loan. When the sub-prime mortgages defaulted, many banks were unable to pay for the money they had borrowed and either defaulted or had to be bailed out by the government.

Moral hazard is another contributing factor to financial crises. Important firms such as big commercial and investment banks and major car manufacturers know that if they fail it will cause a widespread economic crisis. The government views them as Too Big to Fail, and the firms know they will be bailed out if they make risky investments that fail catastrophically. Another form of moral hazard is when managers of firms earn huge bonuses based on short-term profits, which they do not have to pay back if an investment later goes under (this form of moral hazard is referred to in the industry as I.B.G.—I'll be gone). Similarly, in the buildup to the sub-prime mortgage crash, U.S. banks "securitized" sub-prime mortgages—they bundled mortgages up into securities, which were then sold to other investors, which provided the banks with money to make more loans. As long as the securities could be sold, the banks had few incentives to ensure their quality. Under such circumstances, firms and managers win big when gambles pay off while others lose when they fail, leading to far too many highly risky investments.

Information asymmetry, which always exists between lenders and borrowers, may also have a role in some crises. Lenders price loans according to the risks involved, but the borrowers generally have a much clearer understanding of the risks than the lenders. This can lead to **adverse selection.** If lenders raise interest rates to compensate for unknown risks, then only those engaged in the riskiest activities will borrow. The less a lender knows about a borrower, the worse the information asymmetry. As small-town banks are bought up by multinational firms and as more capital crosses international boundaries, we would expect asymmetry to worsen.

National-level crises can occur when countries allow their exchange rates to become overvalued, run current account deficits that are too high, or print too much money. Speculators see these signs and bet money that corrective action will be taken. For example, Soros' Quantum fund bet that the English pound was overvalued by selling it short (see Box 20.2), which actually forced the central bank to devalue the currency in a self-fulfilling prophecy. Pro-cyclical monetary systems, described in Chapter 15, and other pro-cyclical elements of financial institutions also contribute to crises.

The recent sub-prime mortgage crisis provides a good illustration of several of these theories. Rising U.S. real estate prices in 2001 made housing a good place to invest, especially since stock market prices were falling after the collapse of a bubble in information technology stocks (the dot-com bubble) in 2001, and the Fed was keeping interest rates low to stimulate the economy. Conventional mortgages required 20% down payment plus substantial evidence of ability to meet mortgage payments. However, increasing investment drove real estate prices higher. Investment demand for housing increased, and banks increased their faith in the houses themselves as collateral on loans. New types of mortgages arose with lower down payments and lower initial payments, some requiring only interest payments. Such opportunities attracted speculative investors, driving up prices even further. Banks and mortgage companies began offering large numbers of high-interest "sub-prime" loans to increasingly risky borrowers. Some actually earned the name of NINJA loans— No Income, No Job or Assets—and some actually allowed interest to accrue as capital, attracting Ponzi investors (as well as people eager to own their own homes). To aggravate matters, financial experts figured out how to securitize these loans, as we noted above, which were sold, theoretically diffusing their risk throughout the system. With high returns and low (but obviously inaccurate) risk ratings, such securities sold rapidly. This put more money

---

BOX 20-2 SELLING SHORT

Another factor contributing to financial instability is when speculators sell the local currency short. To sell short, a speculator essentially borrows the currency from someone else and sells it at the going price, betting that the price will fall (i.e., the government will be forced to devalue) and the local currency loan can be paid back at a lower dollar cost.[a] Simplifying greatly, selling short increases the supply of domestic currency, putting downward pressure on the price. Governments are forced to sell dollars to cover these short positions. If the banks lack the resources, they cannot cover the positions and are forced to devalue. If enough people sell short, devaluation is inevitable, and the speculators profit. If someone highly respected for his financial acumen makes a big speculative investment to sell short another currency, other speculators will take notice and put their money alongside his. Currency speculators can often outspend national governments, forcing even developed countries to devalue their currencies, as happened to England in 1992. This type of herd behavior can turn selling short into a self-fulfilling prophecy.

The perversity of this type of profit from speculation is that the speculators increase their entitlements to real goods and services by controlling more financial resources, yet the production of real goods and services actually declines. When greater wealth goes to those who produce, we call it earned income. What should we call it when greater wealth goes to those who destroy productive capacity?

---

[a]For example, you borrow pesos and use them to buy dollars. You then hold the dollars, waiting for a peso devaluation. Then you use the dollars to buy pesos. You get a lot more pesos than you originally borrowed, thanks to the devaluation. You pay back your loan and have a lot of pesos left over.

---

in the hands of banks to loan to ever more housing investors, sustaining the price rise even longer. Inevitably, of course, the Ponzi scheme collapsed.

After the collapse of the financial sector, governments around the world spent hundreds of billions of dollars bailing out banks and businesses deemed too big to fail, clearly illustrating the problem of moral hazard. Ironically, the U.S. encouraged the biggest banks to use bailout money to take over smaller banks. By 2009, many of these even bigger bailed-out banks were devising new high-risk financial instruments such as securitized life insurance policies, making huge short-term profits and paying huge bonuses, serious evidence of continuing moral hazard.

■ ECOLOGICAL ECONOMIC EXPLANATIONS OF FINANCIAL CRISIS

Unfortunately, none of the theories described so far fully explains the subprime mortgage crisis, the Latin American debt crisis, the Tequila crisis, or the Asian flu. In particular, these theories fail to explain why there was so little warning of the crises and why they spread as rapidly and extensively as they did. Though decidedly less influential to date than the schools of thought previously described, ecological economic theory extends the Financial Instability Hypothesis to recognize both biophysical constraints on financial sector growth and the inherently complex and unpredictable nature of ecological economic systems.

In terms of biophysical reality, financial assets are abstractions, but they entitle their owners to a share of the real wealth that society produces. They, like IOUs, can be exchanged for real wealth, and like IOUs are a measure of debt. Let's say that, as in Chapter 15, we measure our wealth as some farmers once did, as the number of pigs we own. As abstractions, financial assets can be measured in negative pigs and hence can increase without limit in the short run—that is, they don't depend on the biophysical realities of actual pig production and are unconstrained by food supply, digestive tracts, gestation periods, places to put pigpens, and places to absorb waste. Real wealth, in contrast, is concrete, positive pigs whose rate of increase is limited by those factors. The unlimited negative pigs constitute liens on future positive pigs. There will not be enough positive pigs in the future to redeem the negative pigs. The production of real wealth is limited by biophysical realities—the availability of raw materials provided by nature that can be transformed into market products, the availability of energy to power the transformation, and the availability of ecosystem services, including waste absorption capacity, that are ultimately needed to sustain economic production. When real estate prices boom, as happened in Thailand and the U.S., the flow of services from land and existing houses does not increase, and there is no new source of biophysical value to match the increase in debt. The same is true when stock market prices soar with no underlying increase in productive capacity.

When the value of present real wealth plus biophysically constrained production capacity is no longer sufficient to serve as a lien to guarantee the exploding debt, the debt must implode. To reiterate the words of Frederick Soddy, "You cannot permanently pit an absurd human convention, such as the spontaneous increment of debt [compound interest] against the natural law of the spontaneous decrement of wealth [entropy]."[18] For a time growth in financial assets can sustain the illusion that the economy

---

[18]F. Soddy, *Wealth, Virtual Wealth and Debt*, Sydney: George Allen & Unwin, 1926, p. 20.

continues to grow even as the resources essential to economic production become scarcer, but ultimately financial assets are, for society as a whole, debts to be paid back out of future real growth, and the illusion of growth cannot persist indefinitely.

Financial crises are bound to emerge when biophysical limits become binding. It is no coincidence that when fossil fuel production basically plateaued in 2005 and grain reserves failed to keep pace with growing consumption, growth in the real economy began to fall behind growth in the financial sector. It is of course possible for economic production to keep pace with financial sector growth in the short term by depleting natural capital, but only at the cost of far more catastrophic crises in the future.

Ecological economic systems are inherently complex, exhibiting emergent properties and subject to nonlinear change, surprises, and unpredictable outcomes (see pp. 3–7 and 84–86 of the Workbook). The Efficient Market Hypothesis, in contrast, is built on fairly simple economic models that exhibit none of these characteristics. These models in fact assume the market systems quite predictably move toward equilibrium and are self-correcting. In fact, prior to the 2007 crisis, most major investment firms based their investment decisions on mathematical models that assume normal distributions of investment payoffs exhibiting random, Brownian motion in which risks of market crashes of the type we witness every few years are almost nonexistent. In contrast, agent-based models mimicking actual human behavior show that as credit (i.e., leverage) increases in a financial system, there is an increasing risk of economic collapse, and the collapse can easily spread from industry to industry and country to country. With too much credit in such models, a collapse is almost inevitable.[19] To make matters worse, financial innovators continually come up with new financial instruments that even the experts barely understand and whose impact on the economy no one understands.

In all of the crises mentioned above, the evidence points to positive feedback loops inherent to complex systems leading to self-fulfilling panics rather than smooth adjustments to underlying problems. In the panics of the 1990s, foreign investors (e.g., Americans investing in Mexico or Thailand) begin to fear that national governments and industries would be unable to service their international dollar-denominated debts because of rising interest rates (and hence higher payments in dollars), falling exchange rates (and hence higher payments in national currencies), or economic recession (and hence lower revenues for debt repayment). Holders of short-term debt became reluctant to roll it over and instead demanded repayment, which paradoxically meant that governments had fewer dol-

---

[19]M. Buchanan, Modeling Markets: This Economy Does Not Compute, *New York Times*, October 1, 2008.

lars available for repayment, increasing the risk of default. Holders of assets denominated in local currencies feared devaluation or depreciation and tried to sell their assets for dollars before it occurred.

As some investors withdrew, others became more jittery, and in these panics a chain reaction occurred. Capital fled the countries en masse after being converted from local currencies to dollars. With flexible exchange rates, this leads immediately to depreciation, increasing the quantity of debt as measured in local currencies. With fixed or managed exchange rates, capital flight forces governments to buy local currency and sell dollars. This depletes foreign reserves, depriving governments of the resources needed to pay the foreign debt and maintain exchange rates. While the initial decision to flee may have been irrational, flight quickly became a rational decision for any remaining investors. Investment in the affected countries became very risky, and national bonds were rated as "junk." Desperate to attain capital, governments were forced to offer higher interest rates to attract the dollars needed to meet short-term obligations, increasing the likelihood that they would be unable to repay this new debt. With higher interest rates and no foreign capital available, local businesses collapsed, and the domestic economies spiraled downward. Governments and firms lost the tax and sales revenue necessary to meet debt obligations. Governments were forced to turn to a floating exchange rate, which is almost always accompanied by massive devaluations. Everything the speculators feared came to pass, but largely because investors acted on their fears.

In the more recent sub-prime mortgage crisis, default on mortgages and the securities into which they had been bundled was predictable; real estate bubbles are a regularly recurring phenomenon in market countries. The rate at which the collapse spread to other securities, other countries, and the real economy was an emergent property of the complex ecological economic system.

Many economists treat self-fulfilling panics and widespread collapse of financial systems as examples of multiple equilibriums. If speculators withdraw their capital, the rational thing for others is to withdraw capital also, leading to one equilibrium. If speculators leave their capital in place or invest more, then this also becomes the rational act for others, leading to a different equilibrium.[20] This analysis has more than a grain of truth. However, there is really no such thing as equilibrium in an evolving and growing economy. Eventually, growth in real physical production confronts biophysical limits and must stop. Growth in the monetary value of

---

[20] J. A. Sachs, A. Tornell, and A. Velasco, The Mexican Peso Crisis: Sudden Death or Death Foretold? *Journal of International Economics* 41:265–283 (1996); P. Krugman, Are Currency Crises Self-Fulfilling? *NBER Macroeconomics Annual* 1996:345–378.

financial assets is still possible for a time, but when continued investment of borrowed money drives up the value of financial assets more rapidly than the increase in real goods and services, collapse is inevitable, even if we cannot predict precisely when it will occur.

■ FINANCE AND DISTRIBUTION

Ecological economists also focus on the distributional impacts of financial markets. As anyone who reads newspapers knows, salaries in the financial sector for decades have been much higher than in the rest of the economy as a whole and are often unrelated to performance. But this is barely the tip of the inequity iceberg.

Most financial assets are owned by the already wealthy and grow faster than the economy as a whole over the long run, thus leading to greater concentration of wealth. Furthermore, many financial assets and speculative trades contribute nothing to the growth of real goods and services. For example, many hedge funds use computers that detect minor discrepancies in international exchange rates. They are programmed to buy and sell huge amounts of international currency very rapidly, creating real profits while doing nothing to increase real wealth. This is nothing more than a redistribution of existing wealth.

Furthermore, the financial sector does not use a level playing field. Hedge funds are open only to very wealthy investors. The name *hedge* implies low-risk investment, which implies (as we now know, erroneously) high guaranteed returns to the already wealthy. Large investment banks invest in computers and programs that allow them to act on information 1/30th of a second faster than other investors. Essentially knowing in advance what the market will do, they can make enormous profits with little risk.[21] Average investors compete at a disadvantage, and those with too few resources to invest are not even allowed in the game.

To make matters worse, we have already seen that firms too big to fail in the financial sector can capture the gains from risky gambles while society pays the costs, time and again.

Even when financial assets do contribute to the real growth of market goods and services, for the richest nations, which host the largest financial sectors, marginal costs of economic growth often outweigh the marginal benefits. In this case, the few reap the financial benefits, while the many, even those living in those richest nations, pay the social and environmental costs.

---

[21] C. Duhigg, Stock Traders Find Speed Pays, in Milliseconds, *New York Times,* July 23, 2009.

■ WHAT SHOULD BE DONE ABOUT GLOBAL FINANCIAL CRISES?

Before we ask what should be done about global financial crises, it's worth asking what has been done. In the Latin American debt crisis, Tequila crisis, and Asian flu, the response of conventional economists such as those at the IMF was to impose tight fiscal and monetary policies on the affected countries, then to let affected banks fail or survive on their own. Economists saw this as imposing discipline on those economies. However, the effect of such policies is to plunge economies even further into recession. In response to the sub-prime mortgage crisis, which affected the wealthiest economies, economists (sometimes the very same economists!) called for exactly the opposite approach, favoring extremely loose fiscal and monetary policy, tax breaks, and massive bailouts for the financial sector.

In response to the crises of the 1990s, The IMF also pursued further deregulation of international finance. In an official communiqué issued by the IMF Interim Committee in April 1998, the IMF declared that it would amend its articles to "[make] the liberalization of capital movements one of the purposes of the Fund and [extend], as needed, the Fund's jurisdiction for this purpose."[22] This in spite of the fact that Michel Camdessus, then president of the IMF, predicted that "a number of developing countries may come under speculative attacks after opening their capital account" as a result of their unsound macroeconomic policies.[23] This approach ignores the possibility of self-fulfilling panics and contradicts the original IMF charter to protect the stability of national economies. As we have seen, advocacy of capital mobility also contradicts the IMF's advocacy of comparative advantage-based free trade, although they do not acknowledge it.

How should we respond to a current financial crisis and prevent future ones? The answers are not clear and are contingent on many factors, but we can be guided by some basic rules for adaptive change in complex systems. First, we need to begin from an appropriate paradigm, one defined by what is biophysically possible. Then, we need to pursue appropriate goals, defined by what is socially, psychologically, and ethically desirable. Finally, we need to develop policies that severely dampen positive feedback loops, strengthen negative ones, and increase information flows. [24]

---

[22]International Monetary Fund, Communiqué of the Interim Committee of the Board of Governors of the International Monetary Fund, press release no. 98/14, Washington, D.C., April 16, 1998. Online: http://www.imf.org/external/np/sec/pr/1997/PR9744.HTM.

[23]Communiqué of the IMF Interim Committee, Hong Kong, September 21, 1997. Cited in M. Chossudovsky, Financial Warfare. Online: http://www.corpwatch.org/trac/globalization/financial/warfare.html.

Currently, an economic crisis is defined as any threat to economic growth; a recession is explicitly defined as two consecutive quarters without growth in GNP. Implicit in this definition is the paradigm of the economic system as the whole and the ecosystem as the part, and the goal of endless growth. If we accept the ecological economic paradigm, in which the economy is sustained and contained by the global ecosystem, then continuous growth is of course impossible. An appropriate goal is to enhance quality of life. In the ecological economic paradigm, we therefore redefine a crisis as economic conditions that generate (or will inevitably generate) unemployment, poverty, misery, or instability. In other words, we define crisis as an economic threat to our quality of life.

How we define a crisis determines what we should do to address it. It follows from this new definition that we should not spend trillions bailing out the financial sector but rather spend to create jobs and end poverty and misery for the current generation, while addressing the critical problems such as climate change that threaten the flow of ecosystem services necessary to the well-being of future generations. Among the policies implicit in these goals are public investments in education, the development and deployment of green energy technologies, maintenance of critical public infrastructure, and restoration of depleted natural capital.

Recent policies to forestall crisis, developed within the conventional paradigm, have been to cut taxes and run massive deficits to stimulate growth, which impose the burden of repayment (financial and ecological) on future generations. The ecological economics framework suggests imposing higher but much more progressive taxes. A tax on financial transactions would reduce short term speculative investors and the risks they generate. Higher taxes would unquestionably deter resumed growth in the economy but, when used to fund the policies we suggest above, would avoid collapse. The difference between economic collapse caused by financial crisis and a carefully planned reduction in growth is analogous to the difference between a plane crash and a hovering helicopter, both of which are stationary.

If new taxes prove inadequate to fund government expenditures needed to achieve the goals suggested above, the government can recapture the right to seigniorage and simply spend money into existence. In the midst of financial crises, banks act in a pro-cyclical manner, refusing to lend new money. The ideal time to raise reserve requirements is when banks are voluntarily keeping larger reserves than required. In response to the sub-prime mortgage crisis, national governments are printing up and selling treasury

---

[24]D. Meadows, *Leverage Points: Places to Intervene in a System*, Hartland, VT: The Sustainability Institute, 1999.

bonds, which must be repaid with interest. It makes more sense to simply print up and spend federal reserve notes, which carry no interest.

Policies aimed at developing green technologies and restoring natural capital would help prevent future crises. They would help maintain the productive capacity of our global ecosystems, upon which all economic activity ultimately depends.

In addition to these polices, much tighter regulation of the financial sector, domestic and international, is needed. Financial institutions (and indeed any other private sector firms) that are too big to fail are too big to exist. Not only is moral hazard less of a concern with smaller banks, but loan officers in local institutions are more likely to know their clients, reducing the problem of adverse selection as well. Leverage (the percentage of borrowed money used) in purchasing stocks must be severely restricted. Higher reserve requirements will make banks much less susceptible to overextension and collapse. Types of financial instruments should also be tightly regulated. Before the financial industry is allowed to introduce any new instrument, such as securitized mortgages or securitized life insurance policies, it should be forced to explain how it works, explain how it creates real value for society, and show that it will not increase the risk of crisis.

Why do conventional economists stubbornly pursue such different policies, in the face of so much evidence that they are wrong? Neoclasscial economists are trained in the Efficient Market Hypothesis. Well-informed people make rational choices based on rational expectations. Market prices are rational and lead markets to general equilibrium. Where effective markets don't exist, they must be created in order to achieve these harmonious results. Economic experts routinely bet billions of other people's dollars that this is the case. In contrast, George Soros, a renowned international financier, routinely bet billions of his own dollars that this is not the case. In his own words, he believes that "people act on the basis of imperfect understanding and equilibrium is beyond reach," and as a result, "market prices are always wrong, in the sense that they present a biased view of the future."[25] Soros frequently wins his bets, while the record of conventional economists currently speaks for itself. Finally, Soros further claims that "extending the market mechanism to all domains has the potential of destroying society,"[26] a belief with which we strongly concur.

---

[25]Quoted in W. Greider, *One World, Ready or Not: The Manic Logic of Global Capitalism*, New York: Simon & Schuster, 1997, p. 242.

[26]See Society Under Threat: Soros, *The Guardian*, October 31, 1997. Cited in Chossudovsky, op. cit.

BIG IDEAS to remember

- Balance of payments (BOP)
- Current account
- Capital account
- Fixed vs. flexible exchange rates
- The Efficient Market Hypothesis
- The Financial Instability Hypothesis
- Hedge, speculative, and Ponzi Investors
- Liberalization
- Speculation
- Leverage
- Moral hazard
- Adverse selection

Conclusions to Part V

Globalization has been proposed as a panacea for economic problems, because it is expected to bring economic growth and wealth to all. This conclusion is based on flawed assumptions. With free international capital flows, globalization favors absolute advantage and not comparative advantage. While globalization may bring greater economic growth, the growth will not benefit all countries, and the costs of growth could outweigh the benefits at the margin. Empirical evidence shows that the benefits of growth go to the already wealthy, while the costs, in terms of lost ecosystem services, are borne by all—and perhaps by future generations most of all. Globalization also weakens the policy levers governments have to address issues of scale and distribution on their own. Particularly problematic are unregulated flows of financial capital, often speculative in nature. Such flows allow the financial sector to capture a greater share of global wealth. Not only do speculative flows fail to create real physical wealth, but they also play an important role in generating financial crises that actually reduce its production.

While international trade and finance can be desirable, they are more likely to be so if they take place between independent nations with the ability to make their own policies, and these nations recognize that economic growth brings costs as well as benefits. Nations must also recognize that markets are complex systems subject to biophysical constraints and destabilizing positive feedback loops which lead to crisis and are exacerbated by globalization.

Realizing the potential for widespread gains from internationalization, reducing the likelihood of future crises, and addressing those crises when they do occur demands that we redefine our paradigm of what is biophysically possible and our goals concerning what is desirable. Endless physical growth of the economy is an impossible goal. We must redefine our goal as improving the quality of life for this and future generations. This goal requires ecological sustainability and social justice. We must redefine economic crisis as the presence or imminent threat of poverty, misery, and unemployment. We now turn our attention to some of the policy tools independent nations could utilize to attain the goals of sustainable scale, just distribution, and efficient allocation on a finite planet.

PART VI

Policy

General Policy Design Principles

Listening to many economists and even policy makers, one might often get the impression that the only role of government is to create conditions that allow the market to function. The argument is that not only are markets the best institution available for allocating scarce resources to attain our desired goals, but they are also the best way to determine what our goals are. After all, if the mere act of buying and selling reveals people's preferences, the job of both economist and policy maker is simply to allow the market to satisfy those preferences. But we know that this is not the case. Markets by definition can only reveal preferences for market goods, yet many of the goods and services that enhance human welfare are nonmarket goods. Thus, not only do markets fail to reveal preferences for these resources, they also fail to allocate them effectively.[1] Markets also fail to address the issues of scale and distribution. Therefore, the first point we must make in this chapter is that the market cannot tell us how much clean air, clean water, healthy wetland, or healthy forest we should have or what level of risk is acceptable when the welfare of future generations is at stake. Nor can it tell us what is a desirable initial distribution of resource ownership.

---

[1]Also, it is not strictly true that markets reveal preferences even for market goods; they reveal choices, which are, to be sure, an expression of preferences, but a very conditioned expression under the constraint of existing prices and incomes.

■ THE SIX DESIGN PRINCIPLES

We have already seen the even more fundamental point in Chapter 3 that, at a minimum, policy requires two philosophical presuppositions: first, that there are real alternatives (nondeterminism); second, that some states of the world really are better than others (nonnihilism). We now examine six general design principles for policies, followed by a consideration of their proper sequencing and of how and where policy interventions should first impinge on the market (that is, on quantity or price): at the input or output end of the throughput? We end the chapter with further reflections on property rights. Then, in Chapters 22–24, within the guidelines of the design principles, we will offer some particular policies for promoting a steady-state economy—one that is sustainable, just, and efficient.

1. Economic policy always has more than one goal, and each independent policy goal requires an independent policy instrument.

If there is only one goal, the problem is technical, not economic. For example, building the most powerful engine possible is purely a technical problem. But building the most powerful engine that is not too heavy to power an airplane involves two objectives, power and lightness, that have to be optimized in terms of a higher goal: making an airplane fly. This is already an economic problem of optimizing the combination of conflicting goals, even though we encounter it at an engineering level.

In ecological economic policy, we have three basic goals: sustainable scale, just distribution, and efficient allocation. Nobel laureate Dutch economist Jan Tinbergen set forth the principle that *for every independent policy goal we must have an independent policy instrument.*[2] You have to be very lucky to hit two birds with one stone—it nearly always takes two stones to hit two birds flying independently.[3] For example, should we tax energy and raise its price for the sake of inducing more efficient use, or should we subsidize energy and lower its price to help the poor? This question is endlessly and uselessly debated. One instrument (price of energy) cannot serve two independent goals (increase efficiency, reduce poverty). We need a second instrument, say an income policy. Then we

---

[2]J. Tinbergen, *The Theory of Economic Policy*, Amsterdam: North Holland Publishing, 1952.

[3]This does not mean that a single policy cannot help achieve more than one goal. For example, a tax on land values can reduce land speculation, generate revenue for government coffers, promote more efficient land use, and, by promoting urban in-fill, reduce urban sprawl. However, the optimal tax will change depending on the policy goal; we cannot use a single policy to optimize for different policy goals simultaneously.

can tax energy for the sake of efficiency and distribute income (perhaps from the tax proceeds) to the poor for the sake of alleviating poverty. With two policy instruments, we can serve both efficiency and equity. With only one instrument, we are forced to choose either efficiency or equity.

Since ecological economics insists on three basic goals, it is already clear that we will need three basic policy instruments. The three goals are independent in the sense that attaining one will not bring about attainment of the others. Of course the goals are not isolated in the sense of having nothing to do with each other—they are, after all, parts of a single economic system. Now that we know how many instruments, what other principles will help answer the question: What kind of instruments? The remaining principles attempt to give guidance in this regard.

2. Policies should strive to attain the necessary degree of macro-control with the minimum sacrifice of micro-level freedom and variability.

Consider this example. If what is limited is the capacity of the atmosphere to absorb CO_2, it is important to limit the total CO_2 emissions. Average per-capita emissions times population will have to equal the limited total. But it is not necessary that each and every person emit exactly the per-capita average. There is room for micro-variation around the average in light of particular conditions, as long as the total is fixed.

Let's take another example. Population stability requires the average of 2.1 children per couple. But it is not necessary (or even possible in this case) for each family to have the required average number of children corresponding to generational replacement. Macro-control is compatible with varying degrees of micro-variability around the average. In general we should opt for the least micro-restrictive way of attaining the macro-goal. Markets are useful in providing micro-variability, but by themselves they do not provide macro-control.

3. Policies should leave a margin of error when dealing with the biophysical environment.

Since we are dealing often with staying within biophysical limits, and since those limits are subject to much uncertainty and at times irreversibility, we should leave a considerable safety margin, or slack, between our demands on the system and our best estimate of its capacity. If we go right up to capacity, we cannot afford mistakes because they are too costly. The inability to tolerate mistakes, or sabotage, exacts a large price in reduced individual freedom and civil liberties.

Security issues surrounding the nuclear fuel cycle and the safeguarding of plutonium have already given us a foretaste of the problems involved in living too close to the edge. Our historical experience with small life-support systems operating close to capacity—namely, spaceships, or even ordinary ships or submarines—has thus far not permitted democracy. Military levels of order and discipline are needed on fragile vessels operating near carrying capacity. Only our large spaceship Earth, with lots of slack, can be sufficiently forgiving of error to tolerate democracy. There are political as well as economic costs to excessive scale, a fact that has received too little attention.

4. Policies must recognize that we always start from historically given initial conditions.

Even though our goal may be far from the present state of the world, the latter remains our starting point. We never start from a blank slate. Reshaping and transforming existing institutions is often more effective than abolishing them. This imposes a certain gradualism. Even though gradualism is often a euphemism for doing nothing, it is nevertheless a principle that must be respected.

What are our present institutions? Basically, the market system and private property, but also public property and government regulation. The World Bank and the IMF may be with us for a while, even though they are not nearly as basic an institution as private property or the market. We have neither the wisdom nor the time to start over again without our most fundamental institutions, even if we could imagine alternatives. The considerable stretching and bending of these institutions that we will recommend will be thought radical by some, so it is important to emphasize the conservative principle of starting from where we are, even if the basic idea is not to remain there.

5. Policies must be able to adapt to changed conditions.

Change is an ever-present reality. Human impacts on the ecosystem are enormous and are likely to cause new problems over time. Ecosystems themselves naturally show considerable variation over time—where time can be measured in seasons, years, or eons. Human knowledge is increasing, leading to a new awareness of previously unrecognized problems, as well as new solutions to old ones. The economic system is also continually evolving, and policies that work well now may not work as the system changes.

In addition, we may find that some policies that seem ideal in theory may not be ideal when implemented, and they may even have seriously negative unforeseen side effects. As we apply policies, we will learn how they work in the real world and thus learn how to improve them. The process of developing and implementing policy solutions must respond to this feedback, and real-life outcomes must carry far more weight than stylized theories. **Adaptive management**—changing our policies as conditions change and as we learn more—must be a guiding principle. Indeed, we believe that ecological economics itself is an example of adaptive management to the problems arising from the transition from an empty to a full planet.

6. The domain of the policy-making unit must be congruent with the domain of the causes and effects of the problem with which the policy deals.

This is often called the **principle of subsidiarity**. The idea is to deal with problems at the smallest domain in which they can be solved; problems should be addressed by institutions on the same scale as the problem. Don't seek global solutions for local problems, and don't try to solve global problems with purely local measures.

Consider the example of garbage collection. Garbage collection is largely a municipal problem. Aggregating all municipal garbage collections into a "global garbage problem" is not helpful. Deal with it at the municipal level, at least in the first instance. If local garbage has to be disposed of farther and farther away, or if it contaminates air or water and is thus transported far away, then it becomes a correspondingly larger problem—county, state, region, and so on. By contrast, global warming is fundamentally a global problem, because emissions anywhere affect the climate everywhere. Here we really do need global policy.

■ WHICH POLICY COMES FIRST?

In ecological economics we have three basic goals, so we need three basic policy instruments. The goal of efficient allocation requires the instrument of the market, at least for goods that are private (excludable and rival). For public goods the market will not work. The goal of sustainable scale requires a social or collective limit on aggregate throughput to keep it within the absorptive and regenerative capacities of the ecosystem. The goal of distributive fairness requires some socially limited range of inequality. As we have seen, the market cannot achieve distributive equity or sustainable scale. Furthermore, the market cannot even attain allocative efficiency unless the distribution and scale questions have already been answered. So now we know that in the sequencing of policy instruments the market comes third, after its preconditions have been established.

But what about the sequencing of scale and distribution? Here it is reasonable to put scale first, because limiting scale usually means that previously free natural resources and services have to be declared scarce economic goods.

Once they are scarce, they become valuable assets, and the question of who owns them must be answered. That is an issue of distribution. The logic is that of the cap-and-trade policies. For example, the total scale of SO_2 emitted into an airshed is capped at a scale deemed sustainable. The right to emit SO_2 is no longer a free good. Who owns that right? Previous users? All citizens equally? The state collectively? Some answer to the distributive question must be given before trading in a market can solve the allocative problem. People cannot trade what is not theirs. Some goods that were outside the market can be made marketable goods (i.e., excludable). By limiting the scale of resource use and distributing the ownership of the resource, we can convert nonmarket into marketable goods. But as we saw earlier, not all goods can be converted into market goods. Many good things are inherently nonrival or nonexcludable. We will return to this issue.

The set of prices that corresponds to a Pareto optimal allocation will be different if we set the cap differently or if we distribute ownership differently. This means that we cannot set the cap or distribution according to computations of their social costs and benefits based on existing prices. To do so would be to engage in circular reasoning because the prices depend on the scale or distribution. The ideal scale or distribution, calculated on the basis of existing prices, would, if attained, result in a different set of prices that would invalidate the original calculation. Thus, we can neither set the scale nor determine distribution according to the criterion of efficient allocation.

What, then, is the criterion for scale? Sustainability is the criterion for scale.

And what is the criterion for distribution? Justice is the criterion for distribution.

These, obviously, are not matters of market economics; rather, they are biophysical and cultural. They must be socially and politically determined, and thus, as a matter of policy, these decisions can be made more or less simultaneously. In strict logic, scale comes before distribution, because if there were no cap on total use the resource would be a free good, and the distribution of ownership of a free good would make no sense. Given these prior social decisions on scale and distribution, the market will determine allocatively efficient prices. Indirectly these prices will reflect the scale and distributive limits and therefore may be thought of as internalizing the values of sustainability and justice that have been previously decided politically, independently of prices.

Economists sometimes argue that scale is not an independent consideration—that if we had perfect information and could internalize all external costs and benefits into prices, then the market would automatically stop growth at the optimal scale. In other words, scale would have been subsumed under allocation. This has a certain plausibility if we accept the assumption of "perfect" information. However, if in the name of perfect internalization we insist that prices should incorporate the costs and benefits of different scales, we would also have to insist that prices reflect the costs and benefits of different distributions. But if we tried to use prices based on a given distribution as the means of measuring the costs and benefits of a change in distribution, we are again being circular.

Economics has clearly recognized the circularity and insisted that just distribution is one thing, efficient allocation another. Economists would not, for example, appeal to perfect information and advocate raising the price of things poor people sell or lowering the price of things poor people buy in order to internalize the external cost of poverty into prices. Instead they might advise us to redistribute income directly to attain a more just distribution and let prices adjust. This also makes sense for questions of scale.

The way to get prices to reflect the values of just distribution and sustainable scale is to impose quantitative restrictions on the market that limit the degree of inequality in distribution of income and wealth to a just range and that limit the scale of physical throughput from and back to nature to a sustainable volume. These imposed macro-level distribution and scale limits reflect the social values of justice and sustainability, which are not personal tastes and cannot be reflected in the market by individualistic actions. The market then recalculates allocative prices that are consistent with the *imposed* scale and distribution constraints, thereby in a sense internalizing these social values into prices.

Since it is circular to use prices to calculate optimal scale and optimal distribution, we need some metric of benefit and cost other than price (exchange value). As already suggested, this metric is the value of justice in the case of distribution; it is ecological sustainability, including intergenerational justice, in the case of scale. These are collective values, not individual marginal utilities per dollar equated between different goods in order to maximize satisfaction of individual tastes. If we reduce all dimensions of value to the level of subjective personal taste, then we cannot capture or bring to bear on the market the real weight of objective social values, such as distributive justice and ecological sustainability.

■ CONTROLLING THROUGHPUT

If we are going to impose macro-level constraints on the market to control scale, we must ask: At which end of the throughput flow should we impose these constraints? We could impose restrictions at the output end (pollution), as in the SO_2 example. Or we could limit the input flow from nature (depletion). Since there are fewer mines and wells than there are tailpipes, smokestacks, garbage dumps, and outflow pipes into rivers, lakes, and oceans, it would be easier to put the control on depletion than on pollution. By the law of conservation of matter-energy, if we limit the inflow we will also automatically limit the outflow. Even if it is the outflow that is causing the immediate problem, even if sinks are more limiting than sources, the outflow may be more easily controlled by limiting the flow at its narrowest point, the inflow.

As a general rule it makes sense to control depletion directly, thereby indirectly controlling pollution. As with all general rules, there are exceptions. Although depletion limits provide a quantitative limit on pollution in a gross sense, many qualities of pollution could result from the same quantity of depletion. Depending on how resources are used, the same inputs could be converted into very toxic or very benign pollutants. Therefore, we cannot entirely concentrate on inflows and expect the outflows to take care of themselves. But inputs (depletion) should be our first control point.

■ PRICE VS. QUANTITY AS THE POLICY VARIABLE

In Chapter 22, we will look in detail at two basic approaches to limiting throughput: raising prices through taxation to reduce demand and limiting quantity directly through quotas and letting prices adjust. Before we examine the specifics of each approach in detail, we investigate the effectiveness of the two approaches.

Given that we should intervene mainly on the input side, how should we do it? Should we try to control quantity and let the market determine the resulting price or try to control price and let the market determine the resulting quantity?

THINK ABOUT IT!
Trying to control both price and quantity would be a bad idea. Can you explain why? Think of a demand curve.

If we can, by taxes, set the price where we want it, that will, via the demand curve, determine a corresponding quantity. Alternatively, if by quotas we set the quantity where we want it, that will, via the demand curve, determine a corresponding price. Theoretically, given a demand curve, we

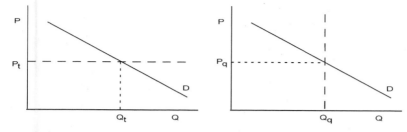

Figure 21.1 • Setting quantity with taxes and quotas. If taxes are set so that the price is equal to P_t, consumers will demand quantity Q_t. Alternatively, quotas could limit output to quantity Q_q, and for this quantity consumers are willing to pay P_q. It is possible, of course, to set Q_q so that $P_t = P_q$ or set P_t so that $Q_t = Q_q$.

could get the same result either by fixing price or by fixing quantity (see Figure 21.1). How do we choose between them?

One point to bear in mind is that demand curves are uncertain and shift. So if we set the price, errors and omissions will result in quantity changes. If we set the quantity, errors and omissions work themselves out in terms of price fluctuations. Therefore, an important criterion by which to choose is: Where is it least painful to experience errors, as price changes or as quantity changes?

Because the ecosystem cares about quantities extracted and absorbed and not the prices people pay, and because prices can adjust more rapidly than ecosystems, ecological economists have a preference for fixing quantity and bearing the adjustment cost of errors in terms of price fluctuations. That is ecologically safer and more in accord with our design principle of leaving a big safety margin. It also follows more strictly the sequencing logic just discussed of setting the scale first outside the market and then allowing prices to be determined by the market.

The superiority of quotas for achieving desirable scale is even more evident when both global population and per-capita levels of resource use continue to grow. Such growth implies increasing demand for both sources and sinks. In the presence of quotas, the increased demand is reflected entirely by increased prices, creating ever-greater incentives to use resources more efficiently. If resource consumption or waste emissions are limited by use of taxes, greater demand will lead to both higher prices and greater consumption, unless taxes are continually raised. In a world of ecological thresholds and irreversible outcomes, taxes lead to a greater risk that we will eventually use up our slack.

We must also recognize that markets do not work as perfectly as we economists might like. In Chapter 20, we discussed financial panics and the tendency for people to follow herd behavior. In Indonesia, the Asian flu crisis led to an 85% depreciation of the currency over the course of weeks. Indonesia has a significant percentage of the planet's rainforests, which supply an abundance of important, nonmarket ecosystem services. Imagine Indonesia sought to internalize the value of these ecosystem services by taxing deforestation at the rate of 10 rupiah per board foot of timber extracted. Most of Indonesia's timber is sold on the global market, so the 85% depreciation in its currency would have implied an 85% decrease in the tax in relation to the trade currency. Even if the tax had been in U.S. dollars, the currency of international trade, most other costs of harvest are in rupiah. The 85% drop in all other costs would still reduce the price and increase demand, leading to greater deforestation. As we know from Chapter 6, greater deforestation can reduce the sustainable yield of the forest and possibly even drive forest stocks below the point of critical depensation. A quota would not be subject to such irrational fluctuations in economic variables.

■ SOURCE VS. SINK

As we have seen, before the market can operate, ownership of the newly scarce asset must be distributed. This raises many difficult issues, and it may cause a retreat from some of our earlier design principles in order to respect the fourth one—namely, that we start from historically given initial conditions. At the input end, most resources are already owned. At the output end, the atmosphere is not privately owned. A **source** is the part of the environment that supplies usable raw materials that constitute the throughput by which the economy produces and that ultimately returns as waste to environmental sinks. A **sink** is the part of the environment that receives the waste flow of the throughput and may, if not overwhelmed, be able to regenerate the waste through biogeochemical cycles back to usable sources. Sources are generally owned, and sinks are generally not owned. Directly controlling sources (depletion) involves more interference with existing property rights than controlling sink access. To socialize all resource ownership after resources are privately owned is revolutionary. To socialize the unowned sink, the atmosphere, and then charge a dumping fee seems less threatening to private property than a direct control over the amount extracted. But to control emissions is to dam the river at its widest point, contrary to the principle that it is easier to dam it at its narrowest point. Should we advocate revolution? Most of us would not, but let's try to stick to our principle for a while at least.

What might be done to reconcile the principle of intervening at the depletion and with the difficulty that sources are private property? We might recognize that property is a "bundle of rights." The resource owner has to give up one stick out of that bundle—namely, the right to decide independently the rate of extraction of his resource. He still owns the resource and receives payment for whatever amount he extracts. But the scale of extraction is no longer a free good. It is socially limited to a national quota, and resource owners must bid at auction for the right to extract a share of the limited total extraction permitted. The scale limit may ultimately be set according to sink limits if they are more binding but still enforced at the source end.

Alternatively, we could have a market in sink permits, capped at an aggregate scale. Suppose, in the case of fossil fuels, all users had to purchase emission permits to burn whatever fossil fuel they purchased. This would indirectly limit demand at the source, and source owners would feel the pinch of scarce sink capacity—the scarcity of one complementary factor reduces the value of the other. This may appear less an infringement on the property rights of the source owners, since they have no claim to the sink, and it is the sink that is being directly limited. But the sink limit on the throughput is surely translated back to the input end. And since that sink limit will be experienced by source owners indirectly, even if we put the limit directly on the sink, why not go ahead and put the limit on the source in the first place? That would be the more efficient place to put it, even if it is the sink that is most scarce.

The other possibility noted is to fix prices through taxes and allow the market to set the corresponding quantity. Once again, this is more efficiently done at the depletion end rather than the pollution end, but both are possible. The tax may be levied at the input end even though its basic motive is to limit the output. The advantage of taxes is administrative simplicity—we already have a tax system, and altering it is less disruptive than setting up a quota system with auctions. This is a significant advantage. On the other hand, taxes really do not limit quantities very strictly, and they maintain the false perception that there are no quantitative limits as long as one pays the price. As long as we pay the price plus a corrective tax, the message conveyed is that we can get as much as we want, individually and collectively. The quota, by contrast, makes it clear that the total quantity will not increase and that all the price is doing is to ration the fixed quantity among competing users. The latter seems a more honest and truthful perception, since we are dealing with a scale-limited physical throughput, not income or welfare.

THINK ABOUT IT!

Does the current U.S. policy on carbon emissions follow the guidelines proposed in this chapter? Should it? Does the policy follow the sequence of scale, distribution, then allocation? To whom does it distribute emission rights?

■ POLICY AND PROPERTY RIGHTS

Before we begin to examine specific policies for achieving a more sustainable, just, and efficient world, we must discuss one of the core features of any policy: property rights. Concern over scale is concern over sustainability, and what is sustainability but the right to resources for future generations? If we believe there is a need for improved distribution, we are basically questioning the existing endowments of property rights. Finally, markets cannot efficiently allocate nonexcludable resources, and excludability is nothing more than a property right. Policy is concerned largely with institutions and laws that create, redefine, and redistribute property rights.

Property rights and excludability are not inherent properties of goods or services. No good is excludable, and no one has property rights unless a social institution exists that makes it excludable and assigns property rights (though we also know that it is not possible to make all goods excludable). A property right for one individual simultaneously imposes a duty or obligation on other individuals to respect those rights. For example, if person A has the right to breathe clean air, then person B has the corresponding duty not to pollute that air. The state ensures that person B will fulfill her duty. Property rights are therefore a three-way relationship between one individual, other individuals, and the state.[4]

In the absence of property rights we have privilege, or presumptive rights. If one person has privilege, he is entitled to behave as he pleases, and others have no rights. If a factory owner has privilege with respect to the atmosphere, he can pollute the air as much as he pleases. If others suffer from this pollution, then they must seek to change the prevailing lack of property rights.

When human populations and impacts were small relative to the sustaining ecosystem, the use of natural capital was appropriately characterized by privilege. Why not allow industries or individuals to pollute if few people lived nearby to be affected by that pollution? Why not allow industries or individuals to harvest fish or harvest trees if they existed in abundance? Why not give away rights to minerals to those who discovered them, as long as seemingly limitless virgin lands remain for future exploration? It makes little sense to establish property rights to super-abundant resources.

However, as we know, the world is no longer so empty. The privilege to extract and pollute now imposes costs on others. This creates pressure to develop environmental policies that assign or modify property rights. Those who have privilege to extract or pollute are likely to defend the sta-

---

[4]D. Bromley, *Environment and Economy: Property Rights and Public Policy*, Oxford, England: Blackwell, 1991.

tus quo, claiming that privilege as a right when in reality it is an absence of defined rights. As we pointed out in Chapter 10, many economists have argued that it does not matter to whom rights are assigned; as long as rights are assigned, the market can efficiently allocate resources. We maintain, in contrast, that while in theory the distribution of rights may not matter in terms of Pareto efficiency (i.e., Pareto efficient outcomes are possible for any distribution of property rights, though it will be a different outcome for different distributions), it matters profoundly for equity. We take the position that property rights belong to the people, as represented by the state, until otherwise assigned, and their distribution should be decided by a democratic process that respects future generations.

There are three important types of property rights, or entitlement rules, and rights to a specific piece of property may be affected by any combination of these.

1. An entitlement known as a **property rule** holds if one person is free to interfere with another, or free to prevent interference. For example, an individual may own a piece of land. If he has the right to build a landfill that destroys the neighbor's view or to prevent the neighbor from walking across the land, he is free to "interfere" with the neighbor. Nor can the neighbor interfere with the landowner's landfill operation. If the neighbor wants to walk across the property or prevent the landfill from being built, the landowner's consent is required.

2. An entitlement known as a **liability rule** holds when one person is free to interfere with another or prevent interference but must pay compensation. For example, the landowner might be free to build the landfill, but by law he is then forced to compensate his neighbor for the smell, loss of view, and other disamenities. At the same time, the state could call on its right of eminent domain to take away the land from the landowner to build a highway and pay compensation at fair market price.

3. An entitlement known as an **inalienability rule** holds if a person is entitled to either the presence or absence of something, and no one is allowed to take away that right for any reason. There may be certain types of chemicals or products that are absolutely not permitted in the landfill, regardless of compensation. The negative impacts of these products are so severe that present and future generations have an inalienable right not to be exposed to them. Dioxins and radioactive waste would fall into this category.

Finally, we must remember that property rights need not be private property rights. Property rights can belong to individuals, communities, the state, the global community, or no one. While many conventional economists favor private property rights, we have already learned that private property rights are not possible in all circumstances (e.g., the ozone layer). In addition, many cultures have successfully managed common property resources for millennia. Almost all nations have certain resources owned by the state; recently, international agreements, such as the Montreal and Kyoto protocols, have recognized the need for ownership and management of some resources by the global community. The search for suitable policies neither can nor should be limited only to those that require private property rights.

Now that we have discussed basic principles of policy, appropriate policy sequence, high-leverage points of intervention, and the relationship between property rights and policies, we turn our attention in the next three chapters to some specific policies. We will generally follow the policy sequence suggested above: scale, distribution, then allocation.

However, while in most cases we cannot hit two birds with one stone, some of the policies we look at are really a bundle of policies affecting all three goals. Other policies may hit one goal squarely while having an impact on another policy goal as well. Our division of the discussion is therefore not exclusive: All three goals will be discussed in each chapter, and policies are grouped only according to their dominant impact. Our three independent goals require three independent policy instruments in the same sense that solving three simultaneous equations for three different variables requires three independent equations; that is, one equation must not be derivable from the other two, and one variable cannot be the same as another, just expressed differently. Three simultaneous equations in three unknowns form a system, so clearly all three variables are interrelated—they are not independent in the sense that a change in one has no effect on the others (i.e., isolated) because they are clearly parts of an interdependent system. But they are independent in the sense in which each independent variable in a set of simultaneous equations requires an independent equation if the system is to be solvable.

BIG IDEAS to remember

- Six policy design principles
- Proper sequence of policies
- Source vs. sink as throughput control point
- Price vs. quantity as control instrument
- Circularity of internalizing scale or distribution in prices
- Property rights

Sustainable Scale

Environmental policies are inherently related to scale. In an empty world, environmental goods and services are not scarce resources, and hence not the focus of policies. The issue is whether environmental policies address scale directly or only tangentially. Just as importantly, to be effective, the policies should square with the six design principles outlined in Chapter 21. We will discuss four different types of policy that affect scale: direct regulation, Pigouvian taxes, Pigouvian subsidies, and tradeable permits. We examine how each is applied in the real world.

■ DIRECT REGULATION

The dominant form of environmental policy affecting scale in most of the world is the regulatory instrument, which can take a variety of forms. Sometimes an activity or substance is considered to have unacceptable costs and is simply banned. For example, many countries no longer allow lead additives to gasoline or the production of DDT, and global bans exist on certain ozone-depleting compounds and persistent organic pollutants. When a substance is sufficiently dangerous, such bans are appropriate.

In other instances, regulation will limit the quantity of a pollutant that can be produced and set emissions levels for the firms or individuals responsible for producing it. For example, individual paper factories may have legal limits to the amount of waste they can discharge into a river, and in many countries, vehicles have to pass emission tests. In yet other instances, regulations will force all firms or individuals to use the best available control technology (BACT) to limit pollution. BACTs may be imposed on all firms or individuals, or only on new entrants to an industry. BACT regulations play an important role in the U.S. clean air laws.

For fisheries, a common regulation has been to limit the fishing season or regulate the type of equipment that may be used in order to reduce the annual catch.[1] Failure to comply with regulations generally involves fines or other penalties. These policies are therefore known as **command-and-control regulations**.

What are the advantages and disadvantages of such policies? Most of them limit the amount of pollution or resource harvest to an acceptable level, thus contributing to the goal of desirable scale. With renewable resources, regulations may be the best way to address biological requirements. Examples are banning harvests during mating seasons, imposing minimum mesh size on fishing nets, forbidding harvest of gravid females, leaving the best and largest of a species as seed stock, or banning certain harvest methods that are particularly destructive of habitat. Regulations can be applied to everyone equally or tailored to meet alternative distributional goals. Finally, policy makers are generally familiar with this approach. It is reasonably easy to understand and can be fairly cheap to monitor and enforce—for example, it is very easy to check whether a given firm is using a mandated technology.

The disadvantage is that in general, regulations fail to meet the criteria for allocative efficiency and thus are often not the most cost-effective way to reach a desired goal. Moreover, they fail to provide incentives for surpassing a goal, such as bringing pollution below the regulated level. These points deserve elaboration.

As shown in Chapter 10, the basic requirement for economic efficiency is that marginal costs equal marginal benefits, at both the individual level and the social level. Ideally, environmental policies should achieve this goal. In practice, however, for our pollution example this would require that we know the marginal social costs of pollution, the marginal net benefits of activities that pollute, and the marginal abatement costs of pollution. Of course, there are really no benefits to pollution per se, but all production causes pollution, and we could not exist with no production at all. In reality, it is virtually impossible to know all the marginal costs of pollution and very difficult for policy makers to know marginal abatement costs. Perfect allocative efficiency therefore is something of a pipe dream.

While we cannot hope for a perfectly efficient solution, we can hope for a cost-effective one. A cost-effective solution achieves a given goal at the lowest cost, even if marginal costs do not exactly equal marginal benefits. It is therefore a very desirable goal but one that is unlikely to be attained by simple regulations. The reason is that command-and-control regulations ignore the second general design principle described in Chap-

---

[1]Regulating inputs to reduce catch is an entirely different issue than regulating inputs to reduce negative externalities, such as excessive bycatch or habitat destruction.

ter 21, that policies should sacrifice the minimum of micro-freedom to attain macro-control.

This point is perhaps most readily illustrated with a specific example. Imagine there are three firms polluting a waterway upstream of the drinking water intake valve for a city. A regulatory agency determines that for health reasons, pollution loads should be cut by 40% and demands that each firm cut its emissions correspondingly. The problem is that different firms may have different marginal abatement costs (MACs) or different operating costs due to a variety of factors, such as manufacturing process or age of manufacturing equipment. It may be very expensive for one firm to cut its emissions by 40% and very cheap for another firm to do so.

Unfortunately, the regulators do not really know the firms' abatement costs. While each firm presumably knows its own abatement costs, gathering this information would be costly, and if the goal were to force firms with lower marginal abatement costs to reduce pollution by a greater extent, each firm would have an incentive to misinform the regulator.[2] Also, it hardly seems fair to make some firms reduce pollution far more than others.

Another problem that arises with regulation is that once regulatory goals have been achieved, there is no incentive to reduce pollution any further and few incentives for new pollution reduction technologies. Similarly, if regulations apply to specific areas (such as U.S. clean air laws), there is no incentive not to increase pollution in areas below the maximum allowed level. Yet we have already seen that pollution has marginal external costs even at very low levels (see Figure 12.7).

What we seek, then, are policies that take advantage of the equimarginal principle of maximization by equalizing MACs across firms, provide incentives to develop new technologies for reducing environmental costs, and keep costs low by allowing firms to act on their private knowledge of their own abatement costs. The ideal policy would also set the marginal benefits of production equal to the marginal environmental costs it imposes, but as we stated earlier, environmental costs are largely unknown.

We will now examine three policies that can theoretically achieve these goals: taxes, subsidies, and tradeable permits, through a cap-and-trade system.

---

[2]When the EPA first proposed tradeable emission permits in SO_2, industry estimates of MACs were as high as $10,000 per ton. EPA estimates were in the neighborhood of $1000 per ton. Permits currently trade for under $100 per ton. Carol Browner, speech, "Public Health and Environmental Protection in the 21st Century," University of Vermont's 2002 Environmental Literacy Seminar Series, March 25, 2002.

■ PIGOUVIAN TAXES

Early in the twentieth century, economist A. C. Pigou began grappling with the problem of internalizing environmental externalities. As we discussed in Chapter 10, externalities occur when one economic agent causes an unintended loss or gain to another agent, and no compensation occurs. In the case of a negative externality, the basic problem is that the economic agent is able to ignore a cost of production (or consumption). Under such circumstances, the market equilibrium of marginal costs equal to marginal benefits does not emerge, and some of the wonderful benefits of markets fail to appear. Pigou came upon the simple solution of imposing a tax equal to the marginal external cost. This would force the economic agent to account for all economic costs, creating an equilibrium in which marginal social costs were equal to marginal social benefits.[3]

Note that this policy requires a change in property rights. When a firm is free to pollute, it has privilege, and those who suffer from the pollution have no rights. A Pigouvian tax essentially creates a property right to the environment for the state, using a liability rule. Firms can still pollute, but they must now pay for the damages from their pollution.

Of course, as we cannot accurately measure marginal environmental costs, the **Pigouvian tax** cannot be set precisely at that level. Even if we did know marginal environmental costs, these costs change with the amount of pollution, and the ideal tax would presumably also have to change. While Pigouvian taxes will not lead to perfectly efficient outcomes, they will reduce environmental costs, and do so cost-effectively. How do they accomplish this?

When abatement costs are less than the tax, it is cheaper for the firms to abate, so that's what they will do. On the other hand, when abatement costs are more than the tax, paying the tax minimizes costs and maximizes profits. This means that after implementation of the tax, the MAC for all firms will be equal to the tax—the equimarginal principle of optimization. Firms for which it is cheap to reduce pollution will therefore make large reductions, and firms for which it is expensive will reduce much less. The latter firms will, of course, pay correspondingly more in taxes than the former. Note that no one but the firm needs to know the firm's marginal abatement costs. Each firm acting on its own preferences and own knowledge with a maximum of micro-freedom generates the cost-effective outcome desirable to society.

Firms continue to pay tax on every unit of pollution that they produce.

---

[3]A. C. Pigou, *The Economics of Welfare*, 4th ed., London: Macmillan, 1932 (originally published in 1920).

This means that there is always an incentive for achieving further reductions in pollution and doing so more cost-effectively. Such incentives are perhaps the most important reason that taxes are superior to command-and-control regulations.

In addition to taxes, firms must also pay abatement costs. It is therefore quite possible that total costs to the firm and industry (all related firms together) under the tax will be higher than they would have been under command-and-control regulations (e.g., forcing each firm to cut emissions by 40%). However, relative to society, the tax is a transfer payment and does not count as a cost. And by ensuring that the firms with the lowest abatement costs make the largest reductions in pollution, the tax ensures that actual costs to society are less under the tax than they would have been under regulations. Nor can the tax really be considered unfair to the firm, as it is simply a payment for the costs the firm is imposing on society. It is possible that the tax would even drive some firms out of business, but as long as the tax were no greater than the marginal external environmental cost, it would simply mean that the costs the firm was imposing on society were greater than the benefits it was providing.

It would be very difficult to predict the decrease in negative externalities that would result from any given tax, and a trial-and-error process might be needed. Yet changing taxes every year, or even every few years, creates a burden for firms, which lose the ability to plan for the future. Perhaps the best approach would be to begin with a fairly low tax but let firms know that the tax will increase over time. This approach would let firms gradually change their practices, reducing overall cost by allowing new technologies to come online before the tax reaches its ultimately desired level.

As long as human populations and the economy are still growing, the demand for activities that impose environmental costs is also likely to grow. This means that to maintain the desired level of environmental amenities or resource depletion, the tax would need to increase over time. As with all environmental policies, the principle of adaptive management is appropriate.

■ PIGOUVIAN SUBSIDIES

A **subsidy** is a bonus or payment for doing something, the opposite of a tax. A **Pigouvian subsidy** is a payment to each firm for each unit by which it reduces environmental costs; it has many of the same attributes as the tax. Ideally, the subsidy will equal the marginal benefit to society of abating pollution. As long as abatement costs are lower than the subsidy, the firm will reduce pollution. Again this will equalize MACs across the industry, the precondition for a cost-effective outcome. Whereas a tax follows the polluter-pays principle, a subsidy basically assumes that the polluter has the privilege to pollute, and society must pay him not to.

One serious problem with subsidies is that they can perversely lead to an increase in pollution. A subsidy increases the profit margin for the polluting industry, possibly attracting new entrants. While each firm pollutes less than in the absence of the subsidy, more firms could still lead to more total pollution. While many people might justifiably resent the notion of paying people not to impose costs on the rest of society, and the potential outcome of greater pollution is entirely undesirable, this does not mean that Pigouvian subsidies are entirely irrelevant. Pigouvian subsidies can be desirable as an incentive to ecosystem restoration. For example, paying farmers to reforest their riparian zones might reduce nutrient runoff and provide a host of other ecological services. In addition, under international law, sovereign nations have the right to do as they choose with their resources, and there is no global government that could impose a Pigouvian tax on the negative environmental costs of deforestation, for example. Under such circumstances, something like a Pigouvian subsidy may be the best option. We will return to this issue in some detail later.

> **THINK ABOUT IT!**
> *Can you explain why we might favor Pigouvian taxes to deal with pollution problems but Pigouvian subsidies to induce forest landowners to reduce timber harvests?*

One final point bears mentioning. While Pigouvian taxes or subsidies may lead to the welfare-maximizing outcome of marginal social costs equal to marginal social benefits, the same does not hold true at the level of the individual. This is a result of the fact that many environmental costs are public bads. Every individual suffers from the same amount of environmental cost, yet each individual has different preferences concerning those costs. A perfect market solution would have to distribute the tax among the afflicted population to exactly compensate for the marginal damage they suffer from the environmental cost. Of course, it would be impossible to determine the marginal cost curve for every individual on the planet, and individuals would have incentives to misinform the agency collecting this information if it would influence how much of the tax they were to receive. Also, if individuals were compensated for the externalities they suffer, they might do less to avoid externalities, and this too could reduce efficiency.[4]

---

[4]E. T. Verhoef, "Externalities." In J. C. J. M. van den Bergh, *Handbook of Environmental and Resource Economics*, Northampton, MA: Edward Elgar, 1999. Again we reiterate that a just distribution should take precedence over efficiency, which would favor compensating individuals for their suffering.

■ CAP AND TRADE

Tradeable permits are another cost-effective mechanism for achieving a specific goal. Rather than increasing prices through a tax to reduce demand, tradeable permits require society to set a **quota**, a maximum amount of pollution or resource depletion that it will allow. This approach, commonly referred to as **cap and trade**, is currently used in the United States to regulate SO_2 emissions, in the European Union to regulate CO_2 emissions, and in several countries to regulate fisheries.

What factors should determine the allowable quota? From the economist's perspective, the ideal quota should be set so that the marginal benefit from one more unit of pollution or harvest is exactly equal to its marginal social and private cost. Uncertainty, ignorance, and price fluctuations make this ideal unattainable. Even if we could accurately estimate marginal costs at the existing scale and price, as we learned in Chapter 21, the very act of setting the quotas (scale) changes the prices used in calculating the costs and benefits.

The sustainable scale for any quota should be determined by biophysical constraints. Harvest quotas for renewable resources must allow for harvest at no greater a rate than that at which the resources can renew themselves, and emission quotas for pollutants must not exceed the waste absorption capacity for the environment. Sustainable quotas must also leave a considerable safety margin between our demands on the system and our best estimate of its capacity (remember design principle #3 from Chapter 21). Quotas must also be congruent with the scale of the problem (remember design principle #6). Quotas for migratory or transboundary species such as Pacific salmon or bluefin tuna must be for total harvests as well as national shares of the total, and the same is true for quotas on transboundary pollutants. Quotas for pollutants must also respect their spatial distribution, not allowing excessive accumulation in any one area.

The desirable scale for any quota may be considerably lower than the sustainable scale, however. In the case of renewable resource harvests, quotas should account for the fact that renewable resource stocks not only provide a flow of harvests but are simultaneously funds that provide a flux of services over time. Where quotas are currently used in fisheries, managers focus almost entirely on the stock-flow aspect of the resource, virtually ignoring the fund-service aspect. While we do not fully understand the role of ecological fund-services in sustaining human well-being, we do know their value is not zero, and they should not be ignored. Certain pollution flows can similarly affect ecosystem services or human well-being even when they are low enough that they do not accumulate as stocks in the environment. The quota process should also respect the principle of adaptive management, allowing adjustment as new information becomes available (design principle #5).

Once quotas have been established according to the criterion of scale, they should be fairly distributed. Most existing cap-and-trade systems have awarded tradeable permits to existing polluters and resource harvesters, transforming a privilege into a right. Permits could alternatively be auctioned off by the government (which essentially assigns resource rights to society as a whole) or distributed to achieve other social goals, such as greater income equality. Permits can be issued annually or once and for all. They can be for set quantities or for a proportion of an adjustable quota.

Finally, making the permits tradeable market goods uses the market mechanism to achieve the overall resource use reduction implicit in the quota as cost-effectively as possible. A firm will abate pollution as long as that is cheaper than the price of a permit, and it will purchase permits when abatement costs are more expensive. Again, this leads to equimarginal abatement costs and maximum micro-freedom, a precondition for cost-effective outcomes. For resource harvesting permits, the firms with the most profitable use of the resource will be able to pay the highest price for the permits. This theoretically ensures that the resource will be allocated toward the most desirable ends, but, we reiterate, only if you believe that individual "votes" concerning the desirable ends should be weighted by individual wealth.

While many variations are possible, sustainability and justice criteria favor annual permit auctions in a cap-and-rent system. Even when regulated by quotas, pollution and resource extraction have negative impacts on the public goods of clean air, clean water, and ecosystem services, and fairness requires that auction revenues captured by governments should be spent on other public goods in compensation. Furthermore, the waste absorption and reproductive capacities of ecosystems are created by nature, not by individual effort. Profits from the use of these capacities are therefore unearned income, or economic rent, and should be captured by society as a whole. In fact, a quota system on resource harvests can actually move a system from the zero profits of open access regimes to a profit-generating (in the sense of a natural dividend) level of harvest, as described in Chapter 12. Since government-imposed quotas create this natural dividend, it is fair that the government should capture it. If instead tradeable permits are awarded once and for all, the owner will capture scarcity rents from any increase in value. Furthermore, renting permits on an annual basis facilitates the annual adjustments that may be necessary in the presence of imperfect information, natural variation, and ecosystem change.

THINK ABOUT IT!

If fishery managers decide to take an ecological economic approach to establishing harvest quotas and explicitly incorporate the fund-service benefits of fish stocks into their decisions, what impact will this have on the quota? (You might want to review Chapter 12.) In determining optimal quotas, do you think policy makers should discount benefits to future generations? Why or why not?

Quotas also require a change in property rights, but whereas taxes impose a liability rule, quotas impose a property rule. The owner of the quota essentially owns a portion of the waste absorption capacity (a rival good made excludable by quotas) of the medium into which they are emitting wastes. The right may initially be awarded to the government, members of the community affected by the pollution, or the polluters. The same principle is true when quotas are used to end privilege in an open access resource harvest regime.

One problem with quotas is that there may be little incentive to reduce total pollution or resource extraction below the quota. If the quota is carefully chosen, this need not be a problem. Within a tradeable quota system, any profit-maximizing firm still has the incentive to reduce emissions or resource harvest so that it may sell a portion of its quota. Thus, while quotas will not drive undesirable activities below the quota level, they provide incentives for reaching quotas ever more cost-effectively. Also, if the economy or population is growing, quotas ensure that resource use will not grow.

Many economists have pointed out that if environmentalists think a quota is too high, they are free to purchase shares of the quota and discard them. Unfortunately, we again run into the problem of public good provision here. The environmentalist would incur the entire cost of purchasing the permit but share the benefits with everyone. In addition, if permits are issued annually in variable amounts, then the government could potentially issue more permits in response to those being purchased and not used. Alternatively, if the regulatory authority decides too many permanent permits were issued to begin with, or new information changes the assessment of how many permits are desirable, the government can readily purchase some back and not use them, as we illustrate with the following case study.

THINK ABOUT IT!

Why will Pigouvian taxes, Pigouvian subsidies, and cap-and-trade policies lead all firms to have equal marginal abatement costs? You may want to review the material on the equimarginal principle of maximization in Chapter 8 to figure this out. The principle is the same.

Tradeable Permits vs. Shorter Seasons

As we have previously discussed, oceanic fisheries have been heavily over-fished, and policies are urgently needed to address this problem. A number have been tried, providing good evidence for the superiority of solutions that maximize micro-freedom. Within this context, we will compare efforts in the United States to reduce unsustainable fish harvests by shortening the season with efforts in New Zealand to implement quotas and tradeable permits.

The halibut fishery is one of the oldest on the Northwest coast of the Americas, and by 1960, open access conditions had led it to be fished almost to extinction. In 1960, the International Pacific Halibut Commission (IPHC) was created to regulate the annual harvest and restore the catch to the maximum sustainable yield. Harvests were limited by imposing a season, which was then gradually reduced as needed. This method proved highly effective at restoring the population and increasing the annual harvest. However, by the early 1990s, the season was as short as one or two 24-hour periods per year (depending on how long it took to reach the annual quota established by the IPHC), during which fishermen engaged in a mad race to maximize their share of the catch.

What are the implications for efficiency and cost-effectiveness of such a short season? First, fishing is already one of the most dangerous industries, and engaging in a mad race just makes it that much more dangerous, especially if the "season" happened to coincide with bad weather. Loss of life was frequent. Boats often captured so much fish they were in danger of sinking and sometimes did. Fishermen were forced to cast as many lines as possible, ensuring that some would be lost. The situation was made worse when the large number of boats caused lines from different boats to get tangled and cut, perhaps leaving already hooked halibut to die. The 2-day open access fishery led fishermen to invest in more equipment to take more fish in a shorter period. In spite of increasing stocks, the season was continually shortened, and the equipment (and labor force) then went unused out of season. Almost all halibut fishermen also take other fish with the same equipment, but the net result was still excess capacity. Demands for rapid harvest led to poorer treatment of the fish and a lower-quality product.

Once landed, all of the fish arrived at the market at the same time. There was therefore a very limited market for fresh halibut, and most had to be frozen. Again, large capital investments were needed to create the infrastructure for freezing all the halibut in such a short period, and there was excess capacity for the remainder of the year. Processing fish by freezing is capital-intensive, and it therefore increased the barriers to market entry, threatening to limit competition. Processing fresh fish, in contrast, is labor-intensive and has far lower capital costs.

In 1990, when Canada modified its system by establishing individual quotas for ships and extending the season, it created a market for fresh halibut. In fact, because Canadian halibut was mostly sold fresh, Canadian fishermen enjoyed a 70% price premium over their Alaskan counterparts. The failures of the U.S. system were so pronounced that in 1995, the U.S. instituted an individual fishing quota system as well. The quotas were assigned to currently active fishermen based on their recent harvests. They were intended to allow fishermen to extend their harvest efforts over the entire season, thereby paying more attention to quality than to speed. Some leasing and trading of the quotas were allowed, but with strict limits on concentration of shares.[5]

New Zealand's fisheries went through the cycles typical of most fisheries. At first the resource was scarcely exploited, with the exception of inshore fisheries. In the 1970s, seeking to exploit a new source of foreign exchange, the government began a program of subsidies to develop the industry. The result was overcapitalization (basically too many boats chasing too few fish) and dramatic declines in fish populations. In 1982, the government forced fishermen who earned less than 80% of their income from fishing out of the market. This had a highly negative impact on Maori fishermen, who traditionally earned their living from a variety of activities, but it did little to reduce pressure on the fishery. Then, in 1986, New Zealand decided to follow the economists' advice and implement a system of transferable fish quotas. Similar systems are also in place in Iceland and the Philippines.

The process is simple. Scientists determine the total allowable catch (TAC) for each species from each of several geographic areas, typically with the goal of achieving maximum sustainable yield. From this number they subtract the expected take by the sport fishery and set aside 20% for the Maori. (An 1840 treaty awarded rights to all New Zealand fisheries to the Maori, but New Zealand chooses not to honor this treaty.) The remainder is the **total allowable commercial catch (TACC)**, which is divided up into **individual transferable quotas (ITQs)**, which may be bought, sold, or leased on the market. The initial ITQs in terms of tons of fish were awarded to fishermen in proportion to their catch history. To make the ITQs more attractive to fishermen, initial awards were close to historic catches and exceeded the TACC. The government then purchased back sufficient ITQs to reach the TACC. Fish populations fluctuate naturally, and so did the TACC. Initially, the government was forced to buy or

---

[5]K. Casey, C. Dewees, et al., The Effects of Individual Vessel Quotas in the British Columbia Halibut Fishery, *Marine Resource Economics* 10(3):211–230 (1995); C. Pautzke and C. Oliver, Development of the Individual Fishing Quota Program for Sablefish and Halibut Longline Fisheries off Alaska. Anchorage, Alaska: North Pacific Management Council, 1997. Online: http://www.fakr.noaa.gov/npfmc/Reports/ifqpaper.htm.

sell ITQs whenever the TACC changed. Then, in 1990, ITQs were changed to represent a proportion of TACC.

In terms of scale and allocation, the policy has been very effective. Fish populations have recovered, though there have been problems with the introduction of new fish species into the market, because typically little is known about their life cycles. For example, considerable evidence suggests that orange roughy has been overexploited in spite of the TACC. In terms of efficiency, fishermen now need invest only enough to capture their share, lowering their capital costs. Harvests are spread out over a longer period, increasing the market for fresh fish. Fishermen can purchase ITQs for different species when they have a large bycatch[6] and sell or lease ITQs when they fail to meet their quota. Less efficient fishermen can sell their quotas to more efficient ones. The value of fisheries in New Zealand has apparently doubled in recent years.

The impact on distribution, however, is far less desirable. ITQs tend to concentrate in the hands of the larger firms, leading to concentration of the wealth in a lucrative industry. Maoris—in spite of the treaty awarding them rights to all the fisheries—have disproportionately been forced out of the market. Part of the problem lies in access to credit. ITQs do not count as collateral for bank loans. When the TACC decreases, small-scale fishermen lack collateral for bank loans to purchase more ITQs, while large firms have other assets they can use as collateral. In part, this problem stems from the initial allocation of ITQs based on catch histories. The firms that played the largest role in overexploiting the fisheries initially were rewarded with more ITQs.[7]

The case of New Zealand fisheries shows the importance of separate policies for achieving separate goals. The TACC (one policy instrument) set the scale, and the ITQs (a separate policy instrument) achieved efficient allocation. But ITQ policies often fail to address distribution issues, which turned out to be problematic for New Zealand. They require a third instrument, perhaps one that could limit the concentration of ITQs to help maintain market competitiveness and avoid forcing poorer fishermen out of the market.

---

[6]Bycatch is the harvest of species other than the target species. Depending on the species and the existing laws, bycatch may be kept or thrown back. Bycatch is often killed in the harvest but is nonetheless thrown back. Dolphins as bycatch for some types of tuna fisheries and sea turtles as bycatch for some types of shrimp fisheries have received considerable attention. For some fisheries such as shrimp, bycatch may be more than 10 times the mass of the target species.

[7]P. Memon and R. Cullen, Fishery Policies and Their Impact on the New Zealand Maori, *Marine Resource Economics* VII(3):153–167 (1992); New Zealand Minister of Fisheries, The Quota Management System, no date. Online: http://www.fish.govt.nz/commercial/quotams.html. R. Bate, The Common Fisheries Policy: A Sinking Ship, *Wall Street Journal*, June 2000. Online: http://www.environmentprobe.org/enviroprobe/evpress/0700_wsj.html.

■ POLICY IN PRACTICE

We see, then, that policies are available that meet environmental goals cost-effectively and that provide incentives for reducing pollution, resource depletion, and so on, even after those goals have been met. Most of these policies are widely accepted by economists as cost-effective solutions, yet regulatory agencies in general seem to prefer the potentially less efficient command-and-control regulations. Why is this so? There are a number of reasons.

Environmental regulations often are administratively simple and may have low monitoring costs. Regulatory agencies have substantial experience with these options, and institutions can be slow to change. Conceptually, regulations are simple and widely perceived as fair, at least when they affect everyone equally. Many regulators pay little attention to cost and may be more concerned with reducing their own transaction costs than with lowering the costs to polluters. Abundant other reasons also exist, but considerable evidence suggests that in many circumstances, the overall costs to society of reaching a given target are higher under regulation than under mechanisms that allow a maximum of micro-level freedom by relying on market allocation, subject to macro-level control.

In the United States, the cap-and-trade systems have had some success on a limited basis (e.g., SO_2), while in Europe tax schemes, referred to as "ecological tax reform," have been more popular. The idea is sold politically under the banner of "revenue neutrality"—the government taxes the same total amount from the public, just in a different way. Following the design principle of gradualism, European governments have sought to impose the most desirable resource tax first and to couple it with the worst existing tax, eliminating the latter to the extent that revenue from the former permits. Thus, one may get a "double dividend"—the environmental benefit of taxing a resource whose price is too low, plus the fiscal benefit of getting rid of a distortionary or regressive tax.[8] Subsequently, one seeks to couple the next most desirable resource tax with the next worse other tax, and so on.

The slogan of ecological tax reform is "Tax bads, not goods." The idea is to shift the tax burden from value added by labor and capital (something we want more of) to "that to which value is added"—namely, the throughput and its associated depletion and pollution (something we want less of). It seems a matter of common sense to tax what you want less of and stop taxing what you want more of. Ever suspicious of common sense, however, neoclassical economists have invented general equilibrium models with particular assumptions (such as the familiar production functions with no resource inputs) that lead to counterintuitive results. We find these models in general to be artificial and unconvincing. In any case the policy, at an incipient level, seems to be working in Europe.[9] The main political dilemma European governments face in trying to implement ecological tax reform seems to be maintaining competitive advantage in international trade by keeping resource prices low, versus internalizing external costs in prices and thereby raising them, to the detriment of competitive advantage—a problem we encountered in our look at globalization.

This latter problem is a severe and general policy difficulty. Our fifth general policy design principle stated that the domain of the policy-making authority should coincide with the domain of actions open to those who cause, or are affected by, the policy (see Chapter 21). If a policy is enacted to limit pollution and a firm can avoid compliance simply by moving across a boundary, then the extent of domains does not coincide. Globalization, as we saw in Chapter 19, expands the domain of actions to the entire world while keeping the domain of public policy confined to the national level. Because national policies are easily evaded in such a situation, we have a general weakening of public policy along with an increase in the relative power of private individuals and corporations. Public efforts at the national level to deal with poverty, environmental degradation, public health, education, and even macroeconomic goals of full employment without inflation are all automatically sacrificed to the overriding goal of growth in the global production of market goods, as stimulated by free trade and free capital mobility.

This is why people are demonstrating in the streets of Seattle, Prague, Genoa, Washington, D.C., and anywhere else the WTO, the IMF, and the World Bank meet. Shortening the length of the meetings and changing the venue to places like Qatar do not address the issues raised by critics. Is it too much to hope that the concepts of ecological economics can provide a framework in which the legitimate claims of both growth and limits can be recognized?

BIG IDEAS to remember

- Direct regulation
- Command-and-control regulations
- Pigouvian taxes and subsidies
- Tradeable permits (quotas)

- Abatement costs
- Total allowable commercial catch (TACC)
- Individual transferable quotas (ITQs)
- Ecological tax reform

---

[8]The existence of a "double-dividend" is a source of frequent dispute among environmental economists but is more accepted by ecological economists.

[9]B. Bosquet, Environmental Tax Reform: Does It Work? A Survey of the Empirical Evidence, *Ecological Economics* 34(1):19–32 (2000).

CHAPTER

23

Just Distribution

The distribution of wealth and income is always a contentious issue. But it is also crucially important. Why?

First, people who are too poor will not care about sustainability. Why should they worry about the welfare of the future when they are not even able to provide for their own basic needs? Throughout the world, the excessively poor are forced to mine soils, clear-cut forests, overgraze grasslands, and tolerate excessive pollution just to survive. And as we have seen, the impacts of these activities are not merely local; they have global consequences.

Second, people who are excessively rich consume large amounts of finite resources, possibly depriving future generations of the basic means of survival. Even the economists most reluctant to make interpersonal comparisons cannot deny that the marginal utility of consumption for those below subsistence is far higher than for those buying increasingly frivolous luxury goods.

Third, if we care about sustainability, we care about intergenerational distribution. We do not want to force the future to live in poverty simply so we can consume more luxuries. Yet what ethical system can justify a concern for the well-being of those yet to be born while not caring for the well-being of those alive today?

Finally, we know that the economic system cannot grow forever on a finite planet. We must limit growth to ensure the well-being of the future, but one cannot ethically tell poor people they must continue to suffer deprivation to ensure that the future does not suffer. If the pie must cease to grow, then we are ethically obliged to redistribute it.

If distribution is so important, then why is it so contentious? Many people believe that in a free market society, people have wealth because they have earned it, and it is unjust to take from people what they have

441

earned with the sweat of their brows. We agree that in general distribution policies should not take away from people what they have earned through their own efforts and abilities. However, people should not be able to capture for themselves values created by nature, by society, or by the work of others. And they should pay a fair price for what they receive from others, including the services provided by government, and for the costs they impose on others. In addition, we must recognize that a less unequal distribution of resources may generate public goods such as economic stability, lower crime rates, stronger communities, and better health (as discussed in Chapter 16), and society should pay for public goods. If we follow these principles, the resulting distribution should be both just and sustainable.[1]

Distribution must focus on both income and wealth and on market goods as well as nonmarket goods. Policies that provide more money for government from higher-income and wealthier individuals can further improve distribution by allowing governments to cut taxes for the less well-off or by funding public goods projects that benefit everyone. Policy makers have devised many plans to achieve distributional goals, both within and between nations. Some have proven successful, some not. We now review some policies designed to achieve a more just distribution.

■ CAPS ON INCOME AND WEALTH

Must we set a maximum individual income? At first glance, many people consider this type of policy an unwarranted intrusion on individual liberties. What right does the state have to take what someone has earned with the sweat of her brow? Income and wealth are the just deserts of hard work. From this viewpoint, income caps are unjust.

However, on a finite planet subject to the laws of thermodynamics, if some people consume too much this generation, they will reduce the resources available to future generations. This means that in the future, society may be worse off than it is today, and people may have to work harder than the current generation to consume even less. In this case, a sense of obligation toward future generations demands that society as a whole reduce consumption so that future generations have the same opportunities to be rewarded for their work as the present, the same opportunities to receive their just deserts. However, to demand that society as a whole reduce consumption yet not to demand that the wealthiest members of society also do so is a difficult position to defend.

---

[1]An unsustainable outcome would be an unacceptable cost for future generations.

Box 23-1	Wealth and Power

Many wealthy people earn far more than they could conceivably con-sume. If Bill Gates invested all his wealth in inflation-indexed govern-ment bonds with real yields of 3%, which is probably as close as one can get to a risk-free investment, he would be earning over $3 million per day.[a] Many of the world's richest people earn more than they or their off-spring could conceivably spend. Why would anyone accumulate wealth if they do not intend to consume it? The only reasonable answer is to amass power and status.

Certainly, it is difficult to argue that wealth does not bring power in existing political systems. While many people argue that the inequitable distribution of wealth is acceptable, in democratic countries, far fewer say the same about inequitable distribution of power.[b] And the power that rewards the accumulation of wealth is readily used to generate yet more wealth and hence more power, in a vicious cycle. For example, it is painfully clear that corporate donations to the political parties in most countries are not made to strengthen democracy but rather to promote legislation that provides greater economic advantage for the contribu-tors. How else can we explain the fact that so many major corporations contribute money simultaneously to two politicians running against each other for the same office? By seeking economic advantage through polit-ical influence, wealth undermines market forces and the beneficial out-comes they are capable of generating.

Strangely enough, most Americans remain opposed to income caps.[c] Americans and citizens of many other capitalist democracies seem to have two completely incompatible core beliefs: We have the right to a democratic government and the right to become richer than Midas. How-ever, as Supreme Court Justice Louis Brandeis reportedly said, "We can have a democratic society, or we can have the concentration of great wealth in the hands of the few. We cannot have both."

[a] In November 2001, Bill Gates topped the Forbes list of the world's richest people for 2003, with a net worth of $40.7 billion.

[b] R. Lane, Market Justice, Political Justice, American Political Science Review 80(2): 383–402 (1986).

[c] Ibid.

Is there any harm in accumulating wealth simply for status? There is, for two reasons. First, people generally exhibit their status through con-spicuous consumption, which increases scale. Second, status is measured relative to others' positions and is thus a zero sum game. Everyone's sta-tus in society cannot increase. Therefore, if I work hard to accumulate wealth and increase my status, I am reducing the status of others relative to me. In order to maintain their status, they will have to work harder as well, sacrificing leisure time, time for community, and time for family. If we all worked twice as hard to increase our status, no one's status would change, we would all have less time to pursue other goals, and we would consume more natural capital. Status through wealth accumulation can turn into a kind of arms race in which we all work harder and become worse off.

Conspicuous consumption is therefore a negative externality, and people should pay for the negative impacts it imposes on others. A progressive consumption tax would help redistribute resources, and by taxing a negative externality, it would lead to a more efficient allocation of resources as well.[2] Empirically, in the wealthier countries there is evidence that people are growing less satisfied with life instead of more satisfied, in spite of continuing dramatic increases in national wealth.[3]

THINK ABOUT IT!
Can you explain how a progressive consumption tax could make even the wealthy better off?

Policies for capping income might also include a highly progressive income tax that asymptotically approaches 100%, more direct limits on how much someone can earn, or relative limits that establish a legal ratio between the highest and lowest income allowed. Progressive income taxes are used worldwide. Many economists claim that such taxes are a deterrent to economic growth, in which case they would help in achieving a steady-state economy at a sustainable scale. However, economic growth in the U.S. was quite high during the 1950s, when the highest marginal federal tax bracket was 90%, compared to less than 40% today.

Policies for capping wealth could include a progressive wealth tax, as currently exists in a number of European nations. People already pay taxes on real estate, which is a form of wealth, so why not extend this to all wealth, particularly the forms that are highly concentrated among the wealthiest? Very high inheritance taxes would also help, as an estimated 46% of accumulated wealth is directly inherited.[4]

Many people would object that progressive taxes take a disproportionate amount from the rich and therefore do not meet the criteria we discussed at the beginning of this chapter. However, governments generally provide most of the infrastructure and institutions that allow businesses to thrive and people to grow wealthy. Would Bill Gates, Warren Buffett,

---

[2] R. Frank, *Luxury Fever: Why Money Fails to Satisfy in an Era of Excess*, New York: Free Press, 1999.

[3] R. Lane, *The Loss of Happiness in Market Democracies*, New Haven, CT: Yale University Press, 2000.

[4] G. Alperovitz, Distributing Our Technological Inheritance, *Technology Review* 97(7):30–36 (October 1994).

and other billionaires be so wealthy if they had been born in sub-Saharan Africa? In addition, political philosophers have long argued that one of the dominant roles of government is to protect private property. Clearly then, the more private property someone owns, the more they benefit from the services of government, and the more they should pay for those services.[5]

Another argument against income caps is that they are also harmful to the poor. From this viewpoint, allowing unlimited accumulation of wealth creates incentives that increase total production and employment opportunities and make the worst off better off than before. Capping income for the wealthiest reduces the opportunities for the poorest to escape poverty. If this is true, then how do we explain the productivity and relative absence of poverty in northern Europe, where taxes are very high?

■ MINIMUM INCOME

Many countries, including the U.S., have instituted policies intended to guarantee a minimum income. These policies can help achieve sustainability by ameliorating poverty, as well as by reducing the gap between a society's richest and poorest members. Moreover, minimum income policies are justified because they can help provide a number of other public goods. In Chapter 17 and 20 we explained how economic recessions can have positive-feedback loops. Something causes consumption to decline. People buy less, so firms produce less and lay off workers. Laid-off workers consume less, so firms again reduce production. In the presence of a minimum income, even when people are laid off, they will continue to consume. Indeed, those with the lowest incomes typically spend the highest percentage of those incomes on consumption. A minimum income helps break the positive-feedback loop that causes economic recessions, and a more stable economy can benefit everyone. In addition, abundant evidence links income disparity to crime, violence, and other public bads. A minimum income may not eliminate these problems, but it can help reduce them, and therefore we favor such a policy.

Neoclassical welfare economics, whose foundations are utilitarian philosophy and diminishing marginal utility, as we saw in Chapter 8, implicitly calls for the elimination of poverty. If the goal of society is to maximize utility summed over individuals, and wealth and income offer diminishing marginal utility, then clearly an additional unit of wealth for a poor person provides more utility than the same unit would provide for a

---

[5]We are glad to report that Warren Buffett and Bill Gates, Sr., are on record as vigorously agreeing with the proposition that society contributes the conditions in which individuals can earn great wealth and that such individuals should be willing to pay significant taxes, especially estate taxes. See W. H. Gates and C. Collins, *Wealth and Our Commonwealth: Why Americans Should Tax Accumulated Fortunes*, Boston: Beacon Press, 2003.

wealthy person. Economists reluctant to accept this conclusion have asserted that different people have immeasurably different capacities to enjoy (or suffer?), and therefore we cannot make interpersonal comparisons of utility. Thus, many economists have focused on maximizing production rather than utility, which effectively skirts the distribution issue.[6]

On the other hand, it is clear that, on average, a unit of additional income would benefit someone living in absolute poverty more than the same amount would benefit a millionaire. People may have different capacities for enjoyment at some level, but we are very alike in our suffering—we are all poisoned by the same toxins, made ill by the same germs—and our biological subsistence needs are the same. The additional utility when one moves from below subsistence needs to above them is obviously immense.

Curiously, most Americans profess to believe that the current distribution of income in the United States is unjust, yet they remain reluctant to provide income to those who have not "earned" it. However, the "just deserts" argument is based on the assumption that people are paid according to their contribution to society. Yet the last two centuries have seen a fairly steady upward trend in real incomes. This is not so much because people make more substantial contributions to society on their own but because they benefit from past contributions to productivity. In other words, the well off are awarded more than their just deserts already, so why not do the same for the worst off?

The specific policy approaches to ensuring a minimum income are more debatable than the need for some policy. The most commonly employed policies are

1. Welfare programs, in which the government provides direct monetary or material aid to the poor

2. Unemployment insurance for the unemployed

3. Minimum wages and negative income taxes for the employed

These approaches can play a role in ensuring minimum incomes, but such simple transfers are probably not the best approach to ending poverty for either society or the recipients of such transfers.

Among traditional approaches to a minimum income, many ecological economists would argue first for equal opportunity in education, job access, and job advancement, followed by guaranteed jobs at a living wage, and direct transfer payments playing a role only when necessary. In addition, we believe that people have equal entitlements to wealth created by nature and by society, independent of the entrepreneurial ability of the individual. Distributing this wealth equally would provide a minimum in-

---

[6] J. Robinson, *Economic Philosophy*, Garden City, NY: Doubleday, 1964.

come. This involves other, less conventional approaches, which we now explore by looking at income as returns from factors of production.

■ DISTRIBUTING RETURNS FROM THE FACTORS OF PRODUCTION

In order to take a systematic look at income distribution, we recall from Chapter 16 that there are four sources of income: wages, profits, interest, and rent. Wages are the returns to labor, profits are the returns to entrepreneurship, interest is the return to capital, and rent is the return to land and other natural resources. Most efforts at distributing income focus on returns to labor, while the greatest disparities in income are actually the result of the other factors of production. We now turn our attention to distributing the returns to capital and the returns to natural capital.

Distributing the Returns to Capital

Financial capital, including equity in productive assets, is highly concentrated both within and between nations. The United States probably offers the most egregious example among the developed countries. As reported in Chapter 16, by the late 1990s, the richest 1% of Americans controlled 95% of the country's financial wealth,[7] up from 48% in 1989. Thus, even though returns to capital (productive and financial) are responsible for less than 30% of income in most developed countries, almost all of that income flows to a small sliver of the population. Between 1997 and 1999, the wealth of the Forbes 400 richest Americans grew by an average of $1,287,671 per day per person. In contrast, between 1985 and 1997, the net worth of the bottom 40% of households declined 80%.[8] Returns to financial wealth, profits, and interest are a major factor in the income disparities seen in the U.S. and many other countries.

Capitalist systems are presumed to be populated by capitalists, and capitalists are the individuals who own the capital. Yet in most so-called capitalist nations, very few people are actually capitalists. Market economies are based on ownership, which is responsible for the impressive productive efficiency of such systems. A broader distribution of capital ownership could enhance the efficiencies of the market economy, and if done correctly, it could actually increase the ability of the system to provide important non-market goods and services. These claims demand some justification.

---

[7] J. Gates, *Democracy at Risk: Rescuing Main Street from Wall Street—A Populist Vision for the 21st Century*, New York: Perseus Books, 2000. Gates cites E. N. Wolff, "Recent Trends in Wealth Ownership," paper for Conference on Benefits and Mechanisms for Spreading Asset Ownership in the United States, New York University, December 10–12, 1998.

[8] Gates, op. cit. Since the net worth of the bottom 40% of households is small, it does not take much of an absolute decline to reduce it by 80%. Nevertheless, the figure is dramatic.

A good place to start might be an analogy with land ownership. Numerous studies have shown that land worked by an owner with secure title is more productive than land worked by sharecroppers or wage laborers.[9] This makes sense. Making land productive requires investments in its productive capacity. A sharecropper or squatter will have little incentive to invest in the productive capacity of his land, and a wage laborer even less. In the case of the sharecropper, returns on the investment must be shared with the landowner, who at any time is able to evict the sharecropper. A squatter also cannot be certain that he will retain control of his land a year hence and will not risk investing resources in the presence of such uncertainty. These points are widely accepted by economists and are considered ample justification for land ownership by the individual.

Yet the labor force in industry may have even fewer incentives to increase productivity than sharecroppers. What incentive do wage laborers have to do any more than the minimum required to keep the job, especially in jobs where managers have little chance to distinguish between the productive capacity of different workers? Workers have the most familiarity with the work they do and in many cases may therefore have the best insights into how to do it faster, better, and cheaper. However, if there are no immediate benefits for the worker from more efficient production, why should she waste her time thinking about how to achieve it?

In addition, as we have pointed out, work is where many of us spend most of our waking hours. Economists typically consider work a disutility to be endured only to gain access to the material goods that provide us with utility, but there is no reason this should be the case. An economic system should not be devoted to the most efficient means of producing material goods but rather to the most efficient means of producing human well-being. Most owners of capital concentrate on maximizing profits. It is rarely the case that profit maximization alone will create working conditions that generate the greatest worker well-being.

Imagine a company in which the workers own significant shares of stock. Such programs, known as Employee Shareholder Ownership Programs (ESOPs), are already widespread throughout the world. In ESOPs, workers do not manage the company, but they do have the same influence over management decisions that shareholders enjoy. Under ESOPs, worker income is composed of wages plus profits on stocks. Workers have much more of an interest in the profitability of the company. If there are mechanisms through which workers can make suggestions, it is in the workers' self-interest to think about ways to improve production. Work-

---

[9]E.g., A. Brandão, P. Salazr, and F. G. Feder, Regulatory Policies and Reform: The Case of Land Markets. In C. Frischtak, ed., *Regulatory Policies and Reform: A Comparative Perspective*, Washington, DC: World Bank, 1995, pp. 191–209.

ers are also concerned about how other workers perform, as it now affects their income as well. The net effect is usually an increase in productivity. If the worker-owners control a large enough share of the stocks, they are likely to work toward making the workplace a more desirable place to be, a place that satisfies a variety of human needs. Rather than an adversarial relationship between workers who want only benefits for themselves and capitalists who desire only profits, worker-owners will strive for a balance between the two. If measured in terms of the ability to satisfy human needs, efficiency is likely to increase under worker ownership, though material production may not.

It is worth mentioning that corporations often offer substantial stock options to CEOs, with a rationale similar to ESOPs. Because CEOs do not own the companies they manage, they may try to manage a company to maximize personal benefit rather than corporate benefits. If stock options form a substantial portion of CEO salaries, then what's good for the corporation (at least in the short term, and as measured by stock value) is also good for the CEO. The problem is, as we have recently seen with the slew of corporate accounting scandals (Enron, WorldCom, and others), some CEOs may focus too hard on the short run and only on the value of stock. Stock values can be inflated through accounting fraud, and CEOs often have enough information to bail out before the crash. In addition, this type of "ESOP" generally aggravates the existing gross inequalities in income and wealth distribution.

Expanded ownership opportunities can also help address externalities. Many industries generate considerable pollution, with highly negative impacts on the local population. If owners live far away, they will seek to maximize profits and in so doing may ignore these negative externalities to the extent allowed by law. What happens if instead sufficient ownership of the industry resides with the local population to give them influence in management? The local population will strive for a balance between the negative externalities and the profits. In effect, the negative externalities have been internalized, a necessary condition for an efficient solution. Transaction costs will be reduced to those of coordination among shareholders, which exist in any publicly owned firm. Such outcomes can be achieved through Community Shareholder Ownership Programs (CSOPs).

Mechanisms for Distributing the Ownership of Capital. Broad-based ownership of capital may be an effective tool for improving distribution, increasing the efficiency of the economy in satisfying human needs, and internalizing externalities. The question is: What policies will help achieve this goal? Simply taking ownership rights away from current owners and handing them over to workers is likely to be unfair and is in any case too radical, departing from our principle that we must pay attention to initial conditions. More feasible alternatives abound, but we wil touch upon only a few of them here. First, productive assets wear out and must continually be replaced. Working toward broader ownership does not require that we directly redistribute existing property, but rather that we change ownership patterns for new capital. Second, not only do mechanisms for achieving this exist, but they have been tested in numerous countries and have received support from across the political spectrum.[10]

ESOPs are perhaps the most widely used system for broadening ownership patterns in capitalist countries. In the U.S., by 1996 some 9 million employees were participating in ESOPs, which controlled about 9% of corporate equity in the country. In the case of United Airlines, pilots led a worker coalition to purchase 55% of existing stock to increase their say in managing the company. (The company is currently in bankruptcy proceedings—a good warning against putting all your eggs in one basket.) In other companies, workers are awarded stock as a benefit, perhaps in place of a profit-sharing plan. In yet others, workers may be awarded stock to defend against a hostile takeover by other corporations (workers often lose their jobs in such takeovers and may therefore be reluctant to sell their shares). Some corporations sell stock to workers to fund expansion or even takeovers of their own. A number of tax incentives and other subsidies have been used to encourage ESOPs.

Given the advantages of more broad-based capitalism, there are a number of other feasible strategies governments could use to encourage this phenomenon. Government contracts, purchases, licenses (e.g., for public airwaves), and privatization programs could all show preference for companies that promote broad-based ownership.[11] Existing loan subsidies, such as loan guarantees many governments offer to purchasers of national exports, could be reconfigured to benefit only companies promoting broad-based ownership. National and international development banks could offer preferential loans to such companies. Innumerable other examples of corporate welfare exist and could similarly be channeled toward creating capitalists. Another requirement would be to train people how to become capitalists and manage capital. Public schools in most countries train people to become workers but not owners.[12] In sum, capitalist societies need more capitalists!

---

[10]E.g., right-wing U.S. President Reagan supported ESOPs, as did Robert Reich, the left-wing secretary of labor under President Clinton. J. Gates, *The Ownership Solution: Towards a Shared Capitalism for the 21st Century*, New York: Perseus Books, 1998.

[11]Privatization, the sale of public (government) assets to the private sector, has been occurring at a breakneck pace throughout the world, often as part of IMF structural adjustment programs.

[12]These and other policies are described in greater detail in Gates, *The Ownership Solution*, op. cit.

Distributing the Returns to Natural Capital

Ownership of land and other natural capital is also quite concentrated throughout the world. While the factor share of rent is generally calculated to be only 2% of income, this calculation ignores two major sources of returns to natural capital. First, returns from the extraction of natural resources are often classified as profit, when in reality most of the returns are actually rent. (Recall that rent is the profit above and beyond what is needed to supply the resource. The supply of nonrenewable resources is fixed, and the sales price of many renewable resources is often higher than would be needed to supply the market.) Second, many of the returns to natural resources are in the form of hidden subsidies. For example, when an industry pollutes water or air and is not required to pay for the costs this imposes, the industry is capturing the returns to the waste absorption capacity of the environment.

Ending Public Subsidies. When the state owns the resources in question, extractive industries are typically required to pay royalties on those resources. In many cases, these royalties are quite small. The state should be able to charge a royalty equal to the scarcity rent.[13] By spending the royalty on public goods, using it to reduce taxes, or distributing it as a citizens' dividend, the state can use rents to improve distribution. In some primary industries, government subsidies to natural resource extracting corporations can be quite blatant. A number of examples from the United States illustrate this point.

Under the Mining Law of 1872, corporations can purchase the surface and mineral rights to federal land for $2.50–$5.00 per acre, depending on the nature of the mineral deposit.[14] This law was originally designed to provide incentives for people of European descent to settle the American West, but now it is little more than a giveaway to large corporations, many of which are not even from the U.S. Publicly owned rangeland is frequently leased to big ranchers at a fraction of the fair market value.[15] Rights to timber in national forests are often sold for less than it costs the government to build the access roads to the resource, or at times even for less than it costs to prepare the bids.[16] As a result, many publicly owned

---

[13]D. M. Roodman, *The Natural Wealth of Nations: Harnessing the Market for the Environment*, New York: Norton, 1998.

[14]M. Humphries and C. Vincent, CRS Issue Brief for Congress: IB89130: Mining on Federal Lands, May 3, 2001. National Council for Science and the Environment. Online: http://www.cnie.org/nle/mine-1.html.

[15]B. Cody, Grazing Fees: An Overview. CRS Report for Congress, 1996. Online: http://cnie.org/NLE/CRSreports/Agriculture/ag-5.cfm.

[16]R. Gorte, Below-Cost Timber Sales: Overview. CRS Report for Congress, 1994. Online: http://www.cnie.org/nle/for-1.html.

forests in the U.S. are logged for timber, even when in private hands logging would not be economically viable.

One of the more controversial pieces of environmental legislation in the 1990s was the timber salvage rider attached to the bill that provided federal assistance to the victims of the Oklahoma City bombing.[17] This legislation sought to "salvage" all trees on national forests that were threatened by insects or fires by chopping them down. The bill suspended environmental regulations for salvage operations and explicitly stated that timber should be sold at a financial loss to the government, if necessary. As most forests are at some risk from insects and fire, the bill was little more than a massive giveaway of public resources. The 2003 Healthy Forests Initiative was essentially a continuation of this policy.

While all of these examples are from the United States, similar policies are in place worldwide. Getting rid of all of these subsidies would reduce the loss of ecosystems and their services, save taxpayers money, and generate abundant new government revenue.

Alaska Permanent Fund and Sky Trust. The state of Alaska has taken a direct step toward distributing the income from natural capital. Alaska charges royalties for extraction of its abundant oil reserves. These royalties go into the Alaska Permanent Fund. Interest on this fund is distributed to all residents of Alaska. The idea is that the natural capital of Alaska, or at least its oil supplies, do not belong to corporations but rather are the common property of all Alaska residents. Putting the money in a trust fund helps ensure that even when the oil is exhausted, future Alaska residents can share in the bounty.

Peter Barnes, an eco-entrepreneur, has proposed a "sky trust" similar to the Alaska Permanent Fund to distribute income from nonmarketed natural capital. He begins by asking: Who owns the sky? His answer is that the sky is a common property resource, owned by all citizens of a country. Yet some people use the waste absorption capacity of the sky more than others. Industries pollute the sky without paying, and some individuals pollute far more than others. Because there are few institutions defending our common property rights to the sky, it is treated as an open access resource, with the well-known results of poor air quality, acid rain, climate change, and other ill effects.

The sky trust is a bundle of policies designed to address scale, distribution, and allocation. The scale and allocation components are achieved by establishing quotas, then auctioning them off in the form of individual tradeable quotas. We have already discussed how these mechanisms

---

[17]A rider is a piece of legislation attached to another, unrelated bill. The bill cannot be passed without the rider.

work. All the returns from the sales of these quotas would go into a trust fund, the returns from which would be distributed equally to all citizens in the form of cash. All citizens would receive equal shares, but those who pollute more would pay more, so redistribution would occur.[18] The same basic idea could be used with all natural capital on the assumption that it is a gift of nature to all humans, and not just to a select few.

While this is a very promising policy for a number of reasons, the re-distribution mechanism also raises some concerns. The sky is a public good, as are many of the other ecosystem services suitable for sky trust–type policies. Cash payments, in contrast, entitle the recipients only to market goods. In effect, this policy would channel the receipts from rationing public goods mainly toward the purchase of private goods. When people spend the money they receive from the trust fund, it will stimulate the consumption of other natural resources and the creation of waste.

We believe another option should be considered—that of using the trust to fund much-needed expenditures on public goods. However, under the current system, with hundreds of billions of dollars spent annually to convince people that the consumption of market goods is the only path to happiness, accompanied by a pervasive public distrust of government, direct cash disbursements might be more politically feasible than spending the trust on public goods, or even on tax reductions. If direct cash disbursements are necessary to make the sky trust politically feasible, the approach has enough to recommend it that it is still well worth pursuing. As people in the future come to better understand and value the importance of nonmarket goods, perhaps the trust could be turned over to the creation of public goods.

Land Tax. Land is another part of the commonwealth, an asset provided by nature that originally belonged to all citizens of a nation. One can do as one likes with many assets, but nations almost always try to retain sovereignty over their territory. Yet land ownership in most countries is highly concentrated, as are the returns to ownership.[19] In addition, as we pointed out in Chapter 11, the value of Ricardian land is almost entirely the result of positive externalities—land is more valuable as a result of proximity to others. In other words, land values are created by society, not by the landowner. Land supply is also fixed. With extremely limited exceptions, no matter how high the price, more land will not be created, and no matter how low the price, the same quantity of land will exist. Therefore, all returns to land are economic rent, payment above the minimum

---

[18]P. Barnes, *Who Owns the Sky? Our Common Assets and the Future of Capitalism,* Washington, DC: Island Press, 2001.

[19]In the U.S., the richest 10% own 60–65% of land by value, and in Brazil, the richest 1% own 50% of the countryside. Roodman, op. cit.

necessary supply price. With supply fixed and demand increasing, both the price of land and the rent on that land will increase, leading to increasing concentration of wealth and income.

A line of thinkers following economist Henry George argue that as society creates the value in land, society should share in the returns to land. While redistributing land itself would be a difficult, disruptive, and politically infeasible policy in many countries, it is much simpler to simply redistribute the rent via land taxes. In the extreme, some proponents of this approach call for a single tax on land, though many interpret this as a tax on all resources that are a gift from nature. In most countries, property taxes fall on both land and the infrastructure on it, and these are two very different types of resources. We should tax that to which value is added, and not the value added. Such a policy has many desirable features.

In terms of distribution, a higher tax on land will drive down the value of land because it drives up the cost of owning it. Theoretically, the price of land should equal the net present value of all future income streams from that land. As a higher tax reduces the income stream, it reduces the price. A land tax also makes land speculation much less profitable. It simply becomes too expensive to pay taxes every year while waiting for land prices to rise. Removing the speculation demand for land reduces land prices even further. Lower prices make land and home ownership more broadly accessible, especially if higher taxes on land are accompanied by reducing or eliminating taxes on buildings on that land. And the entire tax on land will be paid by the landowner—it will not be shifted onto renters because the supply of land is perfectly inelastic. This point is explained in greater detail in Figure 23.1. It is also worth noting that speculative bubbles are a source of economic instability, so a land tax can help stabilize an economy. George argued that almost all business cycles were driven by land speculation.

THINK ABOUT IT!

Can you explain why eliminating profits from speculation will reduce the cost of land? Think in terms of supply and demand.

Land taxes can also help reduce urban sprawl and all of its negative impacts. Those who own high-value land will have greater incentive to either invest in its productive aspects or sell it to someone who will invest. The higher the land value, the higher the pressure to invest or sell. The net result is not necessarily more investment but rather more investment on the most valuable land. Land values are higher the denser the population. This means that land in cities will be more intensively developed, reducing the pressures for urban sprawl.

A land tax accompanied by eliminating the property tax on buildings would reduce the cost of supplying buildings, and more buildings would be

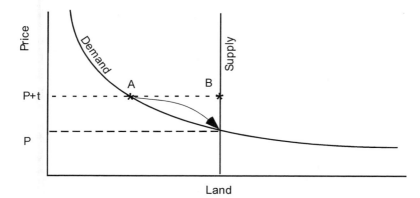

Figure 23.1 • Who pays the land tax? The supply of land is perfectly inelastic with respect to price. This means the same amount will be supplied at any positive price. Demand, in contrast, is sensitive to price, and there is one market price at which supply and demand clear. If land is taxed by amount t, and the land seller tries to pass the tax on to the buyer or renter, raising the price temporarily to P + t, the market will not clear; amount AB will remain unsold. Landowners who were unable to sell or rent at the higher price will therefore be forced to lower their price, driving the price back down to P. This means the government can tax away the entire rent on land, without reducing supply.

built. This approach would improve distribution, because poor people often spend the majority of their income on rent, and a greater supply of buildings would mean lower rent. However, to achieve desirable distribution and scale effects, the land tax (or cap and rent schemes) should extend to all free goods from nature. In this case, new construction would have a higher tax burden (or tradable permit costs) from raw materials than restoration, thereby shifting investments into restoration. Currently, restoring older buildings increases their value and hence the tax burden, and the tax deters the activity. Greater investments in restoration relative to new construction will further favor urban centers over new suburban construction.

In most cases, governments provide the infrastructure for suburban developments, which is basically just a subsidy for the people who move to the suburbs. Cities that have pursued high land taxes accompanied by low or zero taxes on infrastructure on that land include Melbourne, Pittsburgh, Harrisburg, and other cities, and it is well established that the taxes have indeed limited sprawl and led to urban renewal. Land values in rural areas are often a small fraction of those in urban settings, so land taxes may have little impact on land use. Nonetheless, land left in its natural state is already providing ecosystem services and often should not be taxed.

THINK ABOUT IT!

Can you explain why eliminating the tax on buildings will increase the supply and reduce rents?

■ ADDITIONAL POLICIES

Two other distribution policies merit a brief mention. First, if governments recaptured the sole right to seigniorage (the right to print money), they could use that money to improve distribution. Second, we can look at demand-side policies. Where do poor people spend most of their income? On rent and health care.[20] We have already seen how shifting taxes from buildings to land could drive down the cost of housing.

Universal, government-sponsored health care, as most citizens of developed countries already enjoy, would also dramatically increase real income for the poor. It would be likely to decrease national expenditures on health care as well, as the U.S. currently spends a higher percentage of its national income on health care than any other country, and still fails to insure over 40 million Americans. Though it is a highly contentious topic, health care is probably as poorly suited to market provision as ecological services.

BIG IDEAS to remember	
■ Minimum income	■ Rent
■ Caps on income and wealth	■ Public subsidies
■ Wealth, status, and power	■ Alaska Permanent Fund
■ ESOPs and CSOPs	■ Land tax

---

[20]G. Flomenhoft, "The Triumph of Pareto: Does Equity Matter?" Unpublished working paper, presented at the U.S. Society for Ecological Economics 2003 conference, Saratoga Springs, May 23, 2003.

CHAPTER

24

Efficient Allocation

Allocation receives the lion's share of attention in conventional economic thinking, but it plays a tertiary role in the ecological economic
approach to policy. This is not to deny its importance. Efficient or at least
cost-effective allocation is, in fact, a vital component of good policy, and
we have seen how the various policies discussed so far effectively allocate
resources. In this chapter, we turn to some of the "big picture" issues in
allocation.

1. We address the myth that if mechanisms can be developed for internalizing all external costs and valuing all nonmarket goods and
   services, the market alone will lead to efficient allocation.

2. We examine the asymmetric information flows that shape our preferences and how these influence resource allocation. Allocating resources efficiently toward goods that do little to increase our
   well-being is not efficient.

3. We return to policies aimed at macro-allocation, the allocation of resources between private and public goods.

4. We examine some problems confronting the allocation of resources
   under local control and national sovereignty that supply global public goods.

5. We propose an expanded definition of efficiency more compatible
   with the goals of ecological economics.

■ PRICING AND VALUING NONMARKET GOODS AND SERVICES

Recall from Chapter 8 that markets lead to efficient allocation of market
goods by using the price mechanism as a fulcrum on which to balance

supply and demand. Many economists argue that if we could determine monetary values for nonmarket goods, we could then use the market mechanism to efficiently allocate them. As a result, one of the most active research areas in environmental economics is the calculation of "prices" for nonmarket goods. Once prices are established, we need a mechanism to internalize these values into the market system. While it is critically important to establish the value of nonmarket goods and services, it is controversial whether establishing monetary values, which are identical to exchange values, is appropriate or meaningful. Internalizing those values is, in any event, not a simple task.

We already discussed in Chapter 21 the problem of circularity in using prices to determine the optimal scale when those prices were already based on the initial scale. A similar problem arises when we try to put market values on nonmarket ecosystem services, an exercise that is sometimes thought necessary to determine the costs and benefits of changes in scale. However, even if we could overcome that problem, we would still confront a number of serious difficulties.

Recalculating Marginal Values

You will recall the resolution of the diamonds-water paradox from Box 15.1. Water is far more valuable than diamonds, yet it is available at a very low price. Price is exchange value, or the marginal use value of the good or service in question. Use value is the total value, or the value of all units together. The use value of water is infinite, yet where this resource is very abundant, the value of an additional unit approaches zero. However, when water is extremely scarce, an additional unit may mean the difference between life and death, so its marginal value also becomes immeasurably large. The same is true for any essential good or service, such as the life-support functions of ecosystems: When an essential resource is scarce, the marginal value is extremely high, and it increases rapidly with growing scarcity (see the section in Chapter 9 on inelastic demand).

Around 150 years ago, many ecosystem goods and services were so abundant that an extra unit had no appreciable value. As a result, the economic system ignored the value of such goods. Over time, however, these goods and services have become increasingly scarce and their marginal values have soared, which is why economists now attempt to calculate their values. As we approach ecological thresholds, which we may already be doing, the marginal value and hence "price" of these goods and services will increase extremely rapidly.

To internalize these ecosystem values, we would need to continuously recalculate them, centralize the information, then feed it back into the market mechanism via taxes or subsidies. Yet calculating the value of such resources is very expensive, and centralizing the knowledge and feeding it

Box 24-1 PRICING NATURE AND LIFE

In response to a legal demand to assess the value of claims in lawsuits, economists have designed a number of ad hoc methods of estimation. Probably this began with "statistical value of a life" calculations to assess damages in cases involving legal liability for accidental death. The usual procedure is to calculate the present value of expected future lifetime earnings of the deceased. Most people would not give up their life voluntarily for such an amount, primarily because they would no longer be around to receive the payment, but even if the payment went to their heirs, there would be few volunteers. Clearly this sum is in no sense the value of a life. But as a practical procedure for settling legal cases involving accidental death, it is not unreasonable—as long as we always remember that we are valuing human capital as an object, not human beings as subjects. But the question remains: Should compensation be for lost human capital or the lost life of a person?

Question: If you contributed to a fund for the survivors of the victims of September 11, would you prefer that compensation be made according to the differing human capital values of the victims or an equal amount for each victim?

Given this precedent of valuing human capital, it was a short step to price the loss of ecosystem services and natural capital caused by accidental oil spills such as the *Exxon Valdez*, industrial accidents, and so on. As a practical way of making after-the-fact compensation for accidents, this is not unreasonable. Problems arise when we move from after the fact to before the fact, from adjustments to an involuntary occurrence to an imagined voluntary tradeoff. This is what **contingent valuation** estimates do. They present people with imaginary alternatives and ask them, before the fact, and indeed totally hypothetically, how much they would be willing to pay, say, to save 100 grizzly bears, or how much they would accept for the loss of 100 grizzly bears. Interestingly, the two questions usually yield very different answers, although in theory they should be the same for a market transaction. Alternatively, the grizzlies could be valued as a public good. Citizens could be asked how much they would be willing to be taxed along with everyone else to save 100 grizzlies. This makes more sense, but it is still very artificial.

There are other methods besides contingent valuation for estimating hypothetical prices for nontraded goods. Environmental economics texts discuss them in detail. Why we have given little attention to this subject is, we trust, evident from our discussions of the separateness of allocation and scale and the circularity of deciding scale questions by allocative prices.

back into the pricing mechanism would require an enormous and expensive bureaucracy. The paradox is that we love markets precisely because they constantly and almost costlessly recalculate prices on the basis of decentralized information with minimal government intervention. However, this approach to allocation would be expensive and centralized and require large-scale government intervention.

Uncertainty, Ignorance, and Unfamiliarity

In addition, methods for valuing nonmarket goods are fraught with problems. Most rely on artificially constructed markets or ways of inferring nonmarket values through existing markets. Two problems in particular merit discussion: our lack of knowledge of ecosystem function and our lack of familiarity with valuing nonmarket goods.

As an example, the contingent valuation method constructs a hypothetical market basically by asking people what they would be willing to pay for a given nonmarketed good or service. One problem, as we have repeatedly discussed, is that even the experts are ignorant of all the goods and services healthy ecosystems provide, how they provide them, the impacts of human activities on their provision, where critical ecological thresholds lie, and the outcomes when these thresholds are passed.

If we emit a given stream of pollutants into a lake, what will the impact be? What ecosystem services will be lost? Will the waste flow accumulate, causing worse damage over time—perhaps irreversible loss? Will the loss of the system being polluted affect other systems? What is the time scale involved? Even if it were possible for the experts to resolve all these uncertainties (which it is not) and disseminate that information to the population at large, people have no experience with markets in such goods and services and would still have a very difficult time assigning meaningful exchange values.

Time, Distribution, and Valuation

Yet another problem is the time factor. Most ecosystem goods and services are renewable and therefore will provide benefits into the indefinite future. A typical decision is whether to sacrifice a renewable flow from a natural fund-service for a nonrenewable (manmade) fund-service or for a one-time liquidation of stock. This demands that we compare present values with future values. As we discussed in Chapter 16, economists generally do this by discounting future values. The discount rate will typically be one of the most important variables in determining value, and there is no agreed upon objective rule for determining an appropriate rate.

We must also recognize that the question of what should be left to the future is inherently an ethical decision concerning intergenerational *dis–*

| Box 24-2 | METHODS FOR MONETARY VALUATIONS OF ECOSYSTEMS |

Several methods are available for putting dollar values on the nonmarketed goods and services provided by ecosystems. Many of these are appropriate for valuing only a small subset of services. Most textbooks in environmental economics provide an adequate introduction to these methods. We recommend as a good starting point the Web site "Ecosystem Valuation" at http://www.ecosystemvaluation.org, where the following methods are listed:

- *Market Price Method:* Estimates economic values for ecosystem products or services that are bought and sold in commercial markets.

- *Productivity Method:* Estimates economic values for ecosystem products or services that contribute to the production of commercially marketed goods.

- *Hedonic Pricing Method:* Estimates economic values for ecosystem or environmental services that directly affect market prices of some other good.

- *Travel Cost Method:* Estimates economic values associated with ecosystems or sites that are used for recreation. Assumes that the value of a site is reflected in how much people are willing to pay to travel to visit the site.

- *Damage Cost Avoided, Replacement Cost, and Substitute Cost Methods:* Estimate economic values based on the costs of avoided damages resulting from lost ecosystem services, the costs of replacing ecosystem services, or the costs of providing substitute services.

- *Contingent Valuation Method:* Estimates economic values for virtually any ecosystem or environmental service. The most widely used method for estimating nonuse or "passive use" values. Asks people to directly state their willingness to pay for specific environmental services, based on a hypothetical scenario.

- *Contingent Choice Method:* Estimates economic values for virtually any ecosystem or environmental service. Based on asking people to make tradeoffs between sets of ecosystem or environmental services or characteristics. Does not directly ask for willingness to pay; this is inferred from tradeoffs that include cost as an attribute.

- *Benefit Transfer Method:* Estimates economic values by transferring existing benefit estimates from studies already completed for another location or issue.

tribution. Conventional economists argue that the question is not one of distribution but rather one of efficient allocation. If a resource will be sufficiently more valuable in the future than in the present, it should be saved for the future. Therefore, maximizing the net present value (NPV) of resource use will lead to the optimal allocation. However, NPV is the value of present and future resources to this generation. If you recall our discussion of property rights from Chapter 21, this corresponds to a property rule assigning property rights to the current generation, which is free to interfere with the future's access to resources. Under this approach, all that matters is the value of resources to the present generation.

As an alternative, we could assign some resource property rights to future generations. For example, assigning rights via a liability rule would leave this generation free to use resources as long as it compensated the future with an equivalent amount of other resources. Under an inalienability rule, the future would be entitled to a certain share of resources, and the present would be obliged to leave them. These three rules are simply different initial distributions of resources, and each would lead to a different set of prices for both market and nonmarket resources. Which rule to use is an ethical decision, not a matter of allocative efficiency.

THINK ABOUT IT!

Most valuation methods attempt to estimate demand curves for non-market goods. Demand is nothing more than preferences weighted by purchasing power, and decisions based on monetary valuation tend to underestimate the preferences of the poor. Do you think that decisions concerning natural capital should be based on monetary valuation or on some other procedure?

Market vs. Nonmarket Values

Assuming that all these other issues are addressed, would monetary valuation then lead to efficient allocation? Price provides a feedback mechanism used by the market to maximize profit, which economists assume creates the appropriate conditions for maximizing human well-being. Is a single feedback mechanism sufficient for allocating all the resources that contribute to human well-being? Could natural ecosystems in all their incomprehensible complexity function with only one feedback mechanism? Some ecologists might argue that ecological systems do indeed function this way, where maximizing energy consumption is the ultimate feedback system.

Although energy consumption is a useful simplification in some ecological models, it is not the only driving force in nature. And even if it were, as a principle of competitive exclusion, it does not translate well to the human economy, in which maximizing a cost is surely uneconomic. Moreover, human ethical beliefs make the interaction between the human economy and the ecosystem more complex than the functioning of the ecosystem alone. We therefore cannot support the reductionist approach of assuming that the profit motive alone is sufficient to maximize human well-being, much less to guide us in any quest toward an ultimate end.

As a concrete example of the problems with monetary valuation of everything, it would be quite simple to develop a method for calculating a dollar value for democracy. Certainly people in general have a better understanding of democracy than of ecosystem services, and we could readily devise a survey that would tell us how much a voter would be willing to pay for the right to vote (or alternatively the minimum amount for which a voter would sell her vote). We could do the same for human rights, and many people consider the right to live in a healthy environment such a right. But most people would probably agree that politics and human rights are in a different moral sphere than economics, and power in the sphere of economics should not translate to power in these other spheres.[1] (While this, of course, does happen, people do not generally consider it desirable.) Political rights, human rights, and other ethical values are not individual values, but social values. Attempting to estimate social values by aggregating individual tastes suffers from the fallacy of composition and is a categorical mistake.

> **THINK ABOUT IT!**
> *Would you sell your right to vote? If there were such a market for votes, would you expect the price to be high or very low?*

It is true that we are constantly forced to make decisions between mutually exclusive alternatives, such as more forests or more strip malls, which require a comparison between market and nonmarket values. However, many nonmarket goods are fundamentally different from market goods in ways that make "scientific" comparison not only impossible but also undesirable. Putting dollar values on everything does not make the necessary decisions more objective; it simply obscures the ethical decisions needed to make those "objective" valuations.

Most textbooks in environmental economics devote considerable space to discussing methods for valuing nonmarket goods and services. Valuing ecosystems can play an important role in capturing the attention of the public and policy makers and can offer insights into appropriate economic policies. But attempting to calculate an exchange value for all nonmarket goods, then use that value to decide what we will preserve and what we will destroy, is an example of economic imperialism, as discussed in Chapter 3.

Ecological economics takes the broader perspective that such methodologies are inadequate to capture the range of human values and physical

---

[1]M. Walzer, *Spheres of Justice*, New York: Basic Books, 1990.

needs we have for nonmarket goods. Instead of spending time trying to calculate the "correct" price for nonmarket goods, ecological economics stresses that we should act on our knowledge that zero is the incorrect price and spend our time trying to improve upon and implement policies that recognize they have significant, often infinite value, even if we cannot precisely quantify it.

■ MACRO-ALLOCATION

As we discussed in Chapter 17, macro-allocation is the problem of how to allocate resources between the provision of market and nonmarket goods. The government plays an important role in providing nonmarket goods and can also influence demand for market goods through the use of taxes and subsidies. Presumably, in democratic countries, citizens will elect politicians who will make the right choice regarding macro-allocation. One serious problem with this assumption is the discouraging lack of information people have regarding nonmarket goods and services. For people to make appropriate choices, they need appropriate amounts of information. In this section, we will first look at policies addressing unequal information flows concerning the attributes of market and nonmarket goods, then examine possibilities for the government to provide incentives to the private sector for providing public goods.

Asymmetric Information Flows[2]

Asymmetric information is present when the buyer or seller has information that the other does not have, and that information affects the value of the good or service exchanged. Economists have long known that asymmetric information is a market failure, generating serious inefficiencies. For example, if I am selling a car, I know how well it works, but the potential buyer does not. The buyer will adjust the price she is willing to pay based on the risk of purchasing a lemon, and this risk-adjusted price will be less than the value of a good used car. The rational seller will not be willing to sell a good car at the risk-adjusted price, and the market will provide only lemons (at least according to theory). Ackerloff, Spence, and Stiglitz shared the Nobel Prize in economics for such basic insights.

We arguably face a much more serious problem with the asymmetry of information flows that form our preferences. While many economists argue that preferences are innate, businesses are betting an estimated

---

[2]Much of this discussion is adapted from J. Farley, R. Costanza, P. Templet, et al., Synthesis Paper: Quality of Life and the Distribution of Wealth and Resources. In R. Costanza and S. E. Jørgensen, eds., *Understanding and Solving Environmental Problems in the 21st Century: Toward a New, Integrated Hard Problem Science*, Amsterdam: Elsevier, 2002.

$652 billion per year that preferences are heavily influenced by advertising.[3] Advertising costs money, and it can pay for itself only by advertising market goods. Most words we hear today are direct sales pitches for market goods and the programs sponsored by them.[4] In stark contrast to advertising for market goods, very little money is spent convincing people to prefer nonmarket goods. To the extent that advertising alters preferences, it systematically does so in favor of market goods over nonmarket goods.

People have a finite amount of resources to allocate. If advertising convinces us as a society to allocate more resources toward market goods, correspondingly fewer are available to allocate toward nonmarket goods. And as we know, all resources allocated toward consumer goods are extracted from nature and return to nature as waste, destroying public good ecosystem services in the process. Seen in this light, advertising convinces us to degrade or destroy public goods for private gain. It appears that current levels of consumption in the developed countries are incompatible with a sustainable future, yet reducing consumption levels will be exceedingly difficult in the presence of so much advertising.

Nor is this the only market failure associated with advertising. Arguably, human welfare is determined by our ability to satisfy our needs and wants. Advertising creates wants by making us believe we need some product or another, yet it gives us no greater ability to satisfy those wants. In this sense, advertising directly diminishes our welfare. We can make this point no better than B. Earl Puckett, former head of Allied Stores Corporation, who once declared that "it is our job to make women unhappy with what they have."[5] In this line of thinking, advertising is basically a "public bad."

The problem is one of providing symmetric information flows for nonmarket goods. This is a very contentious issue. We briefly present several possibilities here for discussion but welcome new and better ideas for addressing the market failures associated with advertising.

The first involves the recognition that advertising over the airwaves in many countries is subsidized. The airwaves are valuable public property but are often given free of charge or at low cost to communication corporations. Since transmissions beamed over airwaves have properties of public goods in that they are nonexcludable (at least when the transmissions are not scrambled) and nonrival, there is a solid rationale for giving

---

[3]International Advertising Association, 2000. To place this figure in context, only seven countries in the world had a GNP higher than $600 billion in 1997.

[4]A. T. Durning, *How Much Is Enough? The Consumer Society and the Fate of the Earth*, New York: Norton, 1992.

[5]Ibid., pp. 119–120.

away airwaves to those who will beam such transmissions, in spite of the fact that the airwaves themselves are rival, excludable, and scarce. However, if the government charged corporations for the use of airwaves for advertising, it would target only the portion of the airwaves devoted to private profit. Advertising is currently considered a business cost and is tax-exempt. For the reasons listed above, however, it would be more appropriate to tax advertising as a public bad. At a minimum we should not allow advertising to be written off as a cost of production. The rationale is that production is supposed to meet existing demand, not create new demands for whatever happens to be produced.

While taxes would presumably reduce the quantity of ads for market goods, it would not help to generate concern for nonmarket satisfiers of human needs. There are several alternatives for helping to achieve this goal. One approach would be a law mandating "full disclosure" advertising. Just as medicines are labeled with all their potential adverse side effects, advertisements could list all the potential adverse side effects of the products they advertise. This would include, of course, all the negative impacts on the environment and their implications. Another alternative would be to provide free airtime for public service announcements that specifically seek to create demand for environmental services and other nonconsumptive satisfiers of human needs. The media is a phenomenally powerful tool for persuasion, and thus an effective approach to policy would be to make the information flows it provides more symmetric.

A problem with both of these restrictions on advertising is that people will complain that they infringe on the basic right of free speech. However, the right to free speech does have restrictions; it does not include the right to lie or misrepresent. Nor does it include the right to amplification by a powerful megaphone. For example, no one is allowed to shout "fire!" in a crowded theater if there is no fire, because it threatens the well-being of others. Shouting "fire!" may not be fundamentally different from encouraging people to consume when such consumption threatens the well-being of future generations. Many nations already curb advertising on alcohol and tobacco, and the Australian Consumers Association is currently attacking the right to advertise unhealthy foods on children's TV shows.[6] The same rationale also applies to curbing advertising that indirectly encourages destruction of the environment.

Subsidies for Nonmarket Goods

Even if people are well educated about the benefits of nonmarket goods and subsequently elect governments willing to provide them, there remains the problem of how best to do so. Often the best strategy will be

---

[6]Ibid.

for the government to simply supply them outright, or directly pay private sector contractors to do so. In many circumstances, though, the nonmarket goods are positive externalities from the production of market or private goods. For example, when farmers terrace their land, use contour plowing, and retain buffer zones along streams, they may dramatically improve downstream water quality, thereby maintaining the productive capacity of their land for future generations. The problem is that the private sector will supply less of the positive externality than is socially desirable. When the positive externality is in the form of a public good, the best approach may be for the government to subsidize the portion of the private activity that generates the public good.

Several types of subsidies are possible. A direct subsidy can simply compensate the private sector for its provision of the positive externality,[7] at which point the externality is partially internalized and presumably supplied in more adequate amounts. Alternatively, tax relief can be used to subsidize positive externalities. Possibilities range from a decrease in land taxes for farmers who reduce erosion to tax breaks for business investments in training personnel (people these days regularly change jobs, and firms may offer less training than is socially optimal if the worker will move her newly acquired productivity to another firm). As another option, a subsidy can be in the form of subsidized credit. If people underinvest in activities with positive externalities, lower interest payments would stimulate greater investment.

Using Seigniorage

Where would governments find the money for subsidized interest rates? Again we suggest the option of restoring the sole right to seigniorage to the government, as discussed in Chapter 15. When banks create money, they do so through interest-bearing loans. On average, such loans must generate financial returns that can repay the loan plus interest, which means money is loaned for the production (and consumption) of market goods. Unless the economy is growing, it becomes impossible to pay back all loans with positive real interest rates. Governments, in contrast, could use their power to create money to make interest-free loans or even outright grants to activities that best promote the common good, including the provision of nonmarket goods. Not only would this help in the macroallocation of resources toward nonmarket goods, it could lead to a financial system whose viability is not based on unending growth.

---

[7]In most cases, it is difficult to know the exact value of the nonmarket good and hence the optimal level of compensation. We do know that zero is the wrong value, and a reasonable compensation will improve allocation.

■ SPATIAL ASPECTS OF NONMARKET GOODS

In previous chapters we discussed the spatial characteristics of nonmarket goods. Most ecosystems provide services at the local level, the regional level, and the global level. For example, a forest can affect climate stability at each of these levels. Yet the principle of subsidiarity requires that the domain of the policy-making unit be congruent with the domain of the causes and effects of the problem with which the policy deals. The causes of ecosystem degradation are often at the local level; effects are felt at local, regional, and global levels; and policy-making institutions are primarily local and national. This poses serious problems for effective policy.

To make the problem more concrete, we will use the specific example of clearing tropical forests for farmland on the Atherton Tablelands of Australia. On private lands, the decision is local. The net private marginal benefits to farmers of clearing forests for agriculture decreases with area cleared. The first units cleared meet basic needs and are on the very best lands. Additional units cleared meet less important needs and use less adequate lands (e.g., steeper slopes, less fertile soils, and greater distances). Eventually, farmers unaware of the ecosystem services provided by forests cleared to the point that water quality was affected, and shade cover was inadequate, leading to reduced yields—that is, a negative marginal net private benefit (MNPB) to deforestation.

This scenario is depicted in Figure 24.1 by the curve MNPB. With the right to do as they pleased on their land, the rational, well-informed farmers should have cleared the forests until marginal benefits were zero, at point A. Instead, due to ignorance, they cleared to point B. It was not just the farmers who were ignorant; at the time this was occurring, few people knew of the negative impacts caused by deforestation.

Towns downstream of deforested farmlands suffer from irregular water flow and poor quality. The nature tourism industry in the region generates far more income than farming, and it also suffers from deforestation. These local marginal external costs of deforestation (MEC local) are also depicted on the graph. If the local governments had been aware of these negative externalities, they might have implemented policies designed to limit deforestation to point C, perhaps by imposing a local deforestation tax equal to OC′ or issuing tradeable deforestation permits in the quantity OC.

National MEC of deforestation includes local MEC. In addition, deforestation on the tablelands causes erosion, siltation, and nutrient runoff, all of which flows out to sea to be deposited on the coral reefs. This affects fisheries and tourism outside the shire boundaries. The state or national government should have implemented policies to limit deforestation to

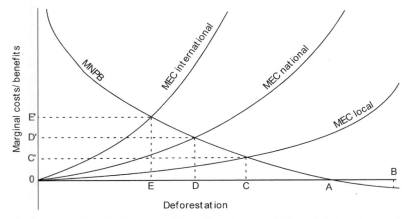

Figure 24.1 • Marginal costs and benefits of deforestation at different spatial levels. Curve MNPB shows the marginal net private benefits of deforestation to farmers on the Atherton Tablelands. This curve accounts for the cost of lost ecosystem services that directly benefit the farmer. Through ignorance, farmers initially cleared amount OB of forest, but had they been better informed, they would have cleared amount OA. Marginal external costs of deforestation are shown for local, national, and international society, along with the corresponding optimal levels of deforestation: C, D, and E, respectively.

point D, perhaps through a national deforestation tax equal to C'D', or by purchasing DC of the OC local level quotas and discarding them.

Global MEC similarly includes national MEC, in addition to deforestation's contribution to global climate change and biodiversity loss. Point E would therefore be the globally optimal level of deforestation, but there is of course no global authority that overrides national sovereignty on issues such as deforestation, so taxation would not be a policy option. Nor is there a global coalition that could purchase deforestation permits (in the event they happened to be internationally tradeable). While individual governments, multilateral institutions, and international NGOs currently play some role in preserving ecosystems and reducing the rate of ecosystem degradation in some countries, it is almost certainly insufficient to achieve globally optimal levels of ecosystem preservation. The insufficiency of these efforts may result from a combination of ignorance and the free-rider effect.

The spatial distribution of ecosystem goods and services thus presents a serious problem. Control over ecosystems is usually at the level of the individual, who retains all the benefits of deforestation (e.g., timber sales and subsequent use of farmland) and shares the costs of lost ecosystem services with society. Where effective institutions and information exist, local and national governments can impose regulations that optimize ecosystem preservation from their points of view. Alternatively, civil society in the form of volunteer community organizations and NGOs can step in when governments fail to act. However, they can rely only on good will, altruism, or other behaviors inconsistent with the assumptions of rational self-interest but thankfully important parts of the human psyche nonetheless. When governments do act, they are unlikely to pursue globally desirable levels of ecosystem preservation, as there are few incentives to do so, and national sovereignty allows them to ignore global benefits. Making optimal outcomes even more elusive, policy makers at all levels are probably unaware of the full range of benefits intact ecosystems actually supply.

While continued conversion of tropical forests may be uneconomic at the global level, in many of the world's remaining tropical forests (and other healthy ecosystems), it may still make economic sense at the local level to continue the deforestation process. If we look at the Amazon rainforest, much of Southeast Asia, and Central Africa, there are still vast tracts of largely untouched forest along with high levels of poverty and landlessness. There is little question that for the individual, the best available alternative under current circumstances is often to clear forest and grow crops. The farmers who do this are not irrational. They may be ignorant of the ecosystem services their activities destroy (though probably considerably less ignorant than city dwellers), but even if they were aware, their personal gains from deforestation almost certainly far outweigh their share of the loss from ecosystem services destroyed.

Much of the forest in these countries occurs in states or administrative districts that are almost entirely forested and have very low population densities. It may thus also make sense at the local and regional level to continue deforesting. There is an important caveat, however. Even if continued deforestation may be appropriate at the local level, in many areas it is carried out in a destructive and inefficient manner. For example, where market access is poor, valuable timber and trees providing important nontimber resources (such as Brazil nuts) may be felled and burnt. Unsustainable production techniques that mine the soil are common, even when small investments could yield far more sustainable and lucrative alternatives. Thus, while deforestation may be appropriate from the local point of view, the way in which it is carried out may at the same time be highly inappropriate.

Even at the national level in countries like Suriname, Guyana, and French Guyana that are up to 90% forested, continued deforestation, if carried out without wanton waste, could help improve well-being for the majority of society (though the indigenous cultures that depend on the forest would almost certainly not be part of that majority). It can also strengthen claims to contested boundaries. In other countries where it may be in the national interest to slow or halt deforestation, it may still not be worth the investment of the resources that would be needed to do so. Vast areas of intact forest make monitoring and enforcement of regulations intended to curb deforestation expensive and difficult. In fact, in many of these countries, large areas of forests are essentially open access resources. For example, an estimated 80% of timber harvests in Brazil are illegal.[8]

Brazil provides an interesting case study of a country in which the national government over the past two decades has moved from policies explicitly and vigorously promoting deforestation toward considerable legislation designed to reduce deforestation. Much of the national legislation designed to preserve forests is poorly enforced, and other national policies will almost certainly dramatically increase deforestation in some areas, but there is nonetheless a real trend toward greater efforts for preservation at the national level. This trend presumably reflects a growing knowledge of ecosystem services provided by the forest, growing marginal utility of remaining forests, and international pressure. Some of the more heavily deforested states have implemented innovative policies for reducing deforestation, while heavily forested states still promote it. In the Atlantic Forest, of which only some 7% remains, some individual landowners are working to protect their remaining forests or are even actively restoring the forest. What are conspicuously absent are adequate resources from the global community to preserve the forest.

International Policies

Problems of global scale must ultimately be solved via global policies. What policies are available to the global community to limit deforestation and other forms of ecosystem destruction to a globally acceptable level? We previously discussed the polluter-pays principle, but when it comes to deforestation and the destruction of coral reefs, wetlands, and other ecosystems (or greenhouse gas emissions), sovereign nations will not pay for the impacts their activities have on the rest of the world. Although all nations in the world benefit from healthy ecosystems in other countries, they do little or nothing to help pay for their preservation. Ecosystem services are global public goods, and most countries are free riders on the provision of those public goods.

One solution is the application of a "beneficiary-pays principle," where those who benefit pay for the benefits they receive. The fact is that less developed countries, which may now contain the bulk of the world's most productive ecosystems, often lack the institutions and resources to limit ecosystem conversion to nationally desirable levels. An effective policy for

---

[8]C. Bright and A. Mattoon, The Restoration of a Hotspot Begins, *World Watch Magazine*, November/December 2001.

preserving ecosystems at a globally desirable level must provide both incentives and resources for doing so.

International Payments for Ecosystem Services. One possibility for implementing a market-based beneficiary-pays principle would be an international Pigouvian subsidy for providing ecosystem services. Take the Brazilian Amazon as a case study. Ideally, to reach point E on Figure 23.1, the global community would need to pay only quantity D′–E′ to get from the nationally optimal to the globally optimal level. The first problem is the transaction cost of getting the wealthy nations to agree to pay Brazil to reduce deforestation and deciding how much each must pay. A potentially larger problem is deciding to whom the subsidy should be paid.

The Amazon is vast, and while deforestation is rapid, it still affects on average less than 1% of the region per year. It would not be efficient to pay all landowners not to deforest, as not all of them plan to deforest in the first place. One possibility would simply be to pay only farmers who are currently deforesting not to do so, but this presents numerous difficulties. Transaction costs of reaching agreements with individual farmers would be enormous, and the monitoring and enforcement costs necessary to ensure landowner compliance would also be substantial. If the farmer agrees not to deforest one area, he may simply deforest another area instead.

Asymmetric information presents another serious problem, as only the landowners themselves know how much they plan to deforest. If payments were made only to farmers in the process of deforesting, other farmers might begin to deforest simply to become entitled to payments. Subsidies directed to the landowners actually involved in deforestation could thus perversely increase the rate of deforestation.

A plausible alternative for a subsidy would be for the international community to pay Brazil for reducing the rate of deforestation below some predetermined baseline. The baseline might be average deforestation over the past several years, or expected deforestation estimated through a more sophisticated model, including variables such as rainfall and economic growth. For example, the most recent available 3-year average of deforestation rates in the Amazon is about 2 million ha per year (average of 1995–1997).[9] The international community could pay a given amount for every hectare deforestation falls short of 2 million. If too much deforestation still occurs, the subsidy can be raised, while if the subsidy is too costly, it can be decreased. Deforesting the full 2 million ha would forgo

---

[9]Instituto Nacional de Pesquisas Espaciais (INPE), Desflorestamento 1995–1997 Amazonia, 1998. Online: http://www.inpe.br/amz-00.htm.

all subsidies, but there still would be no incentive to deforest beyond this point because there is no incentive to do so in the absence of the subsidy.

Another alternative would be to adopt a strategy currently in use in Brazil known as the ICMS *ecológico*. The ICMS is a tax on merchandise and services, and in certain states, some of this money is refunded to municipalities according to the extent to which they meet ecological goals, such as watershed protection and forest conservation. Essentially a payment for the provision of ecological services, it has proven quite effective.

There is no reason a similar approach could not be used at a global level. Initially perhaps, it could be directed toward biodiversity hotspots, 25 areas round the world identified by scientists that contain an unusually large number of species and are seriously threatened, with 70% or more of the area destroyed. Given the importance of biodiversity in maintaining ecosystem resilience and function, it is likely that hotspots offer an unusually large amount of ecosystem services. Similar to the ICMS *ecológico*, a global pool of money could be distributed to the countries that harbor these hotspots according to how well they meet well-defined conservation criteria. It would then be up to the individual countries to determine how best to meet these criteria, thereby allowing micro-freedom to achieve macro-control.

There are several features that contribute toward the feasibility of such international subsidy schemes. First, transaction costs are minimized. Inexpensive satellite photos are capable of providing increasingly accurate estimates of annual deforestation, so monitoring costs would be small.[10] While interpretation of photos may not be an exact science, computer analysis can at least make it a consistent one, in which case quantitative precision is unnecessary. This approach is currently being used to monitor compliance with land use laws in Mato Grosso, Brazil. A subsidy can thus reflect the amount of forest preserved, though an exact dollars-per-hectare figure for the subsidy may not be accurate.

Second, it would not be necessary at the international level to pinpoint who is deforesting. Enforcement and accountability are not major problems, since funds would be disbursed only after conservation occurs; if deforestation is not slowed, no money is spent. Third, national sovereignty would remain intact, as no country would be under any obligation to change behavior. Finally, a major problem in many less developed countries is that they lack the institutions and resources to enforce environmental policy, especially in such vast areas as the Amazon. A subsidy could provide both the incentive and resources for local and national

---

[10]A. Almeida and C. Uhl, Brazil's Rural Land Tax: A Mechanism to Promote Sustainable Land Use in Amazonia, *Land Use and Policy* 12:105–114 (1995).

governments to slow deforestation using some of the previously discussed policies.

A very similar approach would be to pair a national Pigouvian tax with an international subsidy. In this case, the national government would be required to tax deforestation. Such a tax might be costly to implement and administer, both fiscally and politically. While these costs might be less than the revenue gained, they still reduce the incentive to implement such a policy. However, if a global institution matched such a tax or some percentage of such a tax (perhaps greater than 100%), it would effectively combine the polluter-pays and the beneficiary-pays principles in one policy package.

The problem with subsidies for firms to reduce pollution is that it might lead to more firms. If countries decide to grow forests simply to earn a subsidy not to cut them down, that would hardly be a problem.

Although these suggestions make economic sense, they have not been tried at an international level. This is why we call for adaptive management with all policies. When a policy works, use it. When it doesn't, either fix it or replace it with another. What we cannot afford to do is stand by and do nothing when it is obvious that the status quo is not working.

■ REDEFINING EFFICIENCY

In standard economic practice, allocative efficiency is achieved when we put scarce resources to the use that generates most monetary value (which is taken to be a measure of utility). With a central focus on monetary value, allocative efficiency generally ignores nonmarket goods and services. Typically economists defer to Pareto efficiency, which is an allocation such that nobody can be made better off without making someone else worse off. Pareto efficiency does not permit comparisons between individuals, and it accepts the status quo distribution of wealth, however unequal that may be. It ignores the diminishing marginal utility of wealth and the potential for gains from redistribution.

Many economists and policy makers favor potential Pareto efficiency (Hicks-Kaldor welfare criterion, see Chapter 18) as an "objective" decision-making tool that favors any allocations that could potentially create a Pareto improvement after redistribution but that does not require that redistribution. Since wealth clearly generates more wealth in the modern economy, potential Pareto improvements are more likely to benefit the already wealthy than the poor.

Technical efficiency, in contrast, is defined as the maximum amount of physical output one can get from a given amount of resource input. While a desirable goal, it alone does little, if anything, to create a more sustainable society. Greater technical efficiency can reduce the demand for a resource. Alternatively, lowering the quantity of resource needed to make something can lower the cost of the final product. Quite possibly, the result of increased technical efficiency, thanks to lower price, is greater use of resources, not less.

We have already seen that the goal of economics is not to maximize production but rather to provide service. We define service as a psychic flux of satisfaction, which is derived from manmade capital as well as from ecosystem services provided directly by natural capital. Manmade capital can be created only through the transformation of natural capital, so the production of services from manmade capital demands a sacrifice of services from natural capital. We will call this the **comprehensive efficiency identity**. Therefore, an appropriate measure of efficiency is the ratio of services gained from manmade capital stock (MMK) to the services sacrificed from the natural capital stock (NK) as a result. There are several ways to improve this efficiency ratio, as shown in the following identity:

$$\underbrace{\frac{MMK\ services\ gained}{NK\ services\ sacrificed}}_{(1)} = \underbrace{\frac{MMK\ services\ gained}{MMK\ stock}}_{(2)} \times \underbrace{\frac{MMK\ stock}{thruput}}_{(3)} \times \frac{Throughput}{NK\ stock} \times \underbrace{\frac{NK\ stock}{NK\ services\ sacrificed}}_{(4)}$$

Ratio 1 is service efficiency; it is composed of technical design efficiency, allocation efficiency, and distribution efficiency. For example, a well-designed house provides more of the service of shelter than a poorly designed one using the same amount of material; alternatively, glue-laminated beams and laminated I-beam floor joists use less wood to provide the service of structural strength than traditional one-piece solid wood building materials. Allocation efficiency requires that black walnut be used to build fine furniture instead of floor joists. As to distributive efficiency, wood used to provide essential shelter to 50 homeless families provides more service than the same wood used to build a rarely used summer mansion for a billionaire.

Ratio 2 is maintenance efficiency or durability. All MMK stock needs throughput to maintain or replace it, but the less throughput needed, the greater the efficiency. A well-built house lasts longer and needs less maintenance than one sloppily slapped together.

Ratio 3 is growth efficiency of natural capital and harvest efficiency. Well-managed forests and plantations of fast-growing species provide more sustainably harvested timber each year than poorly managed forests or plantations consisting of slowly growing species. For example, studies in the Amazon show that carefully selecting trees to be cut, removing vines on those trees, and carefully planning skid tracks can reduce the time between harvests from 90 to 30 years.

Ratio 4 is increased by creating more natural capital stock or by sacrificing fewer ecosystem services per unit of stock we exploit. Reforestation increases the stock of forest. While a timber plantation may be efficient in terms of growth and timber production, it may provide few other ecosystem services. In contrast, improved management of selective logging of existing forests, as described above, is likely to increase efficiency ratios 3 and 4.

This definition addresses scale by capturing the tradeoff between services gained (numerator) and services lost (denominator), as shown on the left-hand side of the identity. Uneconomic growth invariably reduces efficiency. On the right-hand side of the identity we see the components of overall efficiency—namely, design, distribution, durability, growth, and harvesting.

THINK ABOUT IT!

Analyze the efficiency of burning coal from a strip mine using the comprehensive efficiency identity. Consider each of the four ratios in your answer. (Hint: MMK stock is inventory of mined coal, NK is coal in the ground.)

BIG IDEAS to remember

- Pricing nonmarket goods and services
- Contingent valuation and pricing nature
- Market vs. nonmarket values
- Macro-allocation asymmetric information
- Asymmetric information

- Subsidies for nonmarket goods
- Seigniorage
- International subsidies for ecosystem preservation
- Comprehensive efficiency identity

Looking Ahead

This book has focused on three issues—the allocation of resources, the distribution of income, and the scale of the economy relative to the ecosystem—with special emphasis on the third. A good allocation of resources is efficient; a good distribution of income or wealth is just; a good scale is at least ecologically sustainable.

Allocation and distribution are familiar concepts from standard economics: for any given distribution of income, there is a different optimal allocation of resources with its corresponding optimal set of prices. Standard economics focuses primarily on the allocation issue, paying secondary attention to distribution, first because a given distribution is logically necessary for defining efficient allocation, and second because distributive fairness is important in its own right.

The third issue of scale, the physical size of the economy relative to the containing ecosystem, is not recognized in standard economics and has therefore become the differentiating focus of ecological economics. The preanalytic vision of the economy as an open subsystem of a larger ecosystem that is finite, nongrowing, and materially closed (though open with respect to solar energy) immediately suggests three analytical questions: How large is the economic subsystem relative to the containing ecosystem? How large can it be? How large should it be? Is there an optimal scale beyond which physical growth in the economic subsystem begins to cost more at the margin than it is worth in terms of human welfare? This text has tried to explain the reasons for an affirmative answer to this last question.

If the economy grew into the void, it would encroach on nothing, and its growth would have no opportunity cost. But since the economy in fact grows into and encroaches upon the finite and nongrowing ecosystem, there is an opportunity cost to growth in scale, as well as a benefit. The costs arise from the fact that the physical economy, like an animal, is a "dissipative structure" sustained by a metabolic flow from and back to the environment. This flow, which we have called throughput (adopting the term from engineers) begins with the depletion of low-entropy useful resources from the environment, is followed by the processes of production and consumption (which, despite the connotations of the words, are only physical transformations), and ends with the return of an equal quantity of high-entropy polluting wastes.

Depletion and pollution are costs. Not only does the growing economy encroach spatially and quantitatively on the ecosystem, it also qualitatively degrades the environmental sources and sinks of the metabolic throughput by which it is maintained. This forces a continual co-evolutionary adaptation between the economy and the ecosystem. If that adaptation is made in such a way that the throughput remains within the natural capacity of the ecosystem to absorb wastes and regenerate resources, then the scale of the economy is considered sustainable. Nonrenewable resources, of course, cannot be exploited in a sustained yield manner, but we discussed some rules for "quasi-sustainable" exploitation, along with the analysis of sustained yield exploitation of renewable resources.

From a policy perspective, we have insisted that optimal allocative prices do not guarantee a sustainable scale any more than they guarantee a just distribution of income. Attaining a sustainable scale, a just distribution, and an efficient allocation are three distinct problems. They are certainly not isolated, but solving one does not solve the others. Achieving three different goals generally requires three different policy instruments. This is illustrated by the cap-and-trade systems, a favored policy of ecological economists. Three policy actions are needed in proper sequence. First, a quantitative limit is set, reflecting judgments of sustainable scale—that is, a previously unlimited or free good is recognized as scarce, and the scale of its use is limited. Second, the newly scarce good or right is now a valuable asset—who owns it? Deciding who owns it is a question of distributive justice. Third, once scale and distribution decisions have been politically made, we can have individualistic trading and efficient market allocation. The proper name for such policies should indeed be "cap, distribute, and trade" or "cap and rent," when resource rights are assigned to the state.

As growth pushes us from an empty world to a full world, the limiting factor in production, as we have argued, increasingly becomes natural capital, not manmade capital—for example, the fish catch today is no longer limited by manmade capital of fishing boats but by the complementary natural capital of fish populations in the sea. As we move into a full world, economic logic remains the same: to economize on and invest in the limiting factor. But the identity of the limiting factor changes from manmade to remaining natural capital, and our economizing efforts and policies must change accordingly. Therefore, it becomes more important to study the nature of environmental goods and services in both their stock-flow and fund-service dimensions—are they rival or nonrival, excludable or nonexcludable—in order to know if they are market goods or public goods.

Ecological economics accepts the standard analysis of allocative efficiency, given prior social determination of the distribution and scale questions, and given that the good in question is rival and excludable. Although the main difference has been the focus on scale, that difference has entailed more attention to dimensions of distribution often neglected—namely, intergenerational distribution of the resource base and distribution of places in the sun between humans and all other species (biodiversity). Also, as more vital resources cease being free goods and are allocated by the market, the fairness of the distribution underlying market allocation becomes more critical. Once growth in scale has become uneconomic, it can no longer be appealed to as the solution to poverty. Poverty reduction requires increased sharing. Other issues of debate include whether natural and manmade capital are primarily substitutes or complements, the degree of coupling between physical throughput and GNP, and the degree of coupling between GNP and welfare.

One question not explicitly addressed in the text, but which students are sure to ask, is: What is the relationship between ecological economics and courses in resource economics or environmental economics that are sometimes taught in economics departments? The difference is that the latter are both subfields of neoclassical economics; they do not consider scale an issue, have no concept of throughput, and are focused on efficiency of allocation. Resource economics deals with the efficiency of allocation of labor and capital devoted to extractive industries. It develops many useful concepts that we have covered in this text, such as scarcity rent and user cost. Likewise, environmental economics also focuses on efficiency of allocation and how it is disrupted by pollution externalities. Concepts of internalizing externalities by Pigouvian taxes or well-defined and enforceable property rights (see Coase theorem, Chapter 10) are certainly useful, and we have discussed them. Nevertheless, the aim of both resource and environmental economics is allocative efficiency via right prices, not sustainable scale.

Ecological economics connects resource and environmental economics by recognizing the real-world connection between depletion pollution via the concept of throughput. We have also paid much more attention to impacts on, and feedbacks from, the rest of the ecosystem induced by economic activities that cause depletion, pollution, and entropic degradation. In addition, we have investigated the basic principles (energy flows, material cycles, ecosystem structure and function) governing the containing ecosystem itself, thereby at least partially integrating economics with ecology.

Finally, we have insisted on policy as our guiding philosophical viewpoint. This has led us to recognize and defend the logical presuppositions of policy—namely, nondeterminism and non-nihilism. It really is possible for things to be other than they are, and we really can distinguish better from worse states of the world. If that were not the case, then our effort in writing this book, and your effort in reading it, would both have been in vain.

Glossary

Abiotic resource A nonliving resource that cannot reproduce: fossil fuels, minerals, water, land, and solar energy.

Absolute advantage A country has an absolute advantage if it can produce the good in question at a lower absolute cost than its trading partners. It has a **comparative advantage** if it can produce the good in question more cheaply relative to other goods it produces than can its trading partners, regardless of absolute costs.

Absorptive capacity See "waste absorption capacity."

Adaptive management A basic policy principle whereby we change policies as conditions change or as we gain new information.

Adverse selection

Aggregate macroeconomics The study of the economy from the perspective of key aggregate variables such as the money supply, aggregate price level, the interest rate, aggregate consumption and investment, exports, and imports. The main focus of conventional economics is the rate of growth of GNP. In contrast, ecological economics strives to end physical growth while maintaining or improving social welfare.

Allocation The process of apportioning resources to the production of different goods and services. Neoclassical economics focuses on the market as the mechanism of allocation. Ecological economics recognizes that the market is only one possible mechanism for allocation.

Altruistic punishment

Asymmetric information Occurs when either buyer or seller has information that the other does not have, and that information affects the value of the good or service exchanged.

Balance of payments The sum of the current account (exports minus imports), and the capital account (inflow of capital to the nation minus outflow of capital from the nation).

Barter The direct exchange of goods or services without using money as a medium of exchange. It is very inconvenient in that it requires an unlikely coincidence of reciprocal wants.

Basic market equation $MUxn/MUyn = Px/Py = MPPay/MPPax$, where MU is the marginal utility of good x or good y to person n, and MPP is the marginal physical product of factor a used to produce good x or good y.

Biotic resource A living resource, such as trees, fish, and cattle (elements of ecosystem structure), as well any of the fund-services they provide, such as climate regulation, water regulation, and waste-absorption capacity (ecosystem functions or ecosystem services).

Bretton Woods Institutions (IMF and World Bank) Global financial institutions created in 1945 to finance short-term international trade (International

Monetary Fund) and to lend for long-term investment projects in developing countries (World Bank).

Cap and trade

Capital account A measure of inflows of investments to the nation by foreigners and outflows of investments by nationals to foreign countries.

Carrying capacity Originally the maximum population of cattle that can be sustained on a given area of rangeland. By extension the population of humans that can be sustained by a given ecosystem at a given level of consumption, with a given technology.

Catch-per-unit-effort hypothesis The assumption of a linear relationship between effort, stock, and harvest.

Circular flow The idea that since every expenditure by anyone is at the same time an equal receipt by someone else and receipts in turn become expenditures, money or exchange value flows in a circle. But physical factors and products do not flow in a circle.

Closed system A system that imports and exports energy only; matter circulates within the system but does not flow through it. The Earth closely approximates a closed system.

Coase theorem States that in a perfectly competitive market allocative efficiency will be achieved whether property rights are given to the polluter or the "pollutee." All that is needed is that someone have the property rights and that transaction costs are zero.

Coevolutionary economics The study of the mutual adaptations of economy and environment. Economic activity induces changes in the environment, and changes in the environment in turn induce further changes in the economy in a continuing process of coevolution.

Command-and-control regulation Flat prohibitions, quotas, or standards as opposed to monetary incentives that operate through prices or taxes.

Comparative advantage See "absolute advantage."

Competitive market A market in which there are many small buyers and sellers of an identical product. "Many" means "enough that no single buyer or seller is sufficiently large to affect the market price." Since everyone treats price as a parameter (a given condition) rather than a variable (something one can change), this condition is sometimes called the parametric function of prices.

Complementarity The opposite of substitutability—when goods or factors have to be used together in fairly strict combination with each other rather than instead of each other. Even substitutes have some degree of complementarity, unless they are "perfect" substitutes, in which case they are for all practical purposes identical goods or factors.

Comprehensive efficiency The ratio of services gained from manmade capital stock to the services sacrificed from the natural capital stock.

Conditional cooperators

Congestibility Occurs when a basically nonrival asset is used so heavily that one person's use begins to interfere with or lower the quality of service to other users (e.g., a crowded road or interference from another radio station in a crowded frequency band).

Consumer surplus The maximum that a consumer would be willing to pay for all units of a good he consumes rather than do without the good, minus the amount he actually has to pay for it.

Contingent valuation Hypothetical estimates of prices of nonmarket goods and services based on survey questions asking how much one would be willing to pay for an extra unit of the good or how much one would accept for the loss of a unit of the good.

Critical depensation The population size below which a population will probably go extinct rather than recuperate, even if exploitation ceases. Also known as "minimum viable population."

Crowding out

Current account A measure of the international exchange of real goods and services as well as transfer payments in the current year.

Defensive expenditure An expenditure made to protect one against the unwanted consequences of the production and consumption of other goods by other people. Also called regrettably necessary defensive expenditure.

Deflation

Demand A relationship (inverse) between price of a good and the quantity of the good that consumers would purchase at that price.

Determinism The philosophical doctrine that every event or decision is the inevitable consequence of antecedents, such as physical, psychological, hereditary, or environmental conditions, that are independent of human will or purpose.

Development The improvement in quality of goods and services, as defined by their ability to increase human well-being, provided by a given throughput.

Differential rent See "rent."

Diminishing marginal physical product See "law of diminishing marginal physical product."

Diminishing marginal utility See "law of diminishing marginal utility."

Discount rate The rate at which the present is valued over the future, as a result of uncertainty, or of productivity, or of pure time preference for the present. See "intertemporal discounting."

Disinflation A reduction in the rate of general price increase. See "inflation."

Distribution The apportionment of income or wealth among different people.

Doubling time The fixed time period it takes for a population to double when growing at a constant rate. A handy rule is that doubling time equals 70 divided by the percentage rate of growth (e.g., a population growing at 2% annually will double every 35 years).

Ecological economics The union of economics and ecology, with the economy conceived as a subsystem of the earth ecosystem that is sustained by a metabolic flow or throughput from and back to the larger system. See "throughput."

Ecological reductionism The idea that the human economy is governed entirely by the same laws and forces as the rest of the ecosystem, so there is no necessity to distinguish the human economy as a subsystem.

Economic imperialism The idea that the entire ecosystem can be priced and all values internalized into prices, with the result that price calculations are reliable guides for all decisions.

Ecosystem function An emergent phenomenon in ecosystems, such as energy transfer, nutrient cycling, gas regulation, climate regulation, and the water cycle. As is typical of emergent properties, ecosystem functions cannot be readily explained by even the most extensive knowledge of system components or ecosystem structure.

Ecosystem services Ecosystem functions of value to humans, though given the tightly interconnected nature of ecosystems, it would be difficult to say with certainty that any particular ecosystem function is not of value to humans. See also "fund-service resources."

Ecosystem structure The individuals and communities of plants and animals of which an ecosystem is composed, their age and spatial distribution, and the abiotic resources present. The elements of ecosystem structure interact to create ecosystem functions as emergent properties generated of such a large complex system.

Efficient allocation See "Pareto efficient allocation."

Efficient cause An agent of transformation, such as labor or a machine. See also "fund-service resources" and "material cause."

Elasticity The responsiveness of a change in quantity demanded (supplied) to a change in price, measured as the percentage change in quantity divided by percentage change in price.

Ends–means spectrum

Entropic dissipation The gradual erosion and dispersion into the environment of the matter of which all human artifacts are composed in a one-way flow of low-entropy usefulness to high-entropy waste.

Entropy See "Second Law of Thermodynamics."

Environmental economics The branch of neoclassical economics that addresses environmental problems such as pollution, negative externalities, and valuation of nonmarket environmental services. In general, environmental economics focuses almost exclusively on efficient allocation and accepts the pre-analytic vision of neoclassical economics that the economic system is the whole, not a subsystem of the containing and sustaining global ecosystem.

Equimarginal principle of maximization This is known as the "when to stop" rule. The point at which a consumer reaches an allocation that maximizes her total satisfaction or total utility. That point occurs when the marginal utility per dollar spent on each good is equal. Only when utilities were equal at the margin would it no longer be possible to increase total utility by reallocation of expenditure.

Exchange rate The rate at which one nation's currency is traded for that of another nation. Exchange rates can be fixed by central banks, floating according to daily supply and demand, or some combination of the two.

Exchange value The value of a good in terms of its ability to be traded for other goods, as opposed to its use value.

Excludable resource

Excludability A legal concept that when enforced allows an owner to exclude others from using his asset. An institution is always required to make an asset excludable, but some by their very nature are nonexcludable (e.g., the ozone layer). If the asset is rival (a physical property), then excludability is more or less natural. If the asset is nonrival, then excludability is typically more difficult but sometimes possible, as in the case of patents.

Exponential growth Growth at a constant percentage rate (fixed doubling time). Exponential growth leads to very large numbers surprisingly quickly.

Externality An unintended and uncompensated loss or gain in the welfare of one party resulting from an activity by another party.

Extrinsic incentives

Fallacy of composition The argument that what is true for the part must necessarily be true for the whole, or vice versa.

Fallacy of misplaced concreteness To mistake the map for the territory, to be unmindful of the degree of abstraction in an argument, especially to draw conclusions at a level of abstraction (or concreteness) different from the level of abstraction of the concepts in which the argument is conducted.

Federal Reserve System A coordinated system of district central banks in the U.S. that influences interest rates and money supply by means of open-market operations, discount rate changes, and reserve ratio requirements.

First Law of Thermodynamics Neither matter nor energy can be created or destroyed.

Fiscal policy The attempt to influence GNP, employment, interest rates, and inflation by manipulating government expenditure and taxes.

Fixed exchange rate regime A regime in which the value of one country's currency is pegged to another country's (typically the U.S. dollar).

Flexible exchange rate regime A regime in which exchange rates are determined by the global supply and demand for currencies, and central banks play no direct role.

Floating exchange rate regime See "flexible exchange rate regime."

Fossil fuels Petroleum, coal, natural gas. Fuels formed over geologic ages from biotic materials but now treated as nonrenewable abiotic resources.

Fractional reserve banking The practice of keeping on hand reserves against deposits that are only a small fraction of deposits, so that banks can lend the difference. This practice allows the private banking sector to create money, since demand deposits are counted as money.

Free rider One who enjoys the benefit of a public good without paying a share of the cost of its provision and maintenance.

Fund-service resources Resources not materially transformed into what they produce (efficient cause), which can be used only at a given rate, and their productivity is measured as output per unit of time; cannot be stockpiled; and are worn out, rather than used up. See "stock-flow resources."

General equilibrium model The vision of the economy as a giant system of thousands of simultaneous equations balancing the supply and demand and determining the price and quantity for each commodity in the economy.

Gini coefficient A measure of the inequality of the distribution of wealth or income across a population. A Gini coefficient of one implies perfect inequality (one person owns everything), and a coefficient of zero indicates a perfectly equal distribution.

Globalization The economic integration of the globe by free trade, free capital mobility, and to a lesser extent easy migration. It is the effective erasure of national boundaries for economic purposes. See "internationalization."

Gross national product (GNP) The market value of final goods and services purchased by households, by government, and by foreigners (net of what we purchase from them), in the current year. Alternatively, it is the sum of all value added to raw materials by labor and capital at each stage of production during the given year.

Growth A quantitative increase in size, or an increase in throughput.

Hedge investors

Hotelling rule States that at the margin, the rate of return from holding a resource in the ground (its expected price increase) must be equal to the rate of return from exploiting it now and investing the profits.

Hubbert curve A curve showing the cumulative extraction of a nonrenewable resource over time. The vertical distance for each year is annual extraction for that year. The total area under the curve is total reserves. The typical curve is bell-shaped, rising from zero, reaching a maximum, and falling to zero again with exhaustion.

Human needs assessment A multidimensional concept of welfare that goes beyond income and wealth to include capabilities, capacities, and other existential categories.

Hyperbolic discounting The act of giving more weight to what happens now over what happens in the near future, while being nearly indifferent between the same outcome occurring at different times in the more distant future. Empirical studies suggest that this is a far more accurate representation of the human psyche than exponential discounting.

Hyperinflation Inflation greater than 50% per month.

Inalienability rule An entitlement that holds if a person is entitled to the presence or absence of something, and no one is allowed to take away that right for any reason.

Income The maximum that a community could consume in a given time period and still be able to produce the same amount in the next time period. In other words, the maximum that can be consumed without reducing productive capacity, that is, without reducing capital.

Individual transferable quotas (ITQs)

Inflation An increasing general level of prices (not a state of high prices).

International Bank for Reconstruction and Development (World Bank) An international financial institution composed of member nations and created at Bretton Woods, New Hampshire, in 1945. Originally designed to focus on long-term lending to promote the development of underdeveloped countries, in recent decades it has strayed from its charter.

Internationalization The increasing importance of relations between nations and between citizens of different nations. The nation remains the basic unit of community and policy, controlling to some extent trade, capital flows, and migration. National economies are interdependent but not integrated. See "globalization."

International Monetary Fund (IMF) An international financial institution composed of member nations and created at Bretton Woods, New Hampshire, in 1945. Originally designed to focus on short-term balance of payments financing to promote international economic stability, in recent decades it has strayed from its charter.

Intertemporal allocation The apportionment of resources across different stages in the lifetimes of basically the same set of people (same generation).

Intertemporal discounting The process of systematically weighting future costs and benefits as less valuable than present ones.

Intertemporal distribution The apportionment of resources across different generations (different people).

ISEW Index of Sustainable Economic Welfare, calculated by adjusting personal consumption for various factors that affect sustainability or welfare either positively or negatively, such as depletion of natural capital, increasing inequality in income distribution, or defensive expenditures.

IS-LM model A two-sector general equilibrium model showing how the real and the financial sectors interact to simultaneously determine the national income and the interest rate.

Isolated system An isolated system is one in which neither matter nor energy enters or exits.

Law of diminishing marginal physical product As a producer adds successive units of a variable factor to a production process, other factors constant, the extra output per unit of the variable factor diminishes with each addition (i.e., total output increases at a decreasing rate). This is sometimes called the law of diminishing returns.

Law of diminishing marginal utility As one consumes successive units of a good, the additional satisfaction decreases (i.e., total satisfaction increases, but at a decreasing rate). With a law of constant or increasing marginal utility consumers would spend all their income on only one good.

Law of entropy See "Second Law of Thermodynamics."

Law of increasing marginal cost As one produces more and more of a product, one must use resources (factors of production) that are of lower quality or are more expensive, so that the cost of producing each additional unit is greater than that of the previous unit.

Leverage

Liability rule An entitlement rule that holds if one person is free to interfere with another or prevent interference but must pay compensation.

Liberalize

Linear throughput See "throughput."

Liquidity preference A general preference for holding assets in a form easily convertible into money so as to meet unexpected transaction needs and avoid the inconvenience of barter.

Liquidity trap The failure of lowering interest rates to stimulate economies with low demand.

Lorenz curve A curve plotting the cumulative percentage of the population against the cumulative percentage of total income held by that percentage of the population, which illustrates the degree of equality or inequality in the distribution of income.

Macro-allocation The allocation of resources between market and nonmarket goods and services.

Marginal cost The increment in total cost resulting from producing one more unit of the commodity in question.

Marginal external cost The cost to society of the negative externality that results from one more unit of activity by an economic agent.

Marginal extraction costs The extra total cost needed to extract one more unit of a resource from the ground.

Marginal revenue The increment in total revenue from selling one more unit of the commodity in question.

Marginal user cost The value of one more unit of the resource in its natural state. In a perfectly competitive economy, marginal user cost would in theory equal the price of a resource minus its marginal extraction cost.

Marginal utility The additional pleasure or satisfaction to be gained from consuming one more unit of a good or service.

Material cause A resource that is transformed in the production process. See also "stock-flow resource" and "efficient cause."

Materialism The philosophical doctrine that physical matter (its movements and modifications) is the only reality and that everything in the universe, including thought, feeling, mind, will, and purpose, can be explained in terms of physical laws.

Matrix of human needs (Max-Neef) A cross-classification of basic dimensions of existence with basic human values, yielding a much richer and more detailed description of welfare than the abstract notion of utility.

Maximum sustainable yield Each level of an exploited population has a growth rate that can be harvested, leaving the population undiminished in the next year. There is one level of population for which the sustainable yield is a maximum. In general, however, the biologically maximum sustainable yield is not the economically optimal yield.

Micro-allocation The allocation of resources within the private sector, as opposed to between the private and public sectors. See "macro-allocation."

Mineral resource A useful element or compound, such as copper, iron, petroleum—a class of abiotic resource.

Minimum viable population The population level below which a population is not likely to recuperate but rather will dwindle to extinction. See "critical depensation."

Monetary policy The attempt to influence interest rates, GNP, employment, and inflation by manipulation of the money supply.

Money A unit of account, medium of exchange, and a store of value. Money can be a commodity (gold) whose supply is limited by its real cost of production or a token (fiat money) whose supply is determined by government authority and social conventions.

Monopoly A single seller of a commodity.

Moral hazard A situation in which the existence of insurance against a hazard actually increases the risk of the hazard by making the insured less vigilant in its prevention.

Multiplier effect

Multi-tier pricing The act of charging different prices at different times or for different users.

Natural capital Stocks or funds provided by nature (biotic or abiotic) that yield a valuable flow into the future of either natural resources or natural services.

Natural dividend The unearned income from the harvest of renewable resources. As nature and not human industry produces renewable resources, all profits above "normal" profit (included in the total cost) are unearned, and the natural dividend is equivalent to the total return minus the total cost.

Neoclassical economics The currently dominant school of economics, characterized by its marginal utility theory of value, its devotion to the general equilibrium model stated mathematically, its individualism, and its reliance on free markets and the invisible hand as the best means of allocating resources, with a consequent downplaying of the role of government.

Net present value The amount of money that if available today would generate the future stream of net income in question.

Neuroeconomic

Nihilism In ethics, the rejection of all distinctions in moral value, the rejection of all theories of morality. The view that nothing is better than anything else, or basically "anything goes."

Nonexcludable resource A resource for which no institution or technology exists to make it excludable.

Nonmarket value A value recognized by people but not usually expressed in prices because the valuable thing either is not or cannot be traded in markets.

Non-price adjustments The adjustment of the relative desirability of goods by advertising, or the relative possibility of producing goods by research and development. These are ways to satisfy the basic market equation without adjusting prices.

Nonrenewable resource Low-entropy matter-energy useful to humans and present in fixed stocks whose quantity declines over time. This includes mineral resources, fossil fuels, and fossil aquifers. Because freshwater is naturally recycled through the hydrological process, we do not classify it as a nonrenewable resource.

Nonrival resource A resource whose use by one person does not affect its use by another.

Normal profit The opportunity cost of the time and money the entrepreneur has put into an enterprise (i.e., what she could have earned from her time and money in her next best alternative).

Open access A rival resource that all are free to exploit (no one can be excluded), such as noncoastal fisheries.

Open system An open system takes in and gives out both matter and energy. The economy is such a system.

Opportunity cost The best alternative given up when a choice is made, i.e., if a farmer cuts down a forest to expand his cropland, and if the consequent loss of timber, firewood, and water purification is the next best use of the land, then the value of timber, firewood, and water purification is the opportunity cost of the expanded cropland.

Optimal scale of the macroeconomy Occurs when the increasing marginal social and environmental costs of further expansion are equal to the declining marginal benefits of the extra production. Beyond the optimal scale growth becomes uneconomic, even if we conventionally refer to the expansion of the economy as "economic growth."

Paradox of thrift Occurs when everyone trying to increase their savings rate in the aggregate leads to declining consumption, growing unemployment, and lower aggregate income. People are saving a higher percentage of their income, after increasing their savings rate, but total income has fallen, so that actual savings are less than before.

Pareto efficient allocation Occurs when no other allocation could make at least one person better off without making anyone else worse off. This is also known as a Pareto optimum.

Pareto optimum See "Pareto efficient allocation."

Pigouvian subsidy See "Pigouvian tax."

Pigouvian tax A tax designed to equal the marginal external cost of production of a commodity. It is added to the price, which measures only marginal private costs. The price plus tax now measures marginal social cost, thus internalizing the original external cost. If there originally was an external benefit, then a Pigouvian subsidy would be paid to the producer.

Ponzi investors

Principle of subsidiarity A basic policy principle that the domain of the policy-making unit must be congruent with the domain of the causes and effects of the problem with which the policy deals.

Principle of substitution The assumption that one good or service (or factor of production) can replace another in providing consumer utility (or in the production process). Theoretical utility functions and production functions often exhibit this property, but in reality, many goods, services, and factors of production are complements rather than substitutes. See also "substitutability" and "complementarity."

Prisoner's dilemma

Procedural utility

Producer surplus The difference between the price (equal to marginal cost) at which the producer sells his total output and the lower marginal costs at which all inframarginal units were produced.

Production function A production "recipe" describing how certain quantities of inputs are combined to yield a certain quantity of output or product.

Property rule An entitlement rule that holds if one person is free to interfere with another or free to prevent interference.

Prosocial behavior

Public bad Something that is nonrival, nonexcludable, and undesirable.

Public good A resource that is nonrival, nonexcludable, and desired by the public. Because they are nonexcludable, they will not be produced by profit-seeking firms. Because it is nonrival, the marginal cost of another person using one is zero, so its efficient price should also be zero. A public good should be supplied collectively by the government or other social institution.

Pure time rate of preference The rate at which we prefer goods in the present over the future, independently of considerations of productivity.

Quota

Renewable resource A living resource that is capable of regeneration and growth in perpetuity if exploited in a sustainable manner and that provides raw materials for the economic process.

Rent A payment over and above minimum necessary supply price (cost of production). Since land has zero cost of production, all payment for land is rent. Part of payment for labor may also be rent if the laborer would still do the job for less. If Tiger Woods would still play golf even if he earned only a million dollars a year, then all his earnings over a million dollars is rent. Producer surplus is also an example of rent.

Ricardian land Land as an extension, surface area, and substrate for holding things (i.e., the "indestructible" characteristics of land, excluding its fertility or underground minerals).

Risk The known probability (relative frequency) of occurrence of an event. Risk is insurable. See "uncertainty."

Rival resource

Rivalness An inherent characteristic of certain resources whereby consumption or use by one person reduces the amount available for everyone else.

Royalty The payment to the owner of a resource for the right to exploit that resource. Theoretically, in a competitive market, the per-unit royalty should be equal to the marginal user cost.

Say's law Supply creates its own demand. In production the payments to factors, plus residual profit, generate exactly enough income to purchase, if spent, the total amount produced.

Scale The physical size of the economic subsystem relative to the ecosystem that contains and sustains it. It could be measured in its stock dimension of population and inventory of artifacts, or in its flow dimension of throughput needed to maintain the stocks.

Scarcity rent See "rent."

Second Law of Thermodynamics Entropy never decreases in an isolated system. Although matter and energy are constant in quantity (First Law), they change in quality. The measure of quality is entropy, and basically it is a physical measure of the degree of "used-up-ness" or randomization of the structure or capacity of matter or energy to be useful to us. Entropy increases in an isolated system. We assume the universe is an isolated system, so the Second Law says that the natural, default tendency of the universe is "shuffling" rather than "sorting." In everyday terms, left to themselves, things tend to get mixed up and scattered. Sorting does not occur by itself.

Seigniorage The benefit that accrues to the issuer of token money, resulting from the fact that the issuer receives real goods and services in exchange for a mere token, whereas everyone else has to give up a real asset to get money to exchange for another asset. There is a transfer of real wealth from the public to the issuer of money, equal to the exchange value of money stock minus the cost of production of the tokens (negligible). That amount is seigniorage. Sometimes the term is used to refer only to the interest that could be earned on such an amount.

Sink The part of the environment that receives the waste flow of the throughput and may, if not overwhelmed, be able to regenerate the waste through biogeochemical cycles back to usable sources.

Social discount rate A rate of conversion of future value to present value that reflects society's collective ethical judgment, as opposed to an individualistic judgment such as the market rate of interest.

Solar energy Radiant energy flowing from the sun, our basic long-run source of low entropy that sustains life and wealth.

Source The part of the environment that supplies usable raw materials that constitute the throughput by which the economy produces and which ultimately returns as waste to environmental sinks.

Speculative investors

Steady-state economy The economy viewed as a subsystem in dynamic equilibrium with the parent ecosystem or biosphere that sustains it. Quantitative growth is replaced by qualitative development or improvement as the basic goal.

Steady-state subsystem See "steady-state economy."

Stock-flow resources Resources materially transformed into what they produce (material cause); can be used at virtually any rate desired (subject to the availability of fund-service resources needed for their transformation); their productivity is measured by the number of physical units of the product into which they are transformed; can be stockpiled; are used up rather than worn out. See "fund-service resources."

Subsidy A bonus or payment for doing something, the opposite of a tax.

Substitutability The capacity of a one factor (or good) to be used in the place of another, the opposite of complementarity. Substitutability is never perfect, and the further a substitution is carried, the less satisfactory it becomes (the more the force of complementarity is felt). Goods and factors may be thought of as varying in a continuum from perfect substitutes to perfect complements.

Supply The relationship between the price of a commodity and the quantity that would be supplied at each price.

Sustainable yield The amount of an exploited population that can be harvested, leaving the population undiminished in the next year; the growth rate of the existing stock. See "maximum sustainable yield."

Thermodynamics The branch of physics that tells us that matter and energy can be neither created nor destroyed and that entropy in the total system always increases. This branch of physics is the most relevant to economics because it helps to explain the physical roots of scarcity.

Throughput The flow of raw materials and energy from the global ecosystem's sources of low entropy (mines, wells, fisheries, croplands), through the economy, and back to the global ecosystem's sinks for high-entropy wastes (atmosphere, oceans, dumps).

Total allowable commercial catch (TACC) An aggregate quota limiting the total fish catch per year or season. The quota can be distributed among individuals in many different ways.

Tradeable permits (quotas) Shares of an aggregate quota that are in some way divided up among individuals, who can then buy and sell their quota rights among themselves.

Transaction cost The costs of making a transaction, including legal fees, the cost of gathering information, locating the interested parties, the time costs of bargaining, and so on.

Transaction demand for money The demand for money balances to carry out everyday plus unforeseen transactions and avoid the inconvenience of barter or the delay of converting a nonliquid asset into money.

Ultimate end The vaguely perceived yet logically necessary ordering principle with reference to which we rank our intermediate ends.

Ultimate means The low-entropy matter-energy, consisting of the solar flow and the terrestrial stock—that which we need to serve our ends and which we cannot ourselves create but can receive only from nature.

Uncertainty A situation in which we may know the range of possible outcomes but do not know the probability distribution of outcomes. Uncertainty is uninsurable. See "risk."

Uneconomic growth Growth of the macroeconomy that costs us more than it is worth. A situation in which further expansion entails lost ecosystem services that are worth more than the extra production benefits of the expanded economy.

Unemployment Refers usually to involuntary unemployment, the number or percentage of the workforce without a job who are actively looking for a job.

Use value The actual service or utility from using a commodity for its intended purpose, as opposed to its exchange value, its capacity to purchase another good through exchange.

User cost The opportunity cost of nonavailability of a natural resource at a future date that results from using up the resource today rather than keeping it in its natural state.

Utility function A psychic relationship showing the amount of utility or satisfaction yielded to a consumer by the consumption of differing amounts and combinations of commodities.

Virtual wealth A concept introduced by Frederick Soddy, similar to seigniorage. The total value of real assets that the community voluntarily abstains from holding in order to hold money instead. Since individuals can easily convert their money into real assets, they count their money holdings as wealth. Yet the community as a whole cannot convert money into wealth because someone has to end up holding the money (see "fallacy of composition"). Money wealth is therefore "virtual."

Waste absorption capacity The capacity of an ecosystem to absorb and reconstitute wastes into usable forms through biogeochemical cycles powered by the sun. This capacity is a renewable resource that can be overwhelmed and destroyed, or used within sustainable limits.

Welfare A psychic state of want satisfaction or enjoyment of life—an experience, not a thing—the basic purpose of economic activity.

World Bank See "International Bank for Reconstruction and Development."

World Trade Organization (WTO) The successor organization to the General Agreement on Trade and Tariffs (GATT) that seeks liberalization of international trade and investment and generally promotes globalization.

Suggested Readings

Part I. An Introduction to Ecological Economics

Costanza, Robert (ed.). 1991. *Ecological Economics: The Science and Management of Sustainability*. New York: Columbia University Press.

Costanza, Robert, Charles Perrings, and Cutler J. Cleveland (eds.). 1997. *The Development of Ecological Economics*. Cheltenham, UK: Edward Elgar.

Daly, Herman, and K. Townsend (eds.). 1993. *Valuing the Earth: Economics, Ecology, Ethics*. Cambridge, MA: MIT Press.

Faber, Malte, Reiner Manstetten, and John Proops. 1998. *Ecological Economics: Concepts and Methods*. Cheltenham, UK: Edward Elgar.

Gowdy, John. 1994. *Coevolutionary Economics: Economy, Society, and Environment*. Boston: Kluwer Academic Free Press.

Krishnan, Rajaraman, Jonathan M. Harris, and Neva R. Goodwin (eds.). 1995. *A Survey of Ecological Economics*. Washington, DC: Island Press.

Martinez-Alier, Juan. 1987. *Ecological Economics*. Oxford, UK: Basil Blackwell.

Meadows, Donella H., et al. 1992. *Beyond the Limits: Confronting Global Collapse, Envisioning a Sustainable Future*. Post Mills, VT: Chelsea Green Publishing Co.

Nadeau, Robert L. 2003. *The Wealth of Nature: How Mainstream Economics Has Failed the Environment*. New York: Columbia University Press.

Norgaard, Richard. 1984. "Co-Evolutionary Development Potential." *Land Economics* 60 (May):160–173.

Schumacher, E. F. 1974. *Small Is Beautiful: Economics As If People Mattered*. New York: Harper & Row.

Wackernagel, M. 2002. "Tracking the Ecological Overshoot of the Human Economy." *Proceedings of the National Academy of Sciences* 99(14) (July 2002): 9266–9271.

Part II. The Containing and Sustaining Ecosystem: The Whole

Cohen, Joel H. 1995. *How Many People Can the Earth Support?* New York: W. W. Norton.

Georgescu-Roegen, Nicholas. 1971. *The Entropy Law and the Economic Process*. Cambridge, MA: Harvard University Press.

Georgescu-Roegen, Nicholas. 1976. *Energy and Economic Myths*. New York: Pergamon Press.

Hall, Charles, Cutler Cleveland, and Robert Kaufmann. 1986. *Energy and Resource Quality: The Ecology of the Economic Process*. New York: John Wiley and Sons.

Hay, Peter. 2002. *Main Currents in Western Environmental Thought*. Bloomington: Indiana University Press. (See especially Chapter 8 on economics.)

Hokikian, Jack. 2002. *The Science of Disorder: Understanding the Complexity, Uncertainty, and Pollution in Our World.* Los Angeles: Los Feliz Publishing.

Kauffman, Stuart. 1995. *At Home in the Universe.* New York: Oxford University Press. (A good introduction to complexity theory.)

Meffe, Gary, et al. 2002. *Ecosystem Management: Adaptive Community-Based Conservation.* Washington, DC: Island Press.

Miller, G. Tyler. 1994. *Living in the Environment: Principles, Connections, Solutions,* 8th ed. Belmont, CA: Wadsworth.

Odum, Eugene. 1997. *Ecology: A Bridge Between Science and Society,* 3rd ed. Sunderland, MA: Sinauer Associates.

Part III. Microeconomics

Barnett, Harold, and Chandler Morse. 1963. *Scarcity and Growth: The Economics of Natural Resource Availability.* Baltimore: Johns Hopkins University Press.

Bollier, David. 2002. *Silent Theft: The Private Plunder of Our Common Wealth.* New York: Routledge.

Brekke, Kjell Arne, and Richard B. Howarth. 2002. *Status, Growth, and the Environment: Goods as Symbols in Applied Welfare Economics.* Cheltenham, UK: Edward Elgar.

Devarajan, S., and A. Fisher. 1981. "Hotelling's Economics of Exhaustible Resources 50 Years Later." *Journal of Economic Literature* 1 (March).

Heal, Geoffrey. 2000. *Nature and the Marketplace: Capturing the Value of Ecosystem Services.* Washington, DC: Island Press.

Page, Talbot. 1977. *Conservation and Economic Efficiency.* Baltimore: Johns Hopkins University Press.

Perrings, Charles. 1987. *Economy and Environment: A Theoretical Essay on the Interdependence of Economic and Environmental Systems.* Cambridge, UK: Cambridge University Press.

Price, Colin. 1993. *Time, Discounting, and Value.* Oxford, UK: Blackwell Publishers.

Smith, V. Kerry (ed.). 1979. *Scarcity and Growth Reconsidered.* Baltimore, MD: Johns Hopkins University Press.

Part IV. Macroeconomics

Binswanger, Hans Christoph. 1994. *Money and Magic: A Critique of the Modern Economy in the Light of Goethe's Faust.* Chicago: University of Chicago Press.

Cobb, Clifford, John Cobb, et al. 1994. *The Green National Product: A Proposed Index of Sustainable Economic Welfare.* New York: University Press of America.

Collins, Robert M. 2000. *More: The Politics of Growth in Postwar America.* New York: Oxford University Press.

Ekins, Paul, and Manfred Max-Neef (eds.). 1992. *Real-Life Economics: Understanding Wealth Creation.* New York: Routledge.

Greider, William. 1987. *Secrets of the Temple: How the Federal Reserve Runs the Country.* New York: Simon & Schuster.

Heyck, Denis Lynn Daly. 2002. *Surviving Globalization in Three Latin American Communities*. Toronto: Broadview Press.

Hueting, Roefie. 1980. *New Scarcity and Economic Growth*. North Holland: Netherlands Central Bureau of Statistics.

Socolow, R., et al. (eds.). 1994. *Industrial Ecology and Global Change*. New York: Cambridge University Press.

Part V. International Trade

Culbertson, J. M. 1971. *Economic Development: An Ecological Approach*. New York: Alfred A. Knopf.

Greider, William. 1997. *One World, Ready or Not: The Manic Logic of Global Capitalism*. New York: Simon & Schuster.

Kaul, I., I. Grunberg, and M. A. Stern (eds.). 1999. *Global Public Goods: International Cooperation in the 21st Century*. New York: Oxford University Press.

Korten, David C. 1995. *When Corporations Rule the World*. West Hartford, CT: Kumarian Press.

Mander, Jerry, and Edward Goldsmith (eds.). 1996. *The Case Against the Global Economy*. San Francisco: Sierra Club Books.

Pincus, Jonathan R., and Jeffrey A. Winters (eds.). 2002. *Reinventing the World Bank*. Ithaca, NY: Cornell University Press.

Rich, Bruce. 1994. *Mortgaging the Earth: The World Bank, Environmental Impoverishment, and the Crisis of Development*. Boston: Beacon Press.

Rodrik, Dani. 1997. *Has Globalization Gone Too Far?* Washington, DC: Institute for International Economics.

Sachs, Wolfgang. 1999. *Planet Dialectics: Explorations in Environment and Development*. New York: Zed Books.

Shiva, Vandana. 2000. *Stolen Harvest: The Hijacking of the Global Food Supply*. Cambridge, MA: South End Press.

Stiglitz, Joseph. 2002. *Globalization and Its Discontents*. New York: W. W. Norton.

Part VI. Policy

Barnes, Peter. 2001. *Who Owns the Sky? Our Common Assets and the Future of Capitalism*. Washington, DC: Island Press.

Booth, Douglas E. 1998. *The Environmental Consequences of Growth: Steady-State Economics as an Alternative to Ecological Decline*. New York: Routledge.

Bromley, Daniel. 1991. *Environment and Economy: Property Rights and Public Policy*. Oxford, UK: Blackwell.

Brown, Lester R. 2001. *Eco-Economy: Building an Economy for the Earth*. New York: W. W. Norton.

Costanza, Robert, et al. (eds.). 1996. *Getting Down to Earth: Practical Applications of Ecological Economics*. Washington, DC: Island Press.

Crocker, David, and Toby Linden (eds.). 1998. *The Ethics of Consumption*. Lanham, MD: Rowman and Littlefield.

Daly, Herman. 1996. *Beyond Growth: The Economics of Sustainable Development.* Boston: Beacon Press.

Daly, Herman, and J. Cobb. 1994. *For the Common Good: Redirecting the Economy Toward Community, the Environment, and a Sustainable Future,* 2nd ed. Boston: Beacon Press.

Hamilton, Clive. 2003. *Growth Fetish.* Crows Neck, NSW, Australia: Allen & Unwin.

Jansson, AnnMari, et al. (eds.). 1994. *Investing in Natural Capital: The Ecological Economics Approach to Sustainability.* Washington, DC: Island Press.

Kemmis, Daniel. 1990. *Community and the Politics of Place.* Norman: University of Oklahoma Press.

Munasinghe, Mohan, Osvaldo Sunkel, and Carlos de Miguel (eds.). 2001. *The Sustainability of Long-Term Growth: Socioeconomic and Ecological Perspectives.* Cheltenham, UK: Edward Elgar.

About the Authors

Herman E. Daly is a professor at the University of Maryland, School of Public Affairs. From 1988 to 1994 he was senior economist in the Environment Department of the World Bank. Prior to 1988 he was Alumni Professor of Economics at Louisiana State University, where he taught economics for 20 years. He holds a B.A. from Rice University and a Ph.D. from Vanderbilt University. He has served as Ford Foundation Visiting Professor at the University of Ceará (Brazil), as a research associate at Yale University, as a visiting fellow at the Australian National University, and as a senior Fulbright lecturer in Brazil. He has served on the boards of directors of numerous environmental organizations, and is co-founder and associate editor of the journal *Ecological Economics*. His interest in economic development, population, resources, and environment has resulted in over a hundred articles in professional journals and anthologies, as well as numerous books, including *Toward a Steady-State Economy* (1973), *Steady-State Economics* (1977, 1991), *Valuing the Earth* (1993), *Beyond Growth* (1996), and *Ecological Economics and the Ecology of Economics* (1999). He is co-author with theologian John B. Cobb Jr. of *For the Common Good* (1989, 1994), which received the 1991 Grawemeyer Award for Ideas for Improving World Order. In 1996 he received Sweden's Honorary Right Livelihood Award and the Heineken Prize for Environmental Science, which was awarded by the Royal Netherlands Academy of Arts and Sciences. In 1999 he was awarded the Sophie Prize (Norway) for contributions in the area of environment and development and in 2002 the medal of the Presidency of the Italian Republic for his work in steady-state economics.

Joshua Farley holds an undergraduate degree in biology from Grinnell College, a master's in international affairs with a certificate in Latin American studies from Columbia University, and a Ph.D. in agricultural, resource, and managerial economics from Cornell University. A fellowship at the University of Brasília first exposed him to the transdisciplinary field of ecological economics and, in particular, the work of Herman Daly. Ecological economics complemented his background in biology and international development and his extensive experience working, studying, and traveling in less developed countries. After his studies at Cornell, he spent several years teaching ecological economics at the Centre for Rainforest Studies in Far North Queensland, Australia, first as resident faculty and later as program director. His teaching emphasized transdisciplinary, applied problem solving, while his research provided abundant concrete examples of the inability of the unregulated free-market economic system to efficiently and rationally allocate ecological resources. He has spent the last four years with the Gund Institute for Ecological Economics, first as executive director while the institute was located at the University of Maryland, then as assistant research professor at the University of Vermont. He is now assistant professor at the Department of Community Development and Applied Economics. Combining research and teaching, he enjoys working with nongovernmental organizations, community groups, and local governments to create applied transdisciplinary workshops and field courses in ecological economics.

Index